Mobility, Data Mining and Privacy

Fosca Giannotti · Dino Pedreschi
Editors

Mobility, Data Mining and Privacy

Geographic Knowledge Discovery

With 96 Figures, 12 in color, and 5 Tables

Fosca Giannotti
KDD Laboratory
ISTI-CNR, Istituto di Scienza
e Tecnologie dell'Informazione "A. Faedo"
Via G. Moruzzi, 1
56124 Pisa, Italy
fosca.giannotti@isti.cnr.it

Dino Pedreschi
KDD Laboratory
Dipartimento di Informatica
Università di Pisa
Largo B. Pontecorvo, 3
56127 Pisa, Italy
pedre@di.unipi.it

ISBN 978-3-540-75176-2   e-ISBN 978-3-540-75177-9

Library of Congress Control Number: 2007936014

ACM Classification: C.2, G.3, H.2, H.3, H.4, I.2, I.5, J.1, J.4, K.4

© 2008 Springer-Verlag Berlin Heidelberg

This work is subject to copyright. All rights are reserved, whether the whole or part of the material is concerned, specifically the rights of translation, reprinting, reuse of illustrations, recitation, broadcasting, reproduction on microfilm or in any other way, and storage in data banks. Duplication of this publication or parts thereof is permitted only under the provisions of the German Copyright Law of September 9, 1965, in its current version, and permission for use must always be obtained from Springer. Violations are liable to prosecution under the German Copyright Law.

The use of general descriptive names, registered names, trademarks, etc. in this publication does not imply, even in the absence of a specific statement, that such names are exempt from the relevant protective laws and regulations and therefore free for general use.

Cover Design: KünkelLopka, Heidelberg, based on an original artwork by Salvatore Rinzivillo

Printed on acid-free paper

9 8 7 6 5 4 3 2 1

springer.com

Preface

The technologies of mobile communications and ubiquitous computing are pervading our society. Wireless networks are becoming the nerves of our territory, especially in the urban setting; through these nerves, the movement of people and vehicles may be sensed and possibly recorded, thus producing large volumes of mobility data. This is a scenario of great opportunities and risks. On one side, data mining can be put to work to analyse these data, with the purpose of producing useful knowledge in support of sustainable mobility and intelligent transportation systems. On the other side, individual privacy is at risk, as the mobility data may reveal, if misused, highly sensitive personal information.

In a nutshell, a novel multi-disciplinary research area is emerging within this challenging conflict of opportunities and risks and at the crossroads of three subjects: mobility, data mining and privacy. This book is aimed at shaping up this frontier of research, from a computer science perspective: we investigate the various scientific and technological achievements that are needed to face the challenge, and discuss the current state of the art, the open problems and the expected road-map of research. Hence, this is a book for researchers: first of all for computer science researchers, from any sub-area of the field, and also for researchers from other disciplines (such as geography, statistics, social sciences, law, telecommunication and transportation engineering) who are willing to engage in a multi-disciplinary research area with potential for broad social and economic impact.

This book was made possible by the project GeoPKDD – *Geographic Privacy-Aware Knowledge Discovery and Delivery*[1] – funded by the European Commission under the Sixth Framework Programme, Information Society Technologies, Future Emerging Technologies (project number IST-6FP-014915, started in December 2005). GeoPKDD is a large research initiative, involving more than 40 researchers from eight institutions from seven countries and coordinated by the editors of this book. Its goal is precisely to explore the frontier of research described in this book, and to provide scientific results and practical evidence to demonstrate that it is possible to create useful mobility knowledge out of raw spatiotemporal data by means

---

[1] http://www.geopkdd.eu.

of privacy-preserving data mining techniques. We acknowledge the support of the European Commission, without which neither the project nor the book would have been possible, and we are grateful to the FET project officers Fabrizio Sestini and Paul Hearn for believing in our idea of producing a book in the early stage of the project.

This is a choral book: the community of GeoPKDD researchers cooperated tightly during the first year of the project to produce this book. The structure of the book was agreed upon, and each of the 13 chapters was developed by a team of researchers from at least two, often three, different institutions. The production of the chapters promoted a great many interactions, meetings and follow-ups; the writing of each of the chapters was coordinated by one or two responsible authors, whose names occur first in the author lists. Afterwards, a phase of internal review started, when cross-reviewing among the GeoPKDD researchers was finalised to harmonise content and terminology. Finally, an external round of review took place: each chapter was reviewed by two or three internationally renowned scientists.

We, as editors, are genuinely grateful to all contributors, who were enthusiastic about this book project despite the heavy burden we put on them – a clear sign that the GeoPKDD community is strong and growing. We owe special thanks to the chapter coordinators. Also, the book would not have been possible without the effort of the external reviewers, whom we gratefully acknowledge: Antonio Albano (University of Pisa), Krzysztof R. Apt (CWI, Amsterdam), Toon Calders (University of Antwerp), Christopher Clifton (Purdue University), Cosimo Comella (Italian Data Protection Commission), Elena Ferrari (University of Insubria, Como), Mark Gahegan (Penn State University), Stefano Giordano (University of Pisa), Dimitrios Gunopulos (University of California at Riverside), Ralf Hartmut Güting (University of Hagen), Donato Malerba (University of Bari), Nikos Mamoulis (University of Hong Kong), Yannis Manolopoulos (Aristotle University, Thessaloniki), Stan Matwin (University of Ottawa), Harvey J. Miller (University of Utah), Dimitris Papadias (Hong Kong University of Science and Technology), Christophe Rigotti (INSA, Lyon), Salvatore Ruggieri (University of Pisa), Marius Thériault (Université Laval), Robert Weibel (University of Zurich), Ouri Wolfson (University of Illinois at Chicago), Xiaobai Yao (University of Georgia) and Carlo Zaniolo (University of California at Los Angeles). Finally, we owe special thanks to our colleagues Mirco Nanni and Fabio Pinelli (ISTI-CNR, Pisa) for their help in editing the manuscript.

Pisa, Italy, *Fosca Giannotti*
August 2007 *Dino Pedreschi*

Contents

Mobility, Data Mining and Privacy: A Vision of Convergence 1
F. Giannotti and D. Pedreschi
    1   Mobility Data 2
    2   Data Mining 3
    3   Mobility Data Mining 4
    4   Privacy 8
    5   Purpose of This Book 9
    References 11

Part I Setting the Stage

1 Basic Concepts of Movement Data 15
N. Andrienko, G. Andrienko, N. Pelekis, and S. Spaccapietra
    1.1   Introduction 15
    1.2   Movement Data and Their Characteristics 18
    1.3   Analytical Questions 25
    1.4   Conclusion 38
    References 38

2 Characterising the Next Generation of Mobile Applications Through a Privacy-Aware Geographic Knowledge Discovery Process 39
M. Wachowicz, A. Ligtenberg, C. Renso, and S. Gürses
    2.1   Introduction 39
    2.2   The Privacy-Aware Geographic Knowledge Discovery Process 41
    2.3   The Geographic Knowledge Discovery Process 43
    2.4   Reframing a GKDD Process Using a Multi-tier Ontological Perspective 47
    2.5   The Multi-tier Ontological Framework 51
    2.6   Future Application Domains for a Privacy-Aware GKDD Process.. 60
    2.7   Conclusions 69
    References 70

3 Wireless Network Data Sources: Tracking and Synthesizing Trajectories 73
C. Renso, S. Puntoni, E. Frentzos, A. Mazzoni, B. Moelans, N. Pelekis, and F. Pini

- 3.1 Introduction 73
- 3.2 Categorization of Positioning Technologies 74
- 3.3 Mobile Location Systems 83
- 3.4 From Positioning to Tracking: Collecting User Movements 89
- 3.5 Synthetic Trajectory Generators 91
- 3.6 Conclusions and Open Issues 98
- References 99

4 Privacy Protection: Regulations and Technologies, Opportunities and Threats 101
D. Pedreschi, F. Bonchi, F. Turini, V.S. Verykios, M. Atzori, B. Malin, B. Moelans, and Y. Saygin

- 4.1 Introduction 101
- 4.2 Privacy Regulations 106
- 4.3 Privacy-Preserving Data Analysis 114
- 4.4 The Role of the Observatory 116
- 4.5 Conclusions 117
- References 118

Part II Managing Moving Object and Trajectory Data

5 Trajectory Data Models 123
J. Macedo, C. Vangenot, W. Othman, N. Pelekis, E. Frentzos, B. Kuijpers, I. Ntoutsi, S. Spaccapietra, and Y. Theodoridis

- 5.1 Introduction 123
- 5.2 Basic Concepts: From Raw Data to Trajectory 124
- 5.3 Modelling Approaches for Trajectories 129
- 5.4 Open Issues 141
- References 147

6 Trajectory Database Systems 151
E. Frentzos, N. Pelekis, I. Ntoutsi, and Y. Theodoridis

- 6.1 Introduction 151
- 6.2 Trajectory Database Engines 151
- 6.3 Trajectory Indexing 154
- 6.4 Trajectory Query Processing and Optimization 159
- 6.5 Dealing with Location Uncertainty 165
- 6.6 Handling Trajectory Compression 170
- 6.7 Open Issues: Roadmap 173
- 6.8 Concluding Remarks 183
- References 183

Contents

7 Towards Trajectory Data Warehouses 189
N. Pelekis, A. Raffaetà, M.-L. Damiani, C. Vangenot, G. Marketos,
E. Frentzos, I. Ntoutsi, and Y. Theodoridis
  7.1 Introduction 189
  7.2 Preliminaries and Related Work 191
  7.3 Requirements for Trajectory Data Warehouses 198
  7.4 Modelling and Uncertainty Issues 206
  7.5 Conclusions 209
  References 210

8 Privacy and Security in Spatiotemporal Data and Trajectories 213
V.S. Verykios, M.L. Damiani, and A. Gkoulalas-Divanis
  8.1 Introduction 213
  8.2 State of the Art 215
  8.3 Open Issues, Future Work, and Road Map 231
  8.4 Conclusion 238
  References 238

Part III Mining Spatiotemporal and Trajectory Data

9 Knowledge Discovery from Geographical Data 243
S. Rinzivillo, F. Turini, V. Bogorny, C. Körner, B. Kuijpers, and M. May
  9.1 Introduction 243
  9.2 Geographic Data Representation and Modelling 244
  9.3 Geographic Information Systems 246
  9.4 Spatial Feature Extraction 247
  9.5 Spatial Data Mining 253
  9.6 Example: Frequency Prediction of Inner-City Traffic 260
  9.7 Roadmap to Knowledge Discovery from Spatiotemporal Data ... 261
  9.8 Summary 263
  References 263

10 Spatiotemporal Data Mining 267
M. Nanni, B. Kuijpers, C. Körner, M. May, and D. Pedreschi
  10.1 Introduction 267
  10.2 Challenges for Spatiotemporal Data Mining 268
  10.3 Clustering 270
  10.4 Spatiotemporal Local Patterns 276
  10.5 Prediction 284
  10.6 The Role of Uncertainty in Spatiotemporal Data Mining ... 289
  10.7 Conclusion 289
  References 292

11 Privacy in Spatiotemporal Data Mining 297
F. Bonchi, Y. Saygin, V.S. Verykios, M. Atzori, A. Gkoulalas-Divanis,
S.V. Kaya, and E. Savaş
- 11.1 Introduction 297
- 11.2 Data Perturbation and Obfuscation 300
- 11.3 Knowledge Hiding 304
- 11.4 Distributed Privacy-Preserving Data Mining 312
- 11.5 Privacy-Aware Knowledge Sharing 320
- 11.6 Roadmap Toward Privacy-Aware Mining of Spatiotemporal Data .. 325
- 11.7 Conclusions 328
- References 329

12 Querying and Reasoning for Spatiotemporal Data Mining 335
G. Manco, M. Baglioni, F. Giannotti, B. Kuijpers, A. Raffaetà,
and C. Renso
- 12.1 Introduction 335
- 12.2 Elements of a Data Mining Query Language 337
- 12.3 DMQL Approaches in the Literature 342
- 12.4 Querying Spatiotemporal Data 358
- 12.5 Discussion 369
- 12.6 Conclusions 370
- References 371

13 Visual Analytics Methods for Movement Data 375
G. Andrienko, N. Andrienko, I. Kopanakis, A. Ligtenberg,
and S. Wrobel
- 13.1 Introduction 375
- 13.2 State of the Art 376
- 13.3 Patterns in Movement Data 383
- 13.4 Helping Users to Detect Patterns: A Roadmap 388
- 13.5 Visualization of Patterns 401
- 13.6 Conclusion 407
- References 408

Contributors

Gennady Andrienko
Fraunhofer Institut Intelligente Analyse- und Informationssysteme, Sankt Augustin, Germany, e-mail: gennady.andrienko@iais.fraunhofer.de

Natalia Andrienko
Fraunhofer Institut Intelligente Analyse- und Informationssysteme, Sankt Augustin, Germany, e-mail: natalia.andrienko@iais.fraunhofer.de

Maurizio Atzori
KDD Laboratory, ISTI-CNR, Pisa, Italy, e-mail: maurizio.atzori@isti.cnr.it

Miriam Baglioni
KDD Laboratory, Dipartimento di Informatica, Università di Pisa, Italy, e-mail: baglioni@di.unipi.it

Vania Bogorny
Theoretical Computer Science Group, Hasselt University and Transnational University of Limburg, Belgium, e-mail: vania.bogorny@uhasselt.be

Francesco Bonchi
KDD Laboratory, ISTI-CNR, Pisa, Italy, e-mail: francesco.bonchi@isti.cnr.it

Maria Luisa Damiani
Dipartimento di Informatica e Comunicazione, Università di Milano, Italy, e-mail: damiani@dico.unimi.it

Elias Frentzos
Computer Technology Institute (CTI) and Department of Informatics, University of Piraeus, Greece, e-mail: efrentzo@unipi.gr

Fosca Giannotti
KDD Laboratory, ISTI-CNR, Pisa, Italy, e-mail: fosca.giannotti@isti.cnr.it

Aris Gkoulalas-Divanis
Department of Computer and Communication Engineering, University of Thessaly,
Volos, Greece, e-mail: arisgd@inf.uth.gr

Seda Gürses
Institute of Information Systems, Humboldt University Berlin, Germany,
e-mail: seda@hu-berlin.de

Selim Volkan Kaya
Faculty of Engineering and Natural Sciences, Sabanci University, Istanbul, Turkey,
e-mail: selim.volkan@su.sabanciuniv.edu

Ioannis Kopanakis
Technological Educational Institute of Crete, Greece, e-mail: i.kopanakis@emark.teicrete.gr

Christine Körner
Fraunhofer Institut Intelligente Analyse- und Informationssysteme, Sankt Augustin, Germany, e-mail: christine.koerner@iais.fraunhofer.de

Bart Kuijpers
Theoretical Computer Science Group, Hasselt University and Transnational University of Limburg, Belgium, e-mail: bart.kuijpers@uhasselt.be

Arend Ligtenberg
Wageningen UR, Centre for GeoInformation, Netherlands,
e-mail: arend.ligtenberg@wur.nl

Jose Antonio Fernandes de Macedo
Database Laboratory, École Polytechnique Fédérale de Lausanne, Switzerland,
e-mail: jose.macedo@epfl.ch

Bradley Malin
Department of Biomedical Informatics, Vanderbilt University, Nashville, USA,
e-mail: b.malin@vanderbilt.edu

Giuseppe Manco
ICAR-CNR, Cosenza, Italy, e-mail: manco@icar.cnr.it

Gerasimos Marketos
Computer Technology Institute (CTI) and Department of Informatics, University of Piraeus, Greece, e-mail: marketos@unipi.gr

Michael May
Fraunhofer Institut Intelligente Analyse- und Informationssysteme, Sankt Augustin, Germany, e-mail: michael.may@iais.fraunhofer.de

Andrea Mazzoni
KDD Laboratory, ISTI-CNR, Pisa, Italy, e-mail: andrea.mazzoni@isti.cnr.it

Bart Moelans
Theoretical Computer Science Group, Hasselt University and Transnational University of Limburg, Belgium, e-mail: bart.moelans@uhasselt.be

Mirco Nanni
KDD Laboratory, ISTI-CNR, Pisa, Italy, e-mail: mirco.nanni@isti.cnr.it

Irene Ntoutsi
Computer Technology Institute (CTI) and Department of Informatics, University of Piraeus, Greece, e-mail: ntoutsi@unipi.gr

Walied Othman
Theoretical Computer Science Group, Hasselt University and Transnational University of Limburg, Belgium, e-mail: walied.othman@uhasselt.be

Dino Pedreschi
KDD Laboratory, Dipartimento di Informatica, Università di Pisa, Italy,
e-mail: pedre@di.unipi.it

Nikos Pelekis
Computer Technology Institute (CTI) and Department of Informatics, University of Piraeus, Greece, e-mail: npelekis@unipi.gr

Fabrizio Pini
Wind Telecomunicazioni, Rome, Italy and Department of Electronic Engineering, Università "Tor Vergata", Rome, Italy, e-mail: fabrizio.pini@mail.wind.it

Simone Puntoni
KDD Laboratory, ISTI-CNR, Pisa, Italy, e-mail: simone.puntoni@isti.cnr.it

Alessandra Raffaetà
Dipartimento di Informatica, Università Ca' Foscari di Venezia, Italy,
e-mail: raffaeta@dsi.unive.it

Chiara Renso
KDD Laboratory, ISTI-CNR, Pisa, Italy, e-mail: chiara.renso@isti.cnr.it

Salvatore Rinzivillo
KDD Laboratory, Dipartimento di Informatica, Università di Pisa, Italy,
e-mail: rinziv@di.unipi.it

Erkay Savaş
Faculty of Engineering and Natural Sciences, Sabanci University, Istanbul, Turkey,
e-mail: erkays@sabanciuniv.edu

Yücel Saygin
Faculty of Engineering and Natural Sciences, Sabanci University, Istanbul, Turkey,
e-mail: ysaygin@sabanciuniv.edu

Stefano Spaccapietra
Database Laboratory, École Polytechnique Fédérale de Lausanne, Switzerland,
e-mail: stefano.spaccapietra@epfl.ch

Yannis Theodoridis
Computer Technology Institute (CTI) and Department of Informatics, University of Piraeus, Greece, e-mail: ytheod@unipi.gr

Franco Turini
KDD Laboratory, Dipartimento di Informatica, Università di Pisa, Italy,
e-mail: turini@di.unipi.it

Christelle Vangenot
Database Laboratory, École Polytechnique Fédérale de Lausanne, Switzerland,
e-mail: christelle.vangenot@epfl.ch

Vassilios S. Verykios
Department of Computer and Communication Engineering, University of Thessaly, Volos, Greece, e-mail: verykios@inf.uth.gr

Monica Wachowicz
Wageningen UR, Centre for GeoInformation, Netherlands,
e-mail: monica.wachowicz@wur.nl

Stefan Wrobel
Fraunhofer Institut Intelligente Analyse- und Informationssysteme, Sankt Augustin, Germany, e-mail: stefan.wrobel@iais.fraunhofer.de

Mobility, Data Mining and Privacy: A Vision of Convergence

F. Giannotti and D. Pedreschi

The comprehension of phenomena related to movement – not only of people and vehicles but also of animals and other moving objects – has always been a key issue in many areas of scientific investigation or social analysis. The human geographer, for instance, studies the flows of migrant populations with reference to geography – places that are sources and destinations of migrations – and time. The historian, another example, studies military campaigns and related movements of armies and populations. (A famous instance is the depiction of Napoleon's March on Moscow, published by C.J. Minard in 1861, discussed in Chap. 1 of this book (see Fig. 1.1); this figure represents with eloquence the fate of Napoleon's army in the Russian campaign of 1812–1813, by showing the movement of the army together with its dramatically diminishing size during its advance and subsequent retreat.) The ethologist studies animal behaviour by the analysis of movement patterns, based on field observations or, sometimes, on data from tracking devices.

Today, in the extremely complex social systems of the gigantic metropolitan areas of the twenty-first century, the observation of the movement patterns and behavioural models of people is needed for the traffic engineers and city managers to reason about mobility and its sustainability and to support decision makers with trustable knowledge. The very same knowledge about people movement and behaviour is precious for the urban planner, e.g. to localise new services, to organise logistics systems and for the timely detection of changes that occur in the movement behaviour. At a finer-grained spatial scale, movement in contexts such as a shopping area or a natural park is an interesting subject of investigation, either for commercial purposes, as in geo-marketing, or for improving the quality of service.

In all the above cases, albeit so different from each other, two key problems recur:

- First, how to *collect mobility data* about extremely complex, often chaotic, social or natural systems made of large populations of moving entities.

F. Giannotti
KDD Laboratory, ISTI-CNR, Pisa, Italy, e-mail: fosca.giannotti@isti.cnr.it

- Second, how to turn this data into *mobility knowledge*, i.e. into useful models and patterns that abstract away from the individual and shed light on collective movement behaviour, pertaining to groups of individuals that it is worth putting into evidence.

In other words, by the observation of (many) individual movements – of a migrant, of one of Napoleon's soldiers, of an animal, of a commuting worker in a city, of a tourist in a park – we aim at understanding the general movement patterns or models – a migratory flow, an army's path, a frequently followed trajectory in the savannah, on the urban street network or in a park – that suddenly become usable knowledge, which makes the original system easier to understand by revealing some of its motion laws, hidden in the chaos. Simple and useful mobility knowledge is learned from complex systems of moving entities.

If this has been a long-time dream, never fully realised in practice, a chance to get closer to the dream is offered, today, by the convergence of two factors:

- The *mobility data* made available by the wireless and mobile communication technologies
- *Data mining* – the methods for extracting models and patterns from (large) volumes of data

1 Mobility Data

Our everyday actions, the way people live and move, leave digital traces in the information systems of the organisations that provide services through the wireless networks for mobile communication. The potential value of these traces in recording the human activities in a territory is becoming real, because of the increasing pervasiveness and positioning accuracy. The number of mobile phone users worldwide was estimated as 1.5 billion in 2005, with regions, such as Italy, where the number of mobile phones is exceeding the number of inhabitants; in other regions, especially developing countries, the numbers are still increasing at a high speed. On the other hand, the location technologies, such as GSM and UMTS, currently used by wireless phone operators are capable of providing an increasingly better estimate of a user's location, while the integration of various positioning technologies proceeds: GPS-equipped mobile devices can transmit their trajectories to some service provider (and the European satellite positioning system Galileo may improve precision and pervasiveness in the near future), Wi-Fi and Bluetooth devices may be a source of data for indoor positioning, Wi-Max can become an alternative for outdoor positioning, and so on.

The consequence of this scenario, where communication and computing devices are ubiquitous and carried everywhere and always by people and vehicles, is that human activity in a territory may be *sensed* – not necessarily on purpose, but simply as a side effect of the ubiquitous services provided to mobile users. Thus, the wireless phone network, designed to provide mobile communication, can also be viewed as an infrastructure to gather mobility data, if used to record the location of its users at different times. The wireless networks, whose pervasiveness and localisation precision increase while new location-based and context-based services are offered to mobile users, are becoming the *nerves* of our territory – in particular, our towns – capable of sensing and, possibly, recording our movements.

From this perspective, we have today a chance of collecting and storing mobility data of unprecedented quantity, quality and timeliness at a very low cost: in principle, a dream for traffic engineers and urban planners, compelled until yesterday to gather data of limited size and precision only through highly expensive means such as field experiments, surveys to discover travelling habits of commuting workers and ad hoc sensors placed on streets.

However, there's a long way to go from mobility data to mobility knowledge. In the words of J.H. Poincaré, 'Science is built up with facts, as a house is with stones. But a collection of facts is no more a science than a heap of stones is a house.' Since databases became a mature technology and massive collection and storage of data became feasible at increasingly cheaper costs, a push emerged towards powerful methods for discovering knowledge from those data, capable of going beyond the limitations of traditional statistics, machine learning and database querying. This is what *data mining* is about.

2 Data Mining

Data mining is the process of automatically discovering useful information in large data repositories. Often, traditional data analysis tools and techniques cannot be used because of the massive volume of data gathered by automated collection tools, such as point-of-sale data, Web logs from e-commerce portals, earth observation data from satellites, genomic data. Sometimes, the non-traditional nature of the data implies that ordinary data analysis techniques are not applicable.

The three most popular data mining techniques are predictive modelling, cluster analysis and association analysis.

- In *predictive modelling*, the goal is to develop *classification models*, capable of predicting the value of a class label (or target variable) as a function of other variables (explanatory variables); the model is learnt from historical observations, where the class label of each sample is known: once constructed, a classification model is used to predict the class label of new samples whose class is unknown, as in forecasting whether a patient has a given disease based on the results of medical tests.
- In *association analysis*, also called *pattern discovery*, the goal is precisely to discover patterns that describe strong correlations among features in the data or associations among features that occur frequently in the data. Often, the discovered patterns are presented in the form of association rules: useful applications of association analysis include market basket analysis, i.e. the task of finding items

that are frequently purchased together, based on point-of-sale data collected at cash registers.
- In *cluster analysis*, the goal is to partition a data set into groups of closely related data in such a way that the observations belonging to the same group, or cluster, are similar to each other, while the observations belonging to different clusters are not. Clustering can be used, for instance, to find segments of customers with a similar purchasing behaviour or categories of documents pertaining to related topics.

Data mining is a step of *knowledge discovery in databases*, the so-called KDD process for converting raw data into useful knowledge. The KDD process consists of a series of transformation steps:

- Data *preprocessing*, which transforms the raw source data into an appropriate form for the subsequent analysis
- Actual *data mining*, which transforms the prepared data into patterns or models: classification models, clustering models, association patterns, etc.
- *Postprocessing* of data mining results, which assesses validity and usefulness of the extracted patterns and models, and presents interesting knowledge to the final users – business analysts, scientists, planners, etc. – by using appropriate visual metaphors or integrating knowledge into decision support systems

Today, data mining is both a technology that blends data analysis methods with sophisticated algorithms for processing large data sets, and an active research field that aims at developing new data analysis methods for novel forms of data. On one side, classification, clustering and pattern discovery tools are now part of mature data analysis systems and have been successfully applied to problems in various commercial and scientific domains. On the other side, the increasing heterogeneity and complexity of new forms of data – such as those arriving from medicine, biology, the Web, the Earth observation systems – call for new forms of patterns and models, together with new algorithms to discover such patterns and models efficiently. One of the frontiers of data mining research, today, is precisely represented by spatiotemporal data, i.e., observations of events that occur in a given place at a certain time, such as the mobility data arriving from wireless networks. Here, the challenge is particularly tough: which data mining tools are needed to master the complex dynamics of people in motion and construct concise and useful abstractions out of large volumes of mobility data is, by large, an unanswered question. Good news, hence, for researchers willing to engage in a highly interdisciplinary, highly risky and highly promising area, with a large potential impact on socially and economically relevant problems.

3 Mobility Data Mining

Mobility data mining is, therefore, emerging as a novel area of research, aimed at the analysis of mobility data by means of appropriate patterns and models extracted by efficient algorithms; it also aims at creating a novel knowledge discovery process explicitly tailored to the analysis of mobility with reference to *geography*, at appropriate scales and granularity. In fact, movement always occurs in a given physical space, whose key semantic features are usually represented by geographical maps; as a consequence, the geographical background knowledge about a territory is always essential in understanding and analysing mobility in such territory. Mobility data mining, therefore, is situated in a Geographic Knowledge Discovery process – a term first introduced by Han and Miller in [2] – capable of sustaining the entire chain of production from raw mobility data up to usable knowledge capable of supporting decision making in real applications.

As a prototypical example, assume that source data are positioning logs from mobile cellular phones, reporting user's locations with reference to the cells in the GSM network; these mobility data come as streams of raw log entries recording users entering a cell – *(userID, time, cellID, in)* – users exiting a cell – *(userID, time, cellID, out)* – or, in the near future, user's position within a cell – *(userID, time, cellID, X, Y)* and, in the case of GPS/Galileo equipped devices, user's absolute position. Indeed, each time a mobile phone is used on a given network, the phone company records real-time data about it, including time and cell location. If a call is taking place, the recording data-rate may be higher. Note that if the caller is moving, the call transfers seamlessly from one cell to the next. In this context, a novel geographic knowledge discovery process may be envisaged, composed of three main steps: *trajectories reconstruction*, *knowledge extraction* and *delivery* of the information obtained, described in the following.

(1) *Trajectory reconstruction.* In this basic phase, the stream of raw mobility data has to be processed to obtain trajectories of individual moving objects; the resulting trajectories should be stored into appropriate repositories, such as a trajectory database or data warehouse.

Reconstruction of trajectories is per se a challenging problem. The reconstruction accuracy of trajectories, as well as their level of spatiotemporal granularity, depend on the quality of the log entries, since the precision of the position may range from the granularity of a cell of varying size to the relative (approximated) position within a cell.

Indeed, each moving object trajectory is typically represented as a set of localisation points of the tracked device, called *sampling*. This representation has intrinsic imperfection mainly due to two aspects. The first source of imperfection is the measurement error of the tracking device. For example, a GPS-enabled device introduces a measurement error of a few metres, whereas the imprecision introduced in a GSM/UMTS network is the dimension of a cell, which could be from less than hundred metres in urban settings to a few kilometres in rural areas. The second source of imperfection is related to the sampling rate and involves the trajectory reconstruction process that approximates the movement of the objects between two localisation points. Although some simple approximated reconstruction techniques are sometimes applicable, more sophisticated reconstruction of trajectories from raw mobility data is to be investigated, to take into account the spatial, and possibly temporal, imperfection in the reconstruction process.

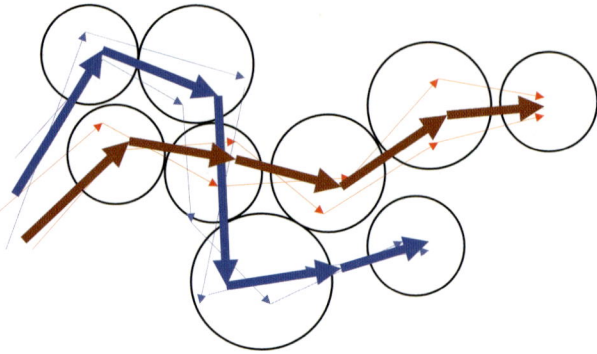

Fig. 1 Trajectory clustering

The management and querying of large volumes of mobility data and reconstructed trajectories also poses specific problems, which are only partly solved by currently available technology, such as moving object databases.

(2) *Knowledge extraction.* Spatiotemporal data mining methods are needed to extract useful patterns out of trajectories. However, spatiotemporal data mining is still in its infancy, and even the most basic questions in this field are still largely unanswered: What kinds of patterns can be extracted from trajectories? Which methods and algorithms should be applied to extract them? The following basic examples give a glimpse of the wide variety of patterns and possible applications it is expected to manage[1]:

- *Clustering*, the discovery of groups of 'similar' trajectories, together with a summary of each group (see Fig. 1). Knowing which are the main routes (represented by clusters) followed by people or vehicles during the day can represent precious information for mobility analysis. For example, trajectory clusters may highlight the presence of important routes not adequately covered by the public transportation service.
- *Frequent patterns*, the discovery of frequently followed (sub)paths (Fig. 2). Such information can be useful in urban planning, e.g. by spotlighting frequently followed inefficient vehicle paths, which can be the result of a mistake in the road planning.
- *Classification*, the discovery of behaviour rules, aimed at explaining the behaviour of current users and predicting that of future ones (Fig. 3). Urban traffic simulations are a straightforward example of application for this kind of knowledge, since a classification model can represent a sophisticated alternative to the simple ad hoc behaviour rules, provided by domain experts, on which actual simulators are based.

---

[1] In the figures, circles represent cells in the wireless network.

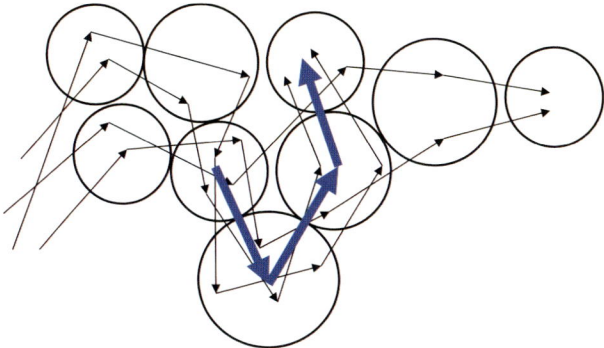

Fig. 2 Trajectory patterns

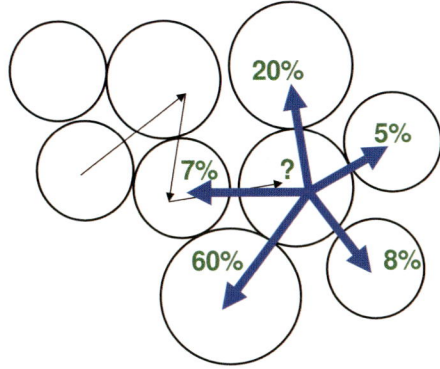

Fig. 3 Trajectory prediction

(3) *Knowledge delivery.* Extracted patterns are very seldom geographic knowledge *prêt-à-porter*: It is necessary to reason on patterns and on pertinent background knowledge, evaluate patterns' interestingness, refer them to geographic information and find out appropriate presentations and visualisations. Once suitable methods for interpreting and delivering geographic knowledge on trajectories are available, several application scenarios become possible. The paradigmatic example is sustainable mobility, namely how to support and improve decision making in mobility-related issues, such as

- Planning traffic and public mobility systems in metropolitan areas
- Planning physical communication networks, such as new roads or railways
- Localising new services in our towns
- Forecasting traffic-related phenomena
- Organising postal and logistics systems
- Timely detecting problems that emerge from the movement behaviour
- Timely detecting changes that occur in the movement behaviour

4 Privacy

Today we are faced with the concrete possibility of pursuing an *archaeology of the present*: discovering from the digital traces of our mobile activity the knowledge that makes us comprehend timely and precisely the way we live, the way we use our time and our land today.

Thus, it is becoming possible, in principle, to understand how to live better by learning from our recent history, i.e. from the traces left behind us yesterday, or a few moments ago, recorded in the information systems and analysed to produce usable, timely and reliable knowledge. In simple words, we advocate that mobility data mining, defined as the collection and extraction of knowledge from mobility data, is the opportunity to construct novel services of great societal and economic impact.

However, there is a little path from opportunities to threats: We are aware that, on the basis of this scenario, there lies a flaw of potentially dramatic impact, namely the fact that the donors of the mobility data are the citizens, and making these data publicly available for the mentioned purposes would put at risk our own privacy, our natural right to keep secret the places we visit, the places we live or work at and the people we meet – all in all, the way we live as individuals. In other words, the personal mobility data, as gathered by the wireless networks, are extremely sensitive information; their disclosure may represent a brutal violation of the privacy protection rights, established in increasingly more laws and regulations internationally.

A genuine positivist researcher, with an unlimited trust in science and progress, may observe that, for the mobility-related analytical purposes, knowing the exact identity of individuals is not needed: anonymous data are enough to reconstruct aggregate movement behaviour, pertaining to whole groups of people, not to individual persons. This line of reasoning is also coherent with existing data protection regulations, such as that of the European Union, which states that personal data, once made anonymous, are not subject any longer to the restrictions of the privacy law. Unfortunately, this is not so easy: the problem is that anonymity means making reasonably impossible the re-identification, i.e. the linkage between the personal data of an individual and the identity of the individual itself. Therefore, transforming the data in such a way to guarantee anonymity is hard: as some realistic examples show, supposedly anonymous data sets can leave unexpected doors open to malicious re-identification attacks. Chapter 4 discusses such examples in different domains such as medical patient data, Web search logs and location and trajectory data; moreover, other possible breaches for privacy violation may be left open by the publication of the mining results, even in the case that the source data are kept secret by a trusted data custodian.

The bottom-line of this discussion is that protecting privacy when disclosing mobility knowledge is a non-trivial problem that, besides socially relevant, is scientifically attractive. As often happens in science, the problem is to find an optimal trade-off between two conflicting goals: from one side, we would like to have precise, fine-grained knowledge about mobility, which is useful for the analytic purposes; from the other side, we would like to have imprecise, coarse-grained knowledge about mobility, which puts us in repair from the attacks to our privacy. It is interesting that the same conflict – essentially between opportunities and risks – can be read either as a mathematical problem or as a social (or ethical or legal) challenge. Indeed, the privacy issues related to the ICTs can only be addressed through an alliance of technology, legal regulations and social norms. In the meanwhile, increasingly sophisticated privacy-preserving techniques are being studied. Their aim is to achieve appropriate levels of anonymity by means of controlled transformation of data and/or patterns – limited distortion that avoids the undesired side effect on privacy while preserving the possibility of discovering useful knowledge. A fascinating array of problems thus emerged, from the point of view of computer scientists and mathematicians, which already stimulated the production of important ideas and tools. Hopefully, in the near future, it will be possible to reach a win–win situation: obtaining the advantages of collective mobility knowledge without divulging inadvertently any individual mobility knowledge. These results, if achieved, may have an impact on laws and jurisprudence, as well as on the social acceptance and dissemination of ubiquitous technologies.

5 Purpose of this Book

Mobility, data mining and privacy: There is a new multi-disciplinary research frontier that is emerging at the crossroads of these three subjects, with plenty of challenging scientific problems to be solved and vast potential impact on real-life problems. This is the conviction that brought us to create a large European project called GeoPKDD – *Geographic Privacy-aware Knowledge Discovery and Delivery* [1] – that, since December 2005, is exploring this frontier of research. The same conviction is the basis of this book, produced by the community of researchers of the GeoPKDD project, which is thoroughly aimed at substantiating the vision advocated above.

The approach that we followed in undertaking this task is twofold: first, in Part I of the book, we set up the stage and make the vision more concrete, by discussing which elements of the three subjects are involved in the convergence: mobility (Which data come from the wireless networks?), data mining (in which classes of applications can be addressed with a geographic knowledge discovery process) and privacy (Which is the interplay between the privacy-preserving technologies and the data protection laws?). Second, in the subsequent parts of the book, we identify the scientific and technological ingredients that, from a computer science perspective, are needed to support a geographic knowledge discovery process; for each such ingredient we discuss the current state of the art and the roadmap of research that we expect.

More precisely, the book is organised as follows.

In Part I (*Setting the stage*), Chap. 1 introduces the basic notions related to the movement of objects and the data that describe the movement; Chap. 2 characterises the next generation of mobility-related applications through a privacy-aware geographic knowledge discovery process; Chap. 3 discusses tracking of mobility data and trajectories from wireless networks and Chap. 4 discusses privacy protection regulations and technologies, together with related opportunities and threats.

In Part II (*Managing moving object and trajectory data*), Chap. 5 discusses data modelling for moving objects and trajectories; Chap. 6 deals with trajectory database management issues and physical aspects of trajectory database systems, such as indexing and query processing; Chap. 7 discusses the first steps towards a trajectory data warehouse providing online analytical tools for trajectory data and Chap. 8 discusses the location privacy problem in spatiotemporal and trajectory data, also taking into account security.

In Part III (*Mining spatiotemporal and trajectory data*), Chap. 9 discusses the knowledge discovery and data mining techniques applied to geographical data, i.e. data referenced to geographic information; Chap. 10 deals with spatiotemporal data mining, i.e. knowledge discovery from mobility data, where the space and time dimensions are inextricably intertwined; Chap. 11 discusses the privacy-preserving methods (and problems) in data mining, with a particular focus on the specific privacy and anonymity issues arising in spatiotemporal data mining; Chap. 12 discusses the quest towards a language framework, capable of supporting the user in specifying and refining mining objectives, combining multiple strategies and defining the quality of the extracted knowledge, in the specific context of movement data and Chap. 13 considers the use of interactive visual techniques for detection of various patterns and relationships in movement data.

This is more a book of questions, rather than a book of answers. It is clearly devoted to shape up a research area, and therefore targeted at researchers that are looking for challenging open problems in an exciting interdisciplinary subject. This is why we tried to speak, as far as possible, a language comprehensible to researchers coming from various subareas of computer science, including databases, data mining, machine learning, algorithms, data modelling, visualisation and geographic information systems. But, more ambitiously, we also tried to speak to researchers from the other disciplines that are needed to fully realise the vision: geography, statistics, social sciences, law, telecommunication engineering and transportation engineering. We believe that at least the material in Part I, and also most of the remaining chapters, can reach the attention of researchers who are interested in the inter-disciplinary dialogue, and perceive the interplay among mobility, the information and communication technologies and privacy as a potential ground for such a dialogue. Most of, if not all, open challenges of the contemporary society are intrinsically multi-disciplinary, and require solutions – hence research – that cross the boundaries of traditional disciplines: we like to think that this book is a little step in this direction.

References

1. GeoPKDD.eu – Geographic Privacy-aware Knowledge Discovery and Delivery. http://www.geopkdd.eu/.
2. H.J. Miller and J. Han (eds). *Geographic Data Mining and Knowledge Discovery*. Taylor & Francis, 2001.

Part I
Setting the Stage

Chapter 1
Basic Concepts of Movement Data

N. Andrienko, G. Andrienko, N. Pelekis, and S. Spaccapietra

1.1 Introduction

From ancient days, people have observed various moving entities, from insects and fishes to planets and stars, and investigated their movement behaviours. Although methods that were used in earlier times for observation, measurement, recording, and analysis of movements are very different from modern technologies, there is still much to learn from past studies. First, this is the thorough attention paid to the multiple aspects of movement. These include not only the trajectory (path) in space, characteristics of motion itself such as speed and direction, and their dynamics over time but also characteristics and activities of the entities that move. Second, this is the striving to relate movements to properties of their surroundings and to various phenomena and events.

As an illustration, let us take the famous depiction of Napoleon's March on Moscow, published by Charles Joseph Minard in 1861 (this representation is reproduced in Fig. 1.1; a detailed description can be found in Tufte [15]). The author engages the readers in the exploration of the fate of Napoleon's army in the Russian campaign of 1812–1813. Beginning at the Polish–Russian border, the thick band shows the size of the army at each position. The path of Napoleon's retreat from Moscow in the cold winter is depicted by the dark lower band, which is tied to temperature and timescales. Tufte [15] identified six separate variables that were shown within Minard's drawing. First, the line width continuously marked the size of the army. Second and third, the line itself showed the latitude and longitude of the army as it moved. Fourth, the lines themselves showed the direction that the army was travelling, both in advance and retreat. Fifth, the location of the army with respect to certain dates was marked. Finally, the temperature along the path of retreat was displayed. It can also be noted that, despite the schematic character of the drawing

---

N. Andrienko
Fraunhofer Institut Intelligente Analyse- und Informationssysteme, Sankt Augustin, Germany,
e-mail: natalia.andrienko@iais.fraunhofer.de

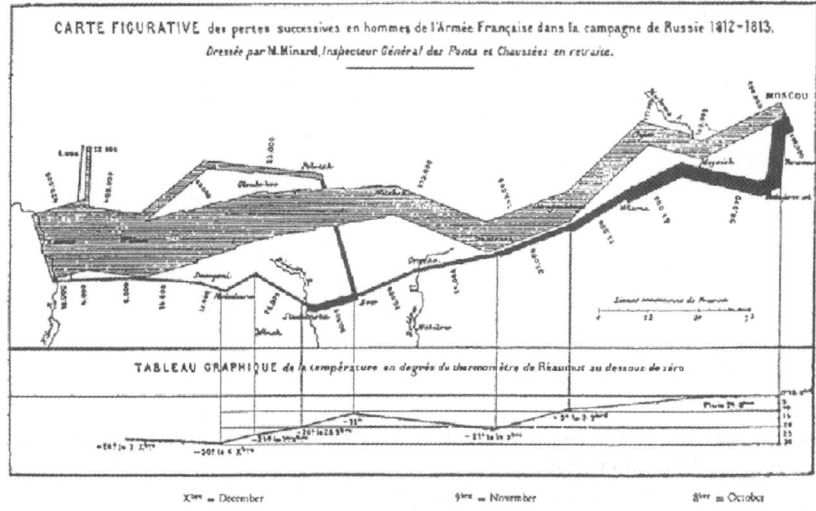

Fig. 1.1 Representation of Napoleon's Russian campaign of 1812, produced by Charles Joseph Minard in 1861

with its rudimentary cartography, Minard depicted some features of the underlying territory (specifically, rivers and towns) he deemed essential for the understanding of the story.

Since the environment in which movements take place and the characteristics of the moving entities may have significant influence on the movements, they need to be considered when the movements are studied. Moreover, movements themselves are not always the main focus of a study. One may analyse movements with the aim to gain knowledge about the entities that move or about the environment of the movements. Thus, in the research area known as time geography, the observation of everyday movements of human individuals was primarily the means of studying activities of different categories of people. On an aggregate level, time geography looks for trends in society.

The ideas of time geography originate from Hagerstrand [5]. A prominent feature of time geography is the view of space and time as inseparable. Hagerstrand's basic idea was to consider space–time paths in a three-dimensional space where horizontal axes represent geographic space and the vertical axis represents time. This representation is known as space–time cube. The idea is illustrated in Fig. 1.2 (left). The line represents the movements of some entity, for example, a working person, who initially was at home, then travelled to his workplace and stayed there for a while, then moved to a supermarket for shopping and, having spent some time there, returned home. Vertical lines stand for stays at a certain location (home, workplace, or supermarket). The workplace is an example of a station, i.e. a place where people meet for a certain activity. The sloped line segments indicate movements. The slower the movement, the steeper will be the line. The straightness of the lines in our drawing assumes that the person travels with constant speed, which is usually

1 Basic Concepts of Movement Data

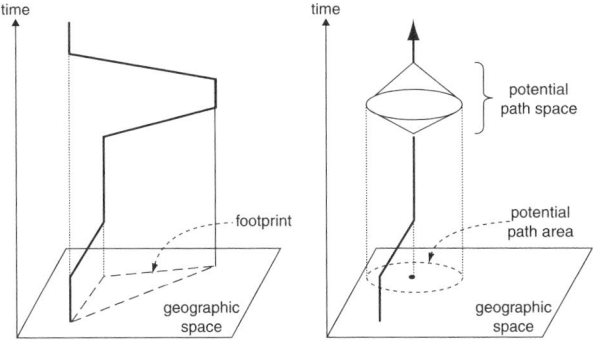

Fig. 1.2 An illustration of the notions of space–time path and space–time prism

just an approximation of the real behaviour. The space–time path can be projected on a map, resulting in the path's footprint.

Another important concept of time geography is the notion of space–time prism, which is schematically illustrated in Fig. 1.2(right). In the three-dimensional representation, this is the volume in space and time a person can reach in a particular time interval starting and returning to the same location (for instance, where a person can get from his workplace during lunch break). The widest extent is called the potential path space and its footprint is called potential path area. In Fig. 1.2(right), it is represented by a circle, assuming it to be possible to reach every location within the circle. In reality, the physical environment will not always allow this. In general, the space–time paths of individuals are influenced by constraints. One can distinguish between capability constraints (for instance, mode of transport and need for sleep), coupling constraints (for instance, being at work or at the sports club) and authority constraints (for instance, accessibility of buildings or parks in space and time).

In the era of pre-computer graphics, it was time consuming and expensive to produce space–time cube visualisations to support the exploration of movement behaviours. However, with the rise of new visualisation technology and interactivity, researchers revisited this concept [7, 13]. Moreover, modern time geography is not entirely based on visual representations and qualitative descriptions. Thus, Miller [10] suggests a measurement theory for its basic entities and relationships, which includes formal definitions of the space–time path, space–time prism, space–time stations as well as fundamental relationships between space–time paths and prisms. This provides foundations for building computational tools for time geographic querying and analysis.

Whatever tools and technologies have been used for the collection, representation, exploration and analysis of movement data, the underlying basic concepts related to the very nature of movement in (geographical) space remain stable and the characteristics of movement examined in past studies do not lose their relevance. In Sect. 1.2, we present a synthesis from existing literature concerning the basic concepts and characteristics of movement. Movement occurs in space and in time, so we discuss the possible ways of spatial and temporal referencing and relevant properties of space and time. We also briefly mention other matters that may have an impact on movement and therefore need attention in analysis. These include properties and activities of moving entities and various space and/or time-related phenomena and events.

Data analysis is seeking answers to various questions about data. In Sect. 1.3, we define the types of questions that can arise in analysis of movement data. For the question types to be independent of any analysis methods and tools, we define them on the basis of an abstract model of movement data, which involves three fundamental components: population of entities, time and space. We distinguish between elementary questions, which refer to individual data items, and synoptic questions, which refer to the data as a whole or to data subsets considered in their entirety. Synoptic questions play the primary role in data analysis. At the end, we relate the tool-independent taxonomy of analytical questions to the established typology of data mining tasks.

1.2 Movement Data and Their Characteristics

This section presents a synthesis from the current literature talking about movement and movement data: what is movement? How can movement be reflected in data? How can movement be characterised? What does it depend on?

1.2.1 Trajectories

A strict definition of *movement* relates this notion to change in the physical position of an entity with respect to some reference system within which one can assess positions. Most frequently, the reference system is geographical space.

A *trajectory* is the path made by the moving entity through the space where it moves. The path is never made instantly but requires a certain amount of time. Therefore, time is an inseparable aspect of a trajectory. This is emphasised in the term 'space–time path' [5, 10, 11], one of the synonyms for 'trajectory'. Another well-known term, 'geospatial lifeline' introduced by Hornsby and Egenhofer [6], also refers to time although less explicitly (through the notion of 'life').

If t_0 is the time moment when the path started and t_{end} is the moment when it ended, for any moment t_i between t_0 and t_{end}, there is a position in space that was occupied by the entity at this moment (although in practice this position is not always known). Hence, a trajectory can be viewed as a function that matches time moments with positions in space. It can also be seen as consisting of pairs (time, location). Since time is continuous, there are an infinite number of such pairs in a trajectory. For practical reasons, however, trajectories have to be represented by finite sequences of time-referenced locations. Such sequences may result from various ways that are used to observe movements and collect movement data:

- Time-based recording: positions of entities are recorded at regularly spaced time moments, e.g. every 5 min

- Change-based recording: a record is made when the position of an entity differs from the previous one
- Location-based recording: records are made when an entity comes close to specific locations, e.g. where sensors are installed
- Event-based recording: positions and times are recorded when certain events occur, in particular, activities performed by the moving entity (e.g. calling by a mobile phone)
- Various combinations of these basic approaches

Typically, positions are measured with uncertainty. Sometimes it is possible to refine the positions by taking into account physical constraints, e.g. the street network.

In studying movements, an analyst attends to a number of characteristics, which can be grouped depending on whether they refer to states at individual moments or to movements over time intervals. *Moment-related characteristics* include the following:

- Time, i.e. position of this moment on the timescale
- Position of the entity in space
- Direction of the entity's movement
- Speed of the movement (which is zero when the entity stays in the same place)
- Change of the direction (turn)
- Change of the speed (acceleration)
- Accumulated travel time and distance

Overall characteristics of a trajectory as a whole or a trajectory fragment made during a subinterval $[t_1, t_2]$ of the entire time span $[t_0, t_{end}]$ include the following:

- Geometric shape of the trajectory (fragment) in the space
- Travelled distance, i.e. the length of the trajectory (fragment) in space
- Duration of the trajectory (fragment) in time
- Movement vector (i.e. from the initial to the final position) or major direction
- Mean, median and maximal speed
- Dynamics (behaviour) of the speed
    - Periods of constant speed, acceleration, deceleration and stillness
    - Characteristics of these periods: start and end times, duration, initial and final positions, initial and final speeds, etc.
    - Arrangement (order) of these periods in time
- Dynamics (behaviour) of the directions
    - Periods of straight, curvilinear, circular movement
    - Characteristics of these periods: start and end times, initial and final positions and directions, major direction, angles and radii of the curves, etc.
    - Major turns ('turning points') with their characteristics: time, position, angle, initial and final directions, and speed of the movement in the moment of the turn
    - Arrangement (order) of the periods and turning points in time

Besides examining a single trajectory, an analyst is typically interested in *comparison* of two or more trajectories. These may be trajectories of different entities (e.g. different persons), trajectories of the same entity made at different times (e.g. trajectories of a person on different days) or different fragments of the same trajectory (e.g. trajectories of a person on the way from home to the workplace and on the way back). Generally, the goal of comparison is to establish *relations* between the objects that are compared. Here are some examples of possible relations:

- Equality or inequality
- Order (less or greater, earlier or later, etc.)
- Distance (in space, in time or on any numeric scale)
- Topological relations (inclusion, overlapping, crossing, touching, etc.)

Many other types of relations may be of interest, depending on the nature of the things being compared. In comparing trajectories, analysts are most often interested in establishing the following types of relations:

- Similarity or difference of the overall characteristics of the trajectories, which have been listed above (shapes, travelled distances, durations, dynamics of speed and directions and so on)
- Spatial and temporal relations
    - Co-location in space, full or partial (i.e. the trajectories consist of the same positions or have some positions in common)
        (a) Ordered co-location: the common positions are attained in the same order
        (b) Unordered co-location: the common positions are attained in different orders
    - Co-existence in time, full or partial (i.e. the trajectories are made during the same time period or the periods overlap)
    - Co-incidence in space and time, full or partial (i.e. same positions are attained at the same time)
    - Lagged co-incidence, i.e. entity e_1 attains the same positions as entity e_0 but after a time delay Δt
    - Distances in space and in time

Most researchers dealing with movement data agree in recognising the necessity to consider not only trajectories with their spatial and temporal characteristics but also the structure and properties of the space and time where the movement takes place as having a great impact upon the movement behaviour. The concepts and characteristics related to space and time are briefly discussed below.

1.2.2 Space

Space can be seen as a set consisting of *locations* or *places*. An important property of space is the existence of distances between its elements. At the same time, space has no natural origin and no natural ordering between the elements. Therefore, in order to distinguish positions in space, one needs to introduce in it some reference system, for example, a system of coordinates. While this may be done, in principle, quite arbitrarily, there are some established reference systems such as geographical coordinates.

Depending on the practical needs, one can treat space as two dimensional (i.e. each position is defined by a pair of coordinates) or as three dimensional (each position is defined by three coordinates). In specific cases, space can be viewed as one dimensional. For example, when movement along a standard route is analysed, one can define positions through the distances from the beginning of the route, i.e. a single coordinate is sufficient.

Theoretically, one can also deal with spaces having more than three dimensions. Such spaces are abstract rather than physical; however, movements of entities in abstract spaces may also be subject to analysis. Thus, Laube et al. [8] explore the movement (evolution) of the districts of Switzerland in the abstract space of politics and ideology involving three dimensions: left vs. right, liberal vs. conservative and ecological vs. technocratic.

The physical space is continuous, which means that it consists of an infinite number of locations and, moreover, for any two different locations there are locations 'in between', i.e. at smaller distances to each of the two locations than the distance between the two locations. However, it may also be useful to treat space as a discrete or even finite set of locations. For example, in studying the movement of tourists over a country or a city, one can 'reduce' space to the set of points of interest visited by the tourists. Space discretisation may be even indispensable, in particular, when positions of entities cannot be precisely measured and specified in terms of areas such as cells of a mobile-phone network, city districts, or countries.

The above-cited examples show that space may be *structured*, in particular, divided into areas. The division may be hierarchical; for instance, a country is divided into provinces, the provinces into municipalities and the municipalities into districts. Areas can also be derived from a geometric decomposition (e.g. 1 km^2 cells), with no semantics associated to the decomposition. A street (road) network is another common way of structuring physical space.

Like coordinate systems, space structuring also provides a reference system, which may be used for distinguishing positions, for instance, by referring to streets or road fragments and relative positions on them (house numbers or distances from the ends). The possible ways of specifying positions in space can be summarised as follows:

- Coordinate-based referencing: positions are specified as tuples of numbers representing linear or angular distances to certain chosen axes or angles
- Division-based referencing: referring to compartments of an accepted geometric or semantic-based division of the space, possibly hierarchical
- Linear referencing: referring to relative positions along linear objects such as streets, roads, rivers, pipelines; for example, street names plus house numbers or road codes plus distances from one of the ends

Since it is often the case that positions of entities cannot be determined accurately, they may be represented in data with uncertainty, for example, as areas instead of points.

Sometimes, an analyst is not so much interested in absolute positions in space as in relative positions with regard to a certain place. For example, the analyst may study where a person travels with regard to his/her home or movements of spectators to and from a cinema or a stadium. In such cases, it is convenient to define positions in terms of distances and directions from the reference place (or, in other words, by means of polar coordinates). The directions can be defined as angles from some base direction or geographically: north, northwest and so on.

Comprehensive analysis may require consideration of the same data within different systems of spatial referencing and, hence, transformation of one reference system to another: geographical coordinates to polar (with various origins), coordinate-based referencing to division based or network based, etc.

It may also be useful to disregard the spatial positions of locations and consider them from the perspective of their domain-specific semantics, e.g. home, workplace, shopping place.

It should be noted that space (in particular, physical space) is not uniform but heterogeneous, and its properties vary from place to place. These properties may have a great impact on movement behaviours and, hence, should be taken into account in analysis. The relevant characteristics of individual locations include the following:

- Altitude, slope, aspect and other characteristics of the terrain
- Accessibility with regard to various constraints (obstacles, availability of roads, etc.)
- Character and properties of the surface: land or water, concrete or soil, forest or field, etc.
- Objects present in a location: buildings, trees, monuments, etc.
- Function or way of use, e.g. housing, shopping, industry, agriculture, or transportation
- Activity-based semantics, e.g. home, work, shopping, leisure

When locations are defined as space compartments (i.e. areas in two-dimensional space or volumes in three-dimensional space) or network elements rather than points, the relevant characteristics also include the following:

- Spatial extent and shape
- Capacity, i.e. the number of entities the location can simultaneously contain
- Homogeneity or heterogeneity of properties (listed above) over the compartment

It should be noted that properties of locations may change over time. For example, a location may be accessible on weekdays and inaccessible on weekends; a town square may be used as a marketplace in the morning hours; a road segment may be blocked or its capacity reduced because of an accident or reparation works.

Similar to space, there are different ways of defining positions in time, and time may also be heterogeneous in terms of properties of time moments and intervals.

1.2.3 Time

Mathematically, time is a continuous set with a linear ordering and distances between the elements, where the elements are moments or positions in time. Analogous to positions in space, some reference system is needed for the specification of moments in data. In most cases, temporal referencing is done on the basis of the standard Gregorian calendar and the standard division of a day into hours, hours into minutes and so on. The time of the day may be specified according to the time zone of the place where the data are collected or as Greenwich Mean Time (GMT). There are cases, however, when data refer to relative time moments, e.g. the time elapsed from the beginning of a process or observation, or abstract time stamps specified as numbers 1, 2 and so on. Unlike the physical time, abstract times are not necessarily continuous.

Like positions in space, moments may be specified imprecisely, i.e. as intervals rather than points in time. But even when data refer to points, they are indispensably imprecise: since time is continuous, the data cannot refer to every possible point. For any two successive moments t_1 and t_2 referred to in the data, there are moments in between for which there are no data. Therefore, one cannot definitely know what happened between t_1 and t_2 but can only estimate this by means of interpolation.

Physical time is not only a linear sequence of moments but includes inherent cycles resulting from the earth's daily rotation and annual revolution. These natural *cycles* are reflected in the standard method of time referencing: the dates are repeated in each year and the times in each day. Besides these natural cycles, there are also cycles related to people's activities, for example, the weekly cycle. Various domain- and problem-specific cycles exist as well, for example, the revolution periods of the planets in astronomy or the cycles of the movement of buses or local trains on standard routes.

Temporal cycles may be nested; in particular, the daily cycle is nested within the annual cycle. Hence, time can be viewed as a hierarchy of nested cycles. Several alternative hierarchies may exist, for example, year/month/day-in-month and year/week-in-year/day-in-week.

It is very important to know which temporal cycles are relevant to the movements under study and to take these cycles properly into account in the analysis. For this purpose, it is necessary that the cycles were reflected in temporal references of the data items. Typically, this is done through specifying the cycle number and the position from the beginning of the cycle. In fact, the standard references to dates and times of the day are built according to this principle. However, besides the standard references to the yearly and daily cycles, references to other (potentially) relevant cycles, e.g. the weekly cycle of people's activities or the cycles of the movement of satellites, may be necessary or useful. Hence, an analyst may need to transform the standard references into references in terms of alternative time hierarchies.

Temporal cycles may have variable periods. For example, the cycle of El Nino and La Nina climatic events, which influences the movement of air and water masses in the Pacific Ocean, has an average return period of four and a half years but can recur as little as two or as much as ten years apart. To make data related to different cycles comparable, one needs to somehow 'standardise' the time references, for example, divide the absolute time counts from the beginning of a cycle by the length of this cycle.

Transformation of absolute time references to relative is also useful when it is needed to compare movements that start at different times and/or proceed with different speeds. The relative time references would in this case be the time counts from the beginning of each movement, possibly, standardised in the way of dividing them by the duration of the movement.

As we have noted, the properties of time moments and intervals may vary, and this variation may have significant influence on movements. For example, the movements of people on weekdays notably differ from the movements on weekends; moreover, the movements on Fridays differ from those on Mondays and the movements on Saturdays differ from those on Sundays. In this example, we have a case of a regular difference between positions within a cycle. Another example of the same kind is the difference between times of a day: morning, midday, evening and night. However, the regularity in the variation of properties of time moments may be disrupted, for example, by an intrusion of public holidays. Not only the intrusions themselves but also the preceding and/or following times may be very different from the 'normal' time; think, for example, of the days before and after Christmas. Such irregular changes should also be taken into account in the analysis of time-dependent phenomena, in particular, movements.

The regularity of changes may itself vary, in particular, owing to interactions between larger and smaller temporal cycles. Thus, the yearly variation of the duration of daylight has an impact on the properties of times of a day, which, in turn, influence movements of people and animals. In the results, movements at the same time of the day in summer and in winter may substantially differ.

Typically, the heterogeneity of properties of time is not explicitly reflected in data and, hence, cannot be automatically taken into account in data analysis. Much depends on the analyst's ability to involve his/her background knowledge. Hence, the methods and tools used for the analysis must allow the analyst to do this.

1.2.4 Moving Entities and Their Activities

Like locations in space and moments in time, the entities that move have their own characteristics, which may influence the movement and, hence, need to be taken into account in the analysis. Thus, the movements of people may greatly depend on their occupation, age, health condition, marital status, and other properties. It is also relevant whether an entity moves by itself or by means of some vehicle. The way and means of the movement pose their constraints on the possible routes and other characteristics of the movement.

People are an example of entities that typically move purposely. The purposes determine the routes and may also influence the other characteristics, in particular, the speed. For other types of entities, for example, tornadoes or elementary particles, one needs to attend to the causes of the movement rather than the purposes.

Movement characteristics may also depend on the activities performed by the entities during their movement. For example, the movement of a person in a shop differs from the movement on a street or in a park. The characteristics of the movement may change when the person starts speaking by a mobile phone.

1.2.5 Related Phenomena and Events

Any movement occurs in some environment and is subject to the influences from various events and phenomena taking place in this environment. Thus, Minard included a graph of winter temperatures in his depiction of Napoleon's Russian campaign since he was sure that the temperatures produced a great influence on the movement and fate of the army. Movements of people are influenced by the climate and current weather, by sport and cultural events, by legal regulations and established customs, by road tolls and oil prices, by shopping actions and traffic accidents and so on. To detect such influences or to take them into account in movement data analysis, the analyst needs to involve additional data and background knowledge.

We have reviewed thus far what characteristics and aspects of movement are considered in the analysis of movement data and what other types information are relevant. However, we did not define what it means, 'to analyse movement data', and for what purposes such an analysis is done. Let us now try to do this.

1.3 Analytical Questions

One can hardly find a strict definition of the term 'data analysis' in handbooks or research literature. However, most of the writers agree with the view of data analysis as an iterative process consisting of the following activities:

- Formulate questions
- Choose analysis methods
- Prepare the data for application of the methods
- Apply the methods to the data
- Interpret and evaluate the results obtained

In short, data analysis is formulating questions and seeking answers. In this section, we try to define the types of questions that can arise in analysis of movement data. Examples of various questions concerning moving entities can be easily found in literature, for instance, in Guting and Schneider [4]:

- How often do animals stop
- Which routes are regularly used by trucks
- Did the trucks with dangerous goods come close to a high-risk facility
- Were any two planes close to a collision
- Find 'strange' movements of ships, indicating illegal dumping of waste

However, we did not find a systematic taxonomy of the *types of questions* relevant to the analysis of movement data. Therefore, we try to build such a taxonomy by applying and adapting the general framework suggested by Bertin [2] and extended by Andrienko and Andrienko [1].

Bertin is a French cartographer and geographer, who was the first in articulating a coherent and reasoned theory for what is now called Information Visualisation. Bertin has developed a comprehensive framework for the design of maps and graphics intended for data analysis, where the function of a graphic is answering questions. Logically, a part of Bertin's theory deals with the types of questions that may need to be answered. The question types, as Bertin defines them, have no specific 'graphical flavour' and no influence of any other method for data representation or analysis. Questions are formulated purely in the 'language' of data, and hence have general relevance. Therefore, we can use Bertin's framework to define the types of questions that arise in analysis of movement data irrespective of what analysis methods are chosen.

To achieve this independence, we define the question types on the basis of an abstract view of the structure of movement data, which is presented next. In our typology, we distinguish between elementary questions, which refer to individual data items, and synoptic questions, which refer to the data as a whole or to data subsets considered in their entirety. Synoptic questions play the primary role in data analysis. We consider various types of elementary and synoptic questions. At the end, we relate the tool-independent taxonomy of analytical questions to the established typology of data mining tasks.

1.3.1 Data Structure

According to the general framework, the types of questions are defined on the basis of the structure of the data under analysis, i.e. what components the data consist of and how they are related. On an abstract level, movement data can be viewed as consisting of the three principal components:

- Time: a set of moments
- Population (this term is used in statistical rather than demographic sense): a set of entities that move
- Space: a set of locations that can be occupied by the entities

As noted above, a trajectory may be viewed as a function mapping time moments onto positions in space. Analogously, movement of multiple entities may be seen as a function mapping pairs ⟨time moment, entity⟩ onto positions. This is a very abstract data model, which is independent of any representative formalism (of course, there may be other models; for example, a database-oriented view would consider the same data as a table of tuples with at least three attributes: entity, time and space). The time and population of entities play the role of 'independent variables', or *referential components*, according to the terminology suggested by

Andrienko and Andrienko [1] and the space plays the role of 'dependent variable', or *characteristic component*.

A combination of values of the referential components is called a *reference*. In our case, a reference is a pair consisting of a time moment and an entity. The set of all possible references is called the *reference set*. Values of the characteristic components corresponding to the references are called *characteristics* of these references.

As it was mentioned in the previous section, the state of a moving entity at a selected time moment can be characterised not only by its position in space but also by additional characteristics such as speed, direction, acceleration. These characteristics can be viewed as secondary, since they can be derived from the values of the principal components. Nevertheless, we can extend our concept of movement data and see it as a function mapping references ⟨time moment, entity⟩ onto combinations of characteristics (position, speed, direction, etc.).

We have also mentioned in the previous section that locations, time moments and entities may have their own characteristics. For example, locations may be characterised by altitude, slope, character of the surface, etc.; entities may be characterised by their kind (people, vehicles, animals, etc.), age, gender, activity and so on. Such characteristics are independent of the movement, that is, do not refer to pairs ⟨time moment, entity⟩ but to individual values of the three principal components, time, population and space. Note that the space plays the role of a referential component for altitude, slope and so on. The characteristics of time moments, entities and locations will be further called *supplementary characteristics*. The characteristics of the pairs ⟨time moment, entity⟩ (including the secondary ones) will be called *characteristics of movement*.

Analytical questions arising in the analysis of movement data, address first of all the references (i.e. times and entities) and the characteristics of movement. However, they may also involve supplementary characteristics.

1.3.2 Elementary and Synoptic Questions

The types of questions are differentiated first of all according to their *level*: whether they address individual references or sets of references. Questions addressing individual references are called *elementary*. The term 'elementary' means that the questions address *elements* of the reference set. Questions addressing sets of references (either the whole reference set or its subsets) are called *synoptic*. The word 'synoptic' is defined in a dictionary (Merriam-Webster [9], p. 1197) as the following:

1. Affording a general view of a whole
2. Manifesting or characterised by comprehensiveness or breadth of view
3. Presenting or taking the same or common view; specifically often capitalised: of or relating to the first three Gospels of the New Testament
4. Relating to or displaying conditions (as of the atmosphere or weather) as they exist simultaneously over a broad area

Table 1.1 Different levels of questions about movement data

		Population	
		Elementary	Synoptic
Time	Elementary	Where was entity e at time moment t?	What was the spatial distribution of all entities at time moment t?
	Synoptic	How did entity e move during the time period from t_1 to t_2?	How did all entities move during the time period from t_1 to t_2?

The first interpretation is the closest to what we mean by synoptic questions, which assume a general view of a reference (sub)set as a whole, as will be clear from the examples given below. Interpretations 2 and 4 are also quite consistent with our usage of the term.

When there are two referential components, like in movement data, a question may be elementary with respect to one of them and synoptic with respect to the other. Examples are given in Table 1.1. Note that these examples are templates rather than specific questions, since they contain slots or variables.

The difference between elementary and synoptic questions is not merely the number of elements involved. It is more fundamental: a synoptic question requires one to deal with a set as a whole, in contrast to elementary questions addressing individual elements. Although an elementary question may address two or more elements, it does not require these elements to be considered all together as a unit. Compare, for instance, the following questions:

- What were the positions of entities $e_1, e_2, \ldots, e_n$ at time moment t?
- What was the spatial distribution of the set of entities $e_1, e_2, \ldots, e_n$ at time moment t?

The first question is elementary with respect to the population, although it addresses multiple entities. However, each entity is addressed individually, and the question about n entities is therefore equivalent to n questions asking about each of the entities separately (i.e. the same answer can be given in both cases: entity e_1 was in place p_1, e_2 was in p_2 and so on). The second question does not ask about the individual positions of all entities but about the spatial distribution of the set of entities as a whole. The possible answers could be 'the entities are distributed evenly' (or randomly, or concentrated in some part of the territory, or aligned, etc.).

In our examples, the elementary questions ask about locations of entities at time moments. They may also ask about the secondary characteristics of movement corresponding to references ⟨time moment, entity⟩, e.g. 'What was the speed of entity e at moment t?' Supplementary characteristics may also be involved, as in the question 'Describe the location where entity e was at moment t'. To answer this question, one needs, first, to determine the spatial position of entity e at moment t and, second, to ascertain the supplementary characteristics of the location thus found.

What do synoptic questions ask about? What is common between 'how did the entity (entities) move?' and 'what was the spatial distribution of the entities' (see Table 1.1)?

1.3.3 Behaviour and Pattern

We introduce the notion of *behaviour*: this is the configuration of characteristics corresponding to a given reference (sub)set. The notion of behaviour is a generalisation of such notions as distribution, variation, trend, dynamics, trajectory. In particular, a trajectory of a single entity is a configuration of locations (possibly, in combination with the secondary characteristics of movement) corresponding to a time interval. We say 'configuration' rather than 'set' meaning that the characteristics are arranged in accordance with the structure and properties of the reference (sub)set and the relations between its elements. Thus, since a time interval is a continuous linearly ordered set, a trajectory is a continuous sequence of locations ordered according to the times they were visited.

The term 'behaviour' is used here in quite a general sense and does not necessarily mean a process going on in time. Thus, the spatial distribution of a set of entities at some time moment is also a kind of behaviour, although it does not involve any temporal variation.

Since a population of entities is a discrete set without natural ordering and distances between the elements, it does not impose any specific arrangement of the corresponding characteristics. Still, the corresponding behaviour is not just a set of characteristics. Thus, one and the same characteristic or combination of characteristics can occur several times, and these occurrences are treated as different, while in a set each element may occur only once. A behaviour over a set of entities may hence be conceptualised as the frequency distribution of the characteristic values over this set of entities.

The absence of natural ordering and distances on a population of entities does not mean that ordering and distances between entities cannot exist at all. Thus, a set of participants of a military parade is spatially ordered and has distances between the elements. However, the ordering and distances are defined in this case on the basis of certain characteristics of the entities, specifically, their spatial positions. The characteristics that define ordering and/or distances between entities can be chosen, in principle, quite arbitrarily. Thus, participants of a parade can also be ordered according to their heights, or weights, or ages. In data analysis, it may be useful to consider different orderings of the entities and the corresponding arrangements of characteristics. In such cases, the behaviours are not just frequency distributions but more complex constructs where characteristic values are positioned according to the ordering and/or distances between the entities they are associated with.

The collective movement behaviour of a population of entities over a time period is a complex configuration built from movement characteristics of all entities at all time moments, which has no arrangement with respect to the population of entities and has a continuous linear arrangement with respect to the time.

Hence, synoptic questions address reference (sub)sets and corresponding behaviours, while elementary questions address individual references and corresponding characteristics. An answer to an elementary question is the value(s) of the characteristic component(s) it is asking about. An answer to a synoptic question is a description of the behaviour or, more generally, a representation of this behaviour in some language, e.g. natural, mathematical, graphical. Such a representation will be called *pattern*. This agrees with the definition of a pattern in the data mining literature: 'a pattern is an expression E in some language L describing facts in a subset F_E of a set of facts F so that E is simpler than the enumeration of all facts in F_E' [3]. Note that the latter definition emphasises the synoptic nature of a pattern: a pattern does not simply enumerate some facts but describes them all together as a whole.

As should be clear from the definition, different patterns (e.g. focusing on different aspects) may represent one and the same behaviour. A pattern may be compound, i.e. composed of other patterns. For example, the description 'most of the people tend to move towards the city centre in the morning and outwards in the evening' is a compound pattern including two simpler patterns, inward and outward movement. Patterns representing movement behaviours of individual entities (i.e. trajectories) and collective movement behaviours of sets of entities base first of all on the characteristics of movement but may also involve supplementary characteristics. Thus, our example pattern concerning the movement of people describes first of all the direction of the movement but also mentions such supplementary characteristics as the character of the moving entities (people), the character of a location (city centre) and the character of the times (morning or evening).

In a pattern describing the movement behaviour on a set of references, one may include various summary values derived from the individual characteristics of the references, for instance, the average speed, prevailing direction or frequency of turns.

1.3.4 Structure of a Question

Any question contains some information that is known to the person who asks the question and aims at gaining some new information, which must be somehow related to the known information. The expected new information will be called the *target* of the question, while the known information will be called the *constraint* (since it sets certain requirements to the content of the new information being sought). Thus, in a question asking about the characteristic corresponding to a given reference, the characteristic is the target, while the reference is the constraint. For example, in the question 'Where was the entity e at time moment t?' the reference, i.e. the pair (e, t), is the constraint and the target is the location corresponding to this pair. There are also inverse questions, which ask about references corresponding to given characteristics, for example, 'What entities visited place p and when?' In this question, the target is the unknown pair consisting of an entity and a time moment that corresponds to the given place p, which is the constraint of this question.

When references consist of two components, as time and entity in the case of movement data, one of the components may be included in the question constraint, with the other being the target:

- What entities were present in place p at time t?
- At what moments (if any) did entity e visit place p?

1 Basic Concepts of Movement Data

These are examples of elementary questions. Synoptic questions, which deal with reference sets and behaviours, have the same structure, i.e. include targets and constraints. In the examples of synoptic questions in Table 1.1, the constraints are (sub)sets of references and the targets are the behaviours corresponding to these (sub)sets. There are also questions where behaviours (described by means of appropriate patterns) are the constraints and reference (sub)sets are the targets, for example, 'What group(s) of entities and in what time period(s) moved as specified by pattern P?'

Like in elementary questions, one of the components defining the references (i.e. set of entities or time interval) may appear in the constraint of a sentence, with the other being the target:

- What entities moved as specified by pattern P during the time interval from t_1 to t_2?
- In what time period(s) did the group of entities $e_1, e_2, \ldots, e_n$ move as specified by pattern P?

Synoptic questions requiring the search for occurrences of specified patterns, as in the above-presented examples, may be called *pattern search tasks*. We highlight this question type and give it a special name since it plays a prominent role in visual data exploration, which is generally viewed as being based on pattern recognition.

1.3.5 Comparison Questions

In the examples considered so far, the questions were targeted at (i.e. asking about) characteristics, or behaviours, or references, or reference sets. Let us give a few examples of a different kind:

1. What were the relative positions of entities e_1 and e_2 at time t?
2. How did the location of entity e change from time t_1 to time t_2?
3. What is the difference in the times when entity e visited places p_1 and p_2?
4. What are the commonalities and differences between the movement behaviours of entities e_1 and e_2 (or groups of entities E_1 and E_2) on the time interval from t_1 to t_2?
5. How does the movement behaviour of entity e (or group of entities E) in time interval from t_1 to t_2 differ from the behaviour in interval from t_3 to t_4? What is in common?
6. Compare the time intervals when entity e (or group of entities E) moved according to pattern P_1 and according to pattern P_2.

These questions are targeted at *relations* between characteristics (questions 1 and 2), between behaviours (questions 4 and 5), between references (question 3), or between reference sets (question 6). Such questions are called comparison questions. Questions 1–3 are *elementary* comparison questions, while questions 4–6 are synoptic comparison questions. The term 'comparison' is used in quite a broad sense as establishing relations between things. The nature of the things determines what relations are possible. Let us list the relations relevant to movement data:

1. Relations between characteristics (including both characteristics of movement and auxiliary characteristics)
   (a) Positions: spatial relations including distance, direction and topological relations such as touch, inside, overlap.
   (b) Numeric characteristics, e.g. speed, acceleration, angle of turn: equality (equal or not equal), order (greater than or less than) and distance (difference)
   (c) Qualitative characteristics, e.g. direction of movement or character of a location: equality
2. Relations between references
   (a) Time moments: equality, order, distance (amount of time between two moments)
       (i) Additionally, relations between the auxiliary characteristics of time moments (numeric or qualitative): see 1(b) and 1(c)
   (b) Entities: equality
       (i) Additionally, relations between the auxiliary characteristics of entities (numeric or qualitative): see 1(b) and 1(c)
3. Relations between behaviours: equality (equal or not equal), similarity (similar or dissimilar) and conformity (conformal or opposite)
   (a) Additionally, relations between the summary characteristics of the behaviours such as the average speed (numeric) or prevailing direction (qualitative): see 1(b) and 1(c)
4. Relations between sets of references
   (a) Time intervals: temporal order, distance, topological relations such as touch, inside, overlap.
       (i) Additionally, relations between the auxiliary characteristics of the intervals such as length (numeric) or character of the times (qualitative): see 1(b) and 1(c)
   (b) Groups (subsets) of entities: equality, inclusion, overlap or absence of overlap
       (i) Additionally, relations between the auxiliary characteristics of the groups such as size (numeric) or character of the entities (qualitative): see 1(b) and 1(c)

1.3.6 Relation Search

Opposite to comparison questions, in which relations are unknown and need to be ascertained, there are questions requiring the search for occurrences of specified relations. In such questions, it is typically necessary to determine and describe the characteristics or behaviours linked by the specified relations and the corresponding references or reference subsets, i.e. where these relations occur. For example,

- Find all cases when two or more entities met in the same location (In what locations? What entities did meet? At what time moments?)
- Find all cases when two or more entities moved together, i.e. simultaneously passed the same locations (What sequences of locations, i.e. paths in space? What entities? On what time intervals?)
- Find all cases when an entity repeatedly made the same path in space (What path in space? What entity? On what time intervals?)
- Finds groups of entities that had similar movement behaviours (What is the common pattern for these behaviours? What entities? On which time intervals?)

From the examples given above, the first is an elementary question, since it addresses individual characteristics (locations) and references (entities and time moments). The remaining examples are synoptic, since they involve behaviours (in particular, paths in space) and reference subsets (in particular, time intervals). It may be noted, however, that both elementary and synoptic questions are not atomic but involve several tasks:

1. Detect an occurrence of the specified relation, i.e. at least two characteristics or at least two behaviours related in this way
2. Find out what references or reference subsets correspond to the characteristics or behaviours thus detected
3. For the answer to be complete, the characteristics or behaviours should also be described, in particular, the behaviours represented by suitable patterns

1.3.7 Building an Overall Pattern

One of the major goals of the analysis of movement data is to characterise the overall movement behaviour of the whole set of entities over the entire time period the data refer to, or, in other words, to build an appropriate pattern representing this overall behaviour (in data mining and statistics, a pattern describing the entire set of facts has the special name *model*). 'Appropriate' means adequate to the further goals, which may be, for example, prediction of the future behaviour or optimisation of the road network. The overall pattern (model) needs to be sufficiently comprehensive and precise. Typically, the required precision cannot be achieved in a simple (atomic) pattern but rather in a compound pattern built from sub-patterns, which, in turn, may also be compound.

Compound patterns result from decomposing the overall behaviour into parts, representing these partial behaviours by sub-patterns, and, finally, bringing the sub-patterns together into an overall pattern. The decomposition is required because the movement behaviour is not uniform throughout the reference set. The decomposition is based on detecting similarities and differences, i.e. involves relation search and comparison questions. The following synthesis of the compound overall pattern involves pattern search and comparison questions (Where else does this sub-pattern occur? What is the relative position of these sub-patterns in time and in space?). Since the data analysis aims first of all at building patterns and models, elementary questions play a marginal role in it when compared to the role played by synoptic questions.

There are several approaches to the decomposition of the overall movement behaviour of a population of entities E over a time period $[t_0, t_{end}]$:

- Divide E into subsets of entities with similar behaviours; build a pattern for each subset; describe the subset each pattern is valid for
- Divide the period $[t_0, t_{end}]$ into intervals where the behaviour can be regarded as homogeneous; build a pattern for each interval; describe the intervals and relations between them; describe the temporal arrangement of the patterns
- Factorise the time into its component parts, i.e. the linear component and one or more cycles (yearly, weekly, daily, or other, domain-specific cycles); build a pattern for the behaviour with respect to each component

In practice, these approaches are usually combined for the resulting model to be more precise. However, a full precision is hardly reachable. First, any pattern is a result of abstraction and simplification; the real data it represents usually slightly deviate from it. Second, extraordinary values and unusual value combinations may occur in a data set or particular entities may behave in an uncommon way. Such outliers usually need to be analysed and described separately.

The division of the set of entities and/or of the time period may be done either on the basis of observed (or somehow else detected, e.g. computed) differences between the respective behaviours or according to expected differences, where the expectations come from the background knowledge. For example, one can expect that children behave differently from adults and elderly people, and that movements in the morning differ from those in midday and evening. Divisions according to expected differences often based on supplementary characteristics of entities and time moments.

1.3.8 Connection Discovery

When studying a phenomenon, an analyst is interested not only in describing or summarising its behaviour but also in explaining it. The analyst wishes to find out the reasons or driving forces that make the phenomenon behave in the way observed. These forces may be internal or external. Internal forces originate from the inherent structure of the phenomenon and interactions between its structural components. External forces originate from interactions between the phenomenon and other phenomena. Hence, the goal is to determine what components and/or phenomena interact and how they interact. Thus, concerning the movement of entities, an analyst may be interested to know whether and how the movement is related to various spatial, temporal and spatiotemporal phenomena such as weather, events (e.g. traffic jams or accidents), opening hours of shops, activities of people. The analyst may also wish to detect interactions between parts of the overall movement behaviour, e.g. between the behaviours of traffic and of pedestrians, or between properties of movement, e.g. direction and speed.

We use the term *connection discovery tasks* to denote seeking for indications of possible interactions between phenomena or between different aspects of the same phenomenon. A result of such a task (or, in other words, an answer to a question about interactions) is a description in some language of the connection that has been discovered. We call such a description a *connection pattern* while a connection, or interaction, may be viewed as a 'mutual behaviour' of two or more phenomena or parts of the same phenomenon.

In data analysis, the following types of connections are typically looked for:

- *Correlation*: An undirected, or symmetrical, connection. This includes not only the statistical correlation between two numeric variables but also all cases of regular co-occurrence of characteristics or behaviours, possibly, with a temporal and/or spatial lag. For example, working in the centre of a city may correlate with using the public transport or a bike for getting to the workplace.
- *Dependency or influence*: A directed connection; for example, the use of a car or a bike for getting to the workplace depends on the weather (or, in other words, the weather influences whether a car or a bike is used).
- *Structural connection*: An observed movement behaviour results from a composition of two or more different movements performed simultaneously, like the observed movement of the planets is the result of a combination of their own movement and the movement of the Earth.

Connection discovery tasks are synoptic, since they require dealing with sets and behaviours rather than with elements and individual characteristics.

1.3.9 Taxonomy as a Whole

Table 1.2 summarises the taxonomy of the analysis questions concerning movement data, except for the connection discovery tasks, which are listed below:

- Detect correlations and dependencies between different characteristics of the movement
- Detect correlations and dependencies between the movement and various supplementary characteristics of the locations, time moments and entities and/or various external phenomena and events
- Represent the observable movement as a composition of several interacting movements of different kinds

We have defined these question types purely by reasoning about movement data, irrespective of any methods of analysis. It may be interesting to see how these types are related to the established typology of the tasks of data mining.

Table 1.2 Types of analytical questions about movement data

		Population	
		Elementary	Synoptic
Time	Elementary	For given references (which include entities and time moments), find the positions and other movement characteristics	Describe the spatial distribution of the set of entities and the spatial and statistical distributions of the movement characteristics at a given moment
		For given movement characteristics, find the corresponding references	Find time moments when the entities and/or their movement characteristics were distributed according to a given pattern (spatial or statistical)
		Compare the movement characteristics of given references	Compare the distributions (spatial or statistical) of the entities and/or movement characteristics at given time moments
		Find occurrences of given relations between movement characteristics and determine the references they correspond to	Find time moments with similar distributions (spatial or statistical) of the entities and/or movement characteristics
	Synoptic	Describe the movement behaviour of a given entity	Describe the collective movement behaviour of a given set of entities during a given time interval
		Find entities with the movement behaviour corresponding to a given pattern	Find the entity subsets and time periods where the collective movement behaviour corresponds to a given pattern
		Compare the movement behaviours of given entities	Compare the collective movement behaviours (a) of different groups of entities during the same time interval and (b) of the same entities during different time intervals
		Find occurrences of similar movement behaviours. Find entities with behaviours similar to the behaviour of a given entity	Divide the data into groups of entities and/or time intervals so that the behaviours are sufficiently homogeneous within the divisions and substantially differ between the divisions; find outstanding behaviours

1.3.10 Relation to the Data Mining Tasks

The first column of Table 1.3 lists the types of data mining tasks as defined in [3, 12]. The corresponding general types of analytical questions are indicated in the second column.

1 Basic Concepts of Movement Data

Table 1.3 Comparison between the general question types and the types of data mining tasks

Data mining tasks	General question types
Clustering: determining a finite set of implicit classes that describe the data	Divide the data into subsets of entities and/or time intervals so that the behaviours are sufficiently homogeneous within the divisions and substantially differ between the divisions
Classification: finding rules to assign data items to pre-existing classes	Detect dependencies between different characteristics of the movement
	Detect dependencies between the movement and various supplementary characteristics of the locations, time moments, and entities and/or various external phenomena and events
	Note: the definition of the classes may be based on movement characteristics (e.g. according to the movement direction), on supplementary characteristics (e.g. according to the activities of the entities), or on the variation of the external phenomena (e.g. according to the weather)
	Detect dependencies between different characteristics of the movement
Dependency analysis: finding rules to predict the value of an attribute on the basis of the values of other attributes	Detect dependencies between the movement and various supplementary characteristics of the locations, time moments, and entities and/or various external phenomena and events
Deviation and outlier analysis: searching for data items that exhibit unexpected deviations or differences from some norm	Find entities or subsets of entities and/or time intervals with behaviours notably differing from the rest of the entities and/or the time
Trend detection: fitting lines and curves to data to summarise the database	Find the entity subsets and time periods with the collective movement behaviours corresponding to given patterns, which are specified through algebraic formulae
Generalisation and characterisation: obtaining a compact description of the database, for example, as a relatively small set of logical statements that condense the information in the database	Describe the collective movement behaviour of the population of entities during the whole time period

As may be seen, the data mining tasks correspond to synoptic questions, in particular, behaviour characterisation (i.e. representation of a behaviour by a pattern), pattern search, relation search and connection discovery. There are no specific data mining tasks for the synoptic comparison questions. This may be an indication of the need in complementing data mining with other methods for data analysis such as visual analytics methods [14].

It is also interesting to note a clear correspondence between most of the elementary question types and database queries; see, for instance, Guting and Schneider [4]. Again, there are no specific query constructs for elementary comparison questions, and hence additional methods are needed.

1.4 Conclusion

In this chapter, we have discussed the structure of movement data, the nature and properties of their component parts and the things that may have an influence on movement and hence need to be accounted for in the analysis of movement data. On the basis of the treatment of data analysis as seeking answers to questions about data and underlying phenomena, we have also tried to define the possible types of questions about movement of a set of entities in space. The question types have been purely defined on the basis of the structure and characteristics of movement data, irrespective of any existing methods and tools for data analysis. The resulting taxonomy of the question types should therefore be seen as a requirement for the set of methods needed for the analysis of movement data. This means that researchers should suggest appropriate methods and tool developers implement tools that will allow analysts to find answers to these types of queries. Such methods and tools are considered in the remainder of the book.

References

1. N. Andrienko and G. Andrienko. *Exploratory Analysis of Spatial and Temporal Data: A Systematic Approach*. Springer, 2006.
2. J. Bertin. *Semiology of Graphics. Diagrams, Networks, Maps*. University of Wisconsin Press, 1983.
3. U.M. Fayyad, G. Piatetsky-Shapiro, and P. Smyth. From data mining to knowledge discovery: An overview. In *Advances in Knowledge Discovery and Data Mining*, pp. 1–34. MIT press, 1996.
4. R.H. Gueting and M. Schneider. *Moving Objects Databases*. Elsevier, 2005.
5. T. Hagerstrand. What about people in regional science? *Papers of the Regional Science Association*, 24:7–21, 1970.
6. K. Hornsby and M.J. Egenhofer. Modeling moving objects over multiple granularities. *Annual Mathematics Artificial Intelligence*, 36(1–2):177–194, 2002.
7. M.-J. Kraak. The space–time cube revisited from a geovisualization perspective. In *Proceedings of the Twenty-First International Cartographic Conference (ICC'03)*, pp. 1988–1995, 2003.
8. P. Laube, S. Imfeld, and R. Weibel. Discovering relative motion patterns in groups of moving point objects. *International Journal of Geographical Information Science*, 19(6):639–668, 2005.
9. *Merriam-Webster's Collegiate Dictionary, 10 edn.*. Merriam-Webster, Incorporated, 1999.
10. H. Miller. A measurement theory for time geography. *Geographical Analysis*, 37:17–45, 2005.
11. H. Miller. Modeling accessibility using space–time prism concepts within geographical information systems: Fourteen years on. In *Classics of International Journal of Geographical Information Science*, pp. 177–182. CRC Press, 2006.
12. H. Miller and J. Han. Geographic data mining and knowledge discovery: An overview. In *Geographic Data Mining and Knowledge Discovery*, pp. 3–32. Taylor and Francis, 2001.
13. A. Moore, P. Whigwham, A. Holt, C. Alridge, and K. Hodge. A time geography approach to the visualization of sport. In *Proceedings of the Seventh International Conference on Geocomputation*, 2003.
14. J. Thomas and K. Cook. *Illuminating the Path. The Research and development Agenda for Visual Analytics*. IEEE Computer Society, 1983.
15. E. Tufte. *Visual Display of Quantitative Information*. Graphics Press, 1983.

Chapter 2
Characterising the Next Generation of Mobile Applications Through a Privacy-Aware Geographic Knowledge Discovery Process

M. Wachowicz, A. Ligtenberg, C. Renso, and S. Gürses

2.1 Introduction

The proliferation of mobile technologies for 'always-on' at 'any-time' and 'any-place' has facilitated the generation of huge volume of positioning data sets containing information about the location and the movement of entities through the geographic environment. In principle, every time an entity moves through space, it creates a trajectory (i.e. track or path) representing the history of its past and current locations. Examples of interesting trajectories of moving entities may range from hurricane and tornado tracks [19] to individual trajectories of animals [26] and planes [5]. Specially designed sensors can provide the location of a mobile entity as well as information about the geographic environment where this entity is moving. Current research on mobile technologies such as sensor web, wireless communication and portable computers has been crucial for the development of multi-sensor systems. Their use to sense a geographic environment and mobile entities can include photodiodes to detect *light* level, accelerometers to provide *tilt* and *vibration* measurements, passive infrared sensors to detect the *proximity of humans*, omni-directional microphones to detect *sound* and other built-in sensors for *temperature, pressure*, and *CO gas* [9].

Moreover, there are many types of mobile applications, which have been developed to meet many of society's needs for economic development, experience and culture. Some examples include the applications based on location-based services (LBS) such as tourism, marketing and transportation management. Existing LBS can already provide tourists with information about their current location in a way that they can find directions, retrieve geographic information and leave comments on an interactive map [1, 35]. A travel diary can be automatically compiled using the history of *where* a tourist has travelled over time. Some systems are also capable of making suggestions on places of interest to visit by visualising the required

---

M. Wachowicz
Wageningen UR, Centre for GeoInformation, Netherlands, e-mail: Monica.Wachowicz@wur.nl

information within an augmented reality environment. The positioning data are usually collected by the global positioning system (GPS) for outdoor tracking and infrared (IR) positioning for indoor tracking.

Although positioning data sets containing information about the location of mobile entities may be available to develop mobile applications, effective usage of that information is still a challenging problem. New methods and tools are needed in the fields of databases, statistics, geography, remote sensing and artificial intelligence that can automatically transform these very large positioning data sets into information about the movement of entities, and furthermore, be the source of geographic knowledge. Extracting new, insightful information embedded within the large heterogeneous databases that contain private information about the location of the mobile entities and their surrounding geographic environment still remains one of the main challenges.

We need to go beyond the collection of positioning data sets to the delivery of information and knowledge derived from these data. A knowledge discovery process empowers the experts of an application domain to extract relevant and useful geographic knowledge from very large positioning data sets. It also supports the development of the next generation of mobile applications through its ability to cope with data warehousing, target data selection, cleaning and pre-processing, as well as data mining, model selection, evaluation and interpretation of the hidden patterns embedded within very large heterogeneous databases. However, one of the main research issues is *privacy*, which is concerned with the protection of data or information that is considered private. It is important to realise that privacy concerns are very important with respect to the social acceptance of the use of a geographic knowledge discovery process for developing mobile applications [49]. The latest developments in privacy-preserving techniques in databases are of primordial importance if the aim is to collect and reconstruct a vast number of individual trajectories that will allow an efficient and effective storage of these trajectories as well as suitable access methods to support analysis and data mining tasks. Such privacy-preserving techniques should support mechanisms that prevent the disclosure of sensitive data, both explicitly (e.g., providing individual's identity) and implicitly (providing non-sensitive data from which sensitive information can be inferred).

In this chapter, we examine the current state of the existing concepts and methods in geographic knowledge discovery, identify the needs for such a process and describe the research challenges and its potential impact on developing new mobile applications. Emphasis is given to the privacy issues on developing such a geographic knowledge discovery process. The overall goal is to identify a framework that can serve as a road map for developing a privacy-preserving knowledge discovery process. The chapter also demonstrates how a privacy-aware geographic knowledge discovery process constructed from a multi-tier ontological perspective can be used to characterise new applications in the domains of transportation management, spatial planning and marketing.

2.2 The Privacy-Aware Geographic Knowledge Discovery Process

2.2.1 The Process of Knowledge Discovery in Databases

The term 'knowledge discovery in databases' was coined in 1989 in an effort to describe the overall process within which data mining is a step in extracting patterns from data. In general, it has been defined as the non-trivial process of identifying valid, novel, potentially useful and ultimately understandable patterns in data [14]. The proliferation of such a process coincides with an exponential increase in disparate data sets being linked together across place, scale, time, theme and discipline and available to science, government and industry. In particular, very large databases that are rich in terms of attribute depth and large in the sense of having many records or objects represented have played a great role in the development of knowledge discovery.

The knowledge discovery process usually involves experimentation, iteration, user interaction and many design decisions and customisations. Different delineations have been proposed for a knowledge discovery process, including the nine-step process described as follows [14]:

1. Developing an understanding of the knowledge domain, the relevant prior knowledge and the goals of the user
2. Creating a target data set, selecting a data set or focusing on a sub-set of variables or data samples, on which discovery is to be performed
3. Data cleaning and pre-processing
4. Data reduction and transformation
5. Choosing the data mining task
6. Choosing the data mining algorithm(s)
7. Data mining for a particular form of representation such as classification rules or trees, regression, clustering, etc.
8. Evaluating and validating the results
9. Consolidating discovered knowledge: incorporating this knowledge into the performance of the system, or simply documenting it and reporting it to users

Although this list might suggest a sequence of steps, a knowledge discovery process is in fact a *random process* in which the steps are often carried out by an unsystematic approach and do not follow a straightforward analysis. Moreover, many people do not realise that these steps have been treated as separate activities, with their own principles, procedures and limitations. One of the main reasons is that a knowledge discovery process is a combination of individual techniques that are built from the fields of databases, pattern recognition, artificial intelligence, machine learning and has strong ties to related efforts in information visualisation and to exploratory data analysis in statistics [8, 10]. Table 2.1 summarises some of the approaches developed by different scientific communities to carry out the steps of mining (step 7), validating (step 8) and reporting the findings (step 9) for

Table 2.1 Some KDD steps and the respective techniques developed by different scientific communities for classification

KDD step	Scientific community			
	Databases	Statistics	Artificial intelligence	Information visualisation
Mining	Classification rules	Local pattern analysis and global inferential tests	Neural networks	Visual data mining
Validating	Computational models for interestingness, confidence, and support measures	Significance tests	Learning followed by verification using a test data set	User suitability tests
Reporting	Rule lists	Significance power	Likelihood estimation and information gain	Visual communication

classification purposes. The aim of this table is not to describe all the techniques already developed for classification, but rather to illustrate how disparate the examples are and how, unfortunately, there is little research integrating and comparing these techniques. The most representative example is given by the database community who has taken different conceptual and implementation techniques for data mining and reporting that are not necessarily compatible with each other. These approaches are joined together merely at the system level that does not guarantee a better understanding of what a knowledge discovery process is for, and as a result, hampers a useful exploration of large databases.

A different delineation has been proposed by Ramakrishman and Grama [41] based on a taxonomy in which a knowledge discovery process is described according to four perspectives on how knowledge is acquired in the process. The first perspective, and actually the most common, is *induction* with its origin in Artificial Intelligence and Machine Learning, in which the process is based on the 'learning-from-examples' concept. This is reflected by the extensive number of existing data mining algorithms that can extract generalised rules from a target data set and summarise the relationships between attributes at higher concept levels. Some examples include the attribute-oriented induction method [6, 22] that has integrated learning-from-examples algorithms with database operations (e.g. group by). Some authors have investigated attribute-oriented induction methods for extracting generalisation hierarchies for spatial data [23, 59].

The *compression perspective* emerges from the work of the fourteenth century philosopher William of Occam, in which the Occam's razor concept is stated as 'entities are not to be multiplied beyond necessity.' The developments in computational learning theory and the feasibility of models based on minimum encoding inference, such as minimum message length (MML) [58], have played an important role in establishing a solid theoretical foundation to this perspective. The Occam's razor is often used as a guiding principle in model selection in data mining, which suggests a 'good' model should use any relevant variable, relationship or behaviour but ignore all irrelevant ones. Models should capture the essence of an application domain under study by searching for simplicity. Some examples of modelling algorithms are projection pursuit, neural networks, decision trees and adaptive splines [14]. All these models assume the availability of a training data set, and the goal is to find a model to predict y from x that will perform well on a new data set.

In contrast to the previously described perspectives, the *querying perspective* is based on discovering knowledge through database query languages. In general, database models have been developed for storing and querying data, and they still need to be proven to be 'good' models for data mining. Most database management systems do not allow the type of data interaction that a knowledge discovery process requires. Nevertheless, several research efforts focus on enhancing query languages such as structured query language (SQL), mainly because most of the data are available from commercial databases and warehouses. Some examples are the semantic query optimisation approach by using semantic rules to reformulate a query [25, 47, 48] and the FOIL [40] approach using Horn-clause definitions in a query.

Finally, the query perspective is closely related to the *approximation perspective*, which relies on the previous knowledge of a model (e.g. a database schema) in order to find some hidden structure in the data. For example, linear algebraic matrix approximations have been developed to identify hidden structures in text data without using a simple keyword matching (e.g. latent semantic indexing, patented by Bellcore).

2.3 The Geographic Knowledge Discovery Process

The term 'geographic knowledge discovery' was coined later, having the acronym of GKDD, and representing a special case of knowledge discovery in databases, since it required specialised tools and provided unique research challenges to deal with space and time. The process has been defined using the previously mentioned steps, but critical research areas have been identified as developing and supporting geographic data warehouses, richer geographic data types, better spatiotemporal representations and user interfaces [16, 24, 57]. In an effort to frame a geographic knowledge process in the context of spatiotemporal environmental data, MacEachren and Wachowicz [31] have proposed a conceptual framework for the integration database and visualisation techniques, emphasising a merger of meta-operations for the GKDD steps. The knowledge discovery system specifically considered location, time and attribute aspects of each data entity during all steps of analysis (from pre-processing, through application of data mining tools, to interpretation).

A different approach was proposed by Aldrige [2] on advocating the concepts of extensional knowledge (i.e. facts) and intensional knowledge (i.e. rules) based on Pawlak's theory on notions of equivalence relations, generalisation, induction, deduction and supervised and unsupervised learning. Empirical knowledge of real-world phenomena was applied to represent extensional knowledge on choropleth maps, and the results show GKDD as a process of inducing non-trivial, potentially useful intentional geographic knowledge from databases.

However, successful applications of GKDD are not common, despite the vast literature on knowledge discovery in databases. The reason is that, although it is relatively straightforward to find patterns in very large spatiotemporal databases, both establishing their relevance and explaining their causes are very complex problems. In practice, most of the patterns found in a GKDD process may already be the background knowledge of an application domain. Large databases may contain a vast number of hidden patterns which are not necessarily novel or useful. At the moment, a geographic knowledge discovery process has no concept of what is known by experts in a way that the patterns make sense within the context of the current application domain. Addressing these issues requires to consider a knowledge discovery process as a human-centred process, not only in the sense that users need to dynamically interact with the system, but also that knowledge can only be inferred from very large and possibly poorly understood databases if the effective form of a *metaphor* is used.

This is even more relevant for geographic knowledge discovery processes where there is a scarce geographic knowledge on the forms of metaphors on inferring knowledge from spatiotemporal databases. Therefore, the need here is for more complex reasoning modes, which could provide the mapping, required for a systematic set of correspondences between metaphors that we try to understand and the patterns found in a geographic knowledge discovery process. This is discussed in more detail in Chap. 3.

2.3.1 Privacy Issues: Involving the Stakeholders in a GKDD Process

Privacy needs to be addressed from the beginning of the geographic knowledge discovery and therefore needs to be integrated into the complex relationship between patterns and information metaphors. In legal frameworks as well as in the database community, the concept of privacy is generally translated into data protection, or more specifically, the protection of personal data. Personal data are 'any information relating to an identified or identifiable natural person [...]; an identifiable person is one who can be identified, directly or indirectly, in particular by reference to an identification number or to one or more factors specific to his physical, physiological, mental, economic, cultural or social identity' (EU Directive 95/46/EC, Art. 2(a)). This definition has two effects: first, it focuses the attention on data (as opposed to people) and second, it focuses the attention on the identification of a natural person as the problem. Thus, it implicitly declares data-processing activities to be 'privacy-preserving' if and only if they do not (better, if they provably cannot) involve the identification of the natural person who is the 'carrier' of a record of attributes. This notion is reflected in the knowledge discovery literature by regarding goals such as k-anonymity as the defining properties of a privacy-preserving data mining algorithms.

However, the privacy literature suggests that people's view of privacy involves not one but many identities, which can be supported by a number of identity management schemes, that there are concerns over profiles independent of individual identification and that context is all important. Any knowledge discovery process should transcend the algorithm-centric and data-centric views. As a result, it should involve a better understanding of an application domain, and the evaluation of the lessons learned from previous cycles of a knowledge discovery process. This means that the complexities of software development and software use must be considered also in terms of privacy. Different mobile entities release trajectory data in different contexts as a result of complex sets of interactions. Moreover, different mobile entities have partially conflicting interests (some of these concerning privacy) in the information processed and are available through such systems, and each system exists in an environment populated by many systems, which also contain information that, when linked together, may breach different privacy interests.

Guidelines on the protection of privacy and transborder flows of personal data, the fair information practices (FIP) notice, choice, access and security that were set down as a recommendation in the US and updated by the Federal Trade Commission, or the principles of the EU Privacy Directives, define privacy not only as a matter of concealment of personal information but also as the ability to control what happens with it [20]. These are sometimes also referred to as the eight principles of 'FIP', which are the following:

- *Collection limitation*: Data collectors should only collect information that is necessary, and should do so by lawful and fair means, i.e. with the knowledge or consent of the data subject.
- *Data quality*: The collected data should be kept up-to-date and stored only as long as it is relevant.
- *Purpose specification*: The purpose for which data is collected should be specified (and announced) ahead of the data collection.
- *Use limitation*: Personal data should only be used for the stated purpose, except with the data subject's consent or as required by law.
- *Security safeguards*: Reasonable security safeguards should protect collected data from unauthorised access, use, modification or disclosure.
- *Openness*: It should be possible for data subjects to learn about the data controller's identity, and how to get in touch with him.
- *Individual participation*: Data subjects should be able to query data controllers whether or not their personal information has been stored, and, if possible, challenge (i.e. erase, rectify or amend) these data.
- *Accountability*: Data controllers should be accountable for complying with these principles.

These principles show that the challenges that privacy brings to geographic knowledge discovery is beyond concealment of identity-related data. For example, the principle of purpose specification can be considered in conflict with the heuristics of most knowledge discovery processes because patterns are usually discovered from data sets that have no purpose specification.

Beyond regulations and legal frameworks, privacy brings a new component to a GKDD process: the stakeholders. There may be different degrees of privacy interests from different stakeholders within a GKDD process. Basically, three main groups of stakeholders may play important roles in dealing with the privacy issues in a GKDD process. They are

- *Sensor carriers*: Those individuals who produce or own objects that produce the positioning data sets. They should authorise the level of privacy expected for collection, use, openness and individual participation in a GKDD process.
- *Data collectors and miners*: Those individuals or organisations interested in collecting the positioning data and developing the data mining algorithms. They must ensure the level of privacy required for data collection, data quality and security safeguards.
- *The experts of an application domain*: The individuals or organisations interested in applying the outcomes of a GKDD process. They should also define the level of privacy required for the collection, purpose and use of the results of a GKDD process.

Figure 2.1 illustrates the relationship among the three groups of stakeholders who are involved in a GKDD process. For example, the collection of data about AIDS patients and their movement may be seen differently from a public health or epidemiology perspective, with the objective of contact tracing, versus an individual perspective, with the objective of protecting against social or workplace discrimination. Such cases point out the necessity of multilateral privacy and security requirements analysis to accompany a GKDD process in which the privacy and security interests of all the stakeholder need to be documented, conflicts identified and negotiated and constraints elicited for the privacy-aware GKDD process.

Further, positioning data, and hence trajectory models, are particularly sensitive information to sensor carriers because of the specific characteristics of a geographic environment. The sensor carriers cannot avoid being at a location at any point of time (one cannot 'opt out of being somewhere'). Therefore, their impression of lacking self-control and comprehensiveness of surveillance are particularly pronounced in comparison to data miners' and experts' concerns about privacy. In addition, positioning data sets in combination with geographic information allow many inferences because of rich social background knowledge. One reason is that because of physical

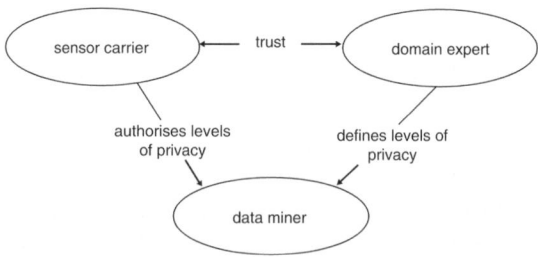

Fig. 2.1 The multilateral relationships of groups of stakeholders in a privacy-aware GKDD process

constraints, the number of interpretations of being at one location is often bounded (e.g. visits to a doctor specialist in the treatment of AIDS victims). Another reason is that there are typical spatiotemporal behaviour patterns (a location where a person habitually spends the night without moving is most likely that person's home). Location data are strong quasi-identifiers. This brings further constraints to the collection, processing and release of positioning data by data miners as well as by experts of an application domain.

The preservation of privacy in ubiquitous environments during data collection is difficult, if not impossible. Multiple channels can be used to collect different kinds of physical information (e.g. radio frequency fingerprints) that can be used to precisely identify the trajectories of a given sensor carrier. Therefore, opting out may not be a possibility during data collection and may bring about legal or social conflicts, especially since trajectories coupled with a priori knowledge on places and social contexts can be analysed by data miners to infer individual identities used to link profiles of individuals, or classify persons into previously defined groups.

Although groups are made up of aggregate patterns, which are not intended to identify single individuals or entities, they may still provoke the problem of group discrimination as a result of the processed data. All members of a group in a specific area or with certain trajectory patterns can be categorised into one group and the discovered category may be attributed to all members when the discovered patterns are applied later. For example, if a specific region is identified as having a high rate of cancer, then all persons linked to this region may be offered insurance at higher rates than in other areas, regardless of how heterogeneous the group members are in reality. Therefore, for an expert of an application domain, the concealment of exact rules, which cover the complete population, or even weaker rules used to classify groups into sensitive categories may also breach privacy of individuals, although they do not identify individuals.

We propose a multi-tier framework as one of the fundamental steps towards framing the complex relationship among movement metaphors and patterns, as well as beginning to understand how privacy issues should be tackled within a GKDD process. Chapter 3 will introduce this multi-tier ontological framework.

2.4 Reframing a GKDD Process Using a Multi-tier Ontological Perspective

2.4.1 The Role of Metaphors in Reasoning to Discover Patterns

Tiezzi [52] stated that knowledge of nature must come from a global and systemic view of *patterns* as well as from a study of the network of information joining the various forms of *metaphors* in time and space. Currently, a geographic knowledge discovery process is supported by computer-based environments that provide a global and systemic view of patterns by allowing users to be interactive and iterative, involving their visual thinking (perceptual-cognitive process) and automated information processing (computer-analytical process) with many decisions made by the users about how to fit models to or how to determine patterns from data.

However, patterns do not explain a metaphor [18]. But on the contrary, it is the *metaphor*, and only after it makes sense, from which an unknown set of patterns from a GKDD process can be interpreted and understood by an expert of an application domain. Why are metaphors important? Metaphors are artefacts of understanding, specifically understanding one kind of conceptual domain in terms of another. They are not just a pattern or a logical form. Johnson [28] defined metaphors as a 'concrete and dynamic, embodied imaginative schemata', which are surely not just logical patterns, images or diagrams. Moreover, Lakoff [29] argues that metaphors are something 'non-propositional', which should not be thought of as if they were commonalities, classes, structures or image schemata, although we might be interested to formulate those.

In a GKDD context, metaphors will help the comprehension of what makes one pattern structurally and meaningfully different from another. Ideally, metaphors would be constructed in the domain of the expert, having a high level or an abstract reason that makes sense within a specific problem context. They will lead to the 'discovery' of higher level entities, relationships or processes within some application domain of interest. Poore and Chrisman [38] have drawn attention to the fact that *information metaphors* do not relate directly to reality, but instead are more successful when they can have the effect of structuring reality to fit. For example, in GIS, the *landscape-as-layer* metaphor has structured the landscape into a set of layers, and nowadays, software packages encourage organisations to collect their data according to layers. Although researchers have proposed new information metaphors such as objects [36, 56] and agents [12, 30], numerous practitioners are locked into the layer form of reasoning.

What will be the information metaphors of a GKDD process? In particular, the *movement-as-trajectories* metaphor is already being used to structure the history of the past and current locations of mobile entities. Pfoser and Jensen [37] employ the metaphor of trajectory as polylines consisting of connected line segments, which can be grouped according to two movement scenarios, termed as unconstrained movement (vessels at sea) and constrained movement (cars and pedestrians). Another example is given by the account of the *movement-as-balance* metaphor that provides an interpretive artefact of a *balance scale* for analysing the traffic flow of cars in the presence of transportation problems [44]. A transportation system that operates under the conditions of free flow will be in balance. On the contrary, if the components such as road and rail are in wrong proportions, they are out of balance, having as a result a traffic that is unbearable with a need to remove the load from the roads.

In a GKDD process, the challenge relies on mapping the discovered patterns with metaphors such as *movement-as-trajectories* or *movement-as-balance*. For example, how discovered patterns, such as clusters or association rules, can be understood as representing those patterns occurred in low-density fringe growth in urban developments that can show the reduced effectiveness of public transport and increased reliance on the private car. It is still to be proven that information metaphors will enhance the likelihood that an expert will not only 'see' the movement patterns, but also will understand their meaning as well. However, it is already clear that information metaphors can generate a chain of commonalities and differences, not a single pattern. A better account of the role of information metaphors in a GKDD process would allow the experts to form and operate on concepts, not on GKDD steps.

The complex relationship between information metaphors and GKDD must remain a topic for further research. In this chapter, we outline our first effort on understanding such a relationship by looking at the reasoning paradigm. Reasoning is the ability of experts to form and operate concepts in abstractions (i.e. metaphors). In our research, reasoning constitutes the 'logic of discovery' as already proposed by the philosopher and logician Peirce. Therefore, three different approaches have been distinguished according to the type of reasoning task. They are

- *Deduction*: A reasoning task by which one infers a consequence from a set of patterns. The consequences are drawn from the general (patterns) down to the specific (metaphor). In this case, the metaphor is already known by the experts, and it usually forms the empirical basis of a GKDD process, because the relationship between the metaphor and the patterns can be verified straightforwardly. The metaphor C is known if patterns A and B are revealed in a GKDD process. One example is the *movement-as-journey* metaphor, where movement is conceptualised as a journey that begins and ends at home and can include one or more stops. The GKDD process might reveal cluster A showing that people make more journeys and spend more time travelling on weekdays, rather than weekends. Cluster B might reveal the patterns of a large neighbourhood shopping centre, where people spend less time travelling and travel less kilometres on weekdays. The *journey* metaphor underlies the expert's understanding of people's behaviour based on properties such as travel distance (from home to destination) and cyclic time (e.g. weekdays). Movement metaphors are needed for deductive reasoning, since they underlie an assumption that represents the expert's understanding of the patterns revealed by a GKDD process.
- *Induction*: A reasoning task by which one infers a generalisation from a set of patterns. It implies reasoning from detailed facts (examples) to general principles (conclusions). This approach of reasoning supports 'learning by example', where the example is the metaphor that contains more information than what was contained in the patterns themselves. The challenge is to uncover what metaphors can explain the causes for the observed patterns. Movement metaphors are needed for inductive reasoning, since they rise from generalisation. For example, the *movement-as-activity* metaphor explains how people organise their movement in a geographic environment by defining a sequence of activities that comprise a person's existence at any temporal scale (daily, monthly, lifetime) and social extent. For example, after the discovery of distinct linear patterns of a set of trajectories between mornings, afternoons and evenings, it would be possible to infer that leisure and work activities are the most common activities

conducted outside the home, closely followed by grocery shopping and bring/get activities such as bringing/getting a child to/from school. If activity is used to explain the linear patterns extracted via inductive reasoning, then a generalised form of 'activities' metaphors needs to be known a priori in order to explain the discovered patterns in the forms of rules, clusters, or classes.

- *Abduction*: A reasoning task by which one infers to the best cause for the occurrence of a set of patterns. An explanation is a relation between one or more hypothesis and the patterns they account for. It is flexible because it is not restricted to using existing metaphors of an application domain, but is instead free to create new metaphors that help to explain the patterns presented. If some theory states that if pattern A causes pattern B, and the GKDD process reveals the occurrence of pattern B, then by abduction an expert can infer A. However, data mining methods do not operate in this way, most either attempt to locate pre-defined patterns (*deduction*) or else learn from examples that are presented or selected (*induction*). Ideally, the new metaphor would be unravelled by the expert, mapping the discovered patterns into a new hypothesis in an application domain.

Figure 2.2 illustrates the relationship between metaphors and the reasoning tasks of a GKDD process. It is important to emphasise the role of metaphors in clarifying, naming and structuring what might otherwise be vague and inapplicable patterns within the context of an application domain. Therefore, reasoning is an integral part of the discovery process, and we propose that discovery and reasoning should be studied together. This will facilitate not only the extraction of patterns from very large databases but also to infer knowledge from these patterns.

Previous research on spatiotemporal reasoning has primarily dealt with hierarchical metaphors based on static and well-defined closed environments, and unfortunately, without having them associated to a geographic knowledge discovery process. Some examples include the spatiotemporal granularity description of

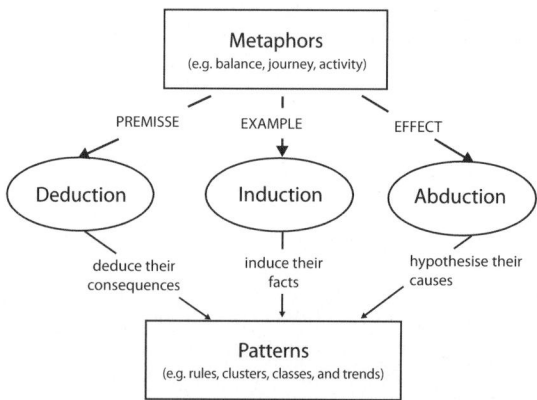

Fig. 2.2 The role of metaphors in different reasoning tasks

spatial regions and the concept of perceptual hierarchical spatial units for representing people behaviour in urban environments [43]. The dominant view has been that these representations are hierarchically organised [32], and the locations, objects, circumstances and factors may be perceived and understood in separate representations, which are required according to a particular situation or task [27]. Most of the studies have been conducted at a specific scale level by building scenarios on the variations in urban form characteristics such as urban morphology, transportation network, availability of facilities and density of a city and the relative location of neighbourhoods. The reasoning task involved has been of deriving the most likely explanations of the known facts and assumptions about urban form characteristics, and their influence on travel behaviour. Such explanations have usually pointed out to four major factors that have explained such an influence on a specific scale. They are *density of development, land use mix, transport networks and layout development* [51].

2.5 The Multi-Tier Ontological Framework

A GKDD process constructed from a multi-tier ontological perspective aims to integrate different reasoning tasks in a unified system by mapping the complex relationship between movement metaphors and patterns. Knowledge discovery is not a trivial process and it requires the examination of metaphors of characteristics, similarities and differences, interrelations, behaviour and evolution of what experts believe the world is like. This will lead to uncovering new and innovative hypothesis of distributions, patterns and structures across very large databases. Therefore, these metaphors will not rely on similar reasoning backgrounds but will be derived from the integration of different inference modes (i.e. abductive, inductive, and deductive).

This section describes the multi-tier ontological framework that has been developed from two previous fundamental research works: first, the work on a set of tiers of ontology previously proposed by Frank [15] for defining consistency constraints, data interoperability and more recently data quality in Geographical Information Systems (GIS) [33], and second, our multi-tier framework largely based on the three 'spaces' paradigm that has been proposed by Ernst Cassirer (1874–1945), a philosopher of the Marburg school, who describes a learning process as a truly dynamic activity of the mind of the human experience of spaces and time. The spaces are from an observed space through sensors and senses (interpretation), to an abstract model of space (guide), to a higher level of concepts incorporated in an internal and cognitive space (synthesis).

Our aim is to describe a GKDD process using ontological tiers that will provide the common base for the organisation of different nature and sources of knowledge of the movement metaphors used by experts of application domains. The tiers also establish the movement metaphors for the integration of different reasoning tasks in a unified system. This can only be achieved because each

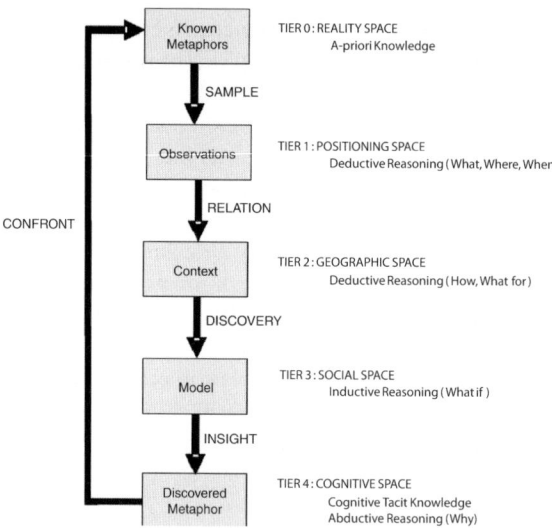

Fig. 2.3 The multi-tier privacy-aware geographic knowledge discovery process

tier instantiates the metaphors of the previous tier, enabling the understanding of interesting, meaningful and previously unknown patterns. Figure 2.3 illustrates the proposed multi-tier ontological framework, in which a successive set of tiers refine the steps of GKDD process, which are named sampling, relating to a geographic context, discovering patterns, generating new insights and confronting them with previous background knowledge. Therefore, five tiers have been defined. They are reality space, positioning space, geographic space, social space and cognitive space.

From a privacy perspective, the multi-tier framework allows a number of legal frameworks, often specific to application domains, to be adhered to throughout the GKDD process. Therefore, one of the initial steps of this process requires understanding which of these laws or regulations apply to each one of the tiers. Often such frameworks require the sensor carriers and domain experts to get the consent of those whose data is being collected to the primary or secondary use of that data. This is particularly challenging in a GKDD process, since the metaphors and the context of their use may not yet be determined during the tier 1 when the collection of data is carried out. Furthermore, the other tiers may also have privacy constraints on the results of the GKDD process that are unknown by the data miner as well as the expert of an application domain.

2.5.1 Tier 0: The Reality Space

Tier 0 of the ontology represents the '*reality space*', which recognises the existence of a known world as a four-dimensional continuous field in space and time. Usually, natural language is used to formalise the background knowledge that is derived from metaphors formulated by experts within an application domain. The process of geographic knowledge discovery may use this type of knowledge for generating a priori knowledge as pre-determined hypothesis, training examples or rules. Several known movement metaphors are currently being used, including the *movement-as-journey* metaphor, which was already mentioned in the previous section. The existing a priori knowledge might formulate that one or two journeys on a day are most common. Over three quarters of all journeys are usually a single-stop tour, while combining more than three stops in a journey is very rare. In contrast, the *movement-as-activity* metaphor also mentioned in the previous section can generate a priori knowledge statements such as most out-of-home activities have a considerable duration. More than 50% of all out-of-home activities take more than an hour and over 30% take even more than 2 h [3].

In tier 0, it is important to establish whether there are any privacy concerns from any of the sensor carriers and the experts of an application domain. This means that it is necessary to define a level of privacy according to who are the sensor carriers from whom data will be collected and who are the involved experts who will define the purposes of collecting these data. In the case of applications for transportation management, the sensor carriers might be those traveling from home to work, and the experts might be the company managers who have privacy goals towards the collected data. Company managers may not want it to be known where their employees travel during work hours, since this could point out to information about the activities of that company and those who are interested in using the data, for example, the supermarkets in the area may be interested in the trajectories relevant for better advertising. Once the stakeholders are identified in tier 0, it is also necessary to identify what their privacy requirements are, which could there be stated in terms of hierarchical levels of privacy.

2.5.2 Tier 1: The Positioning Space

The tier 1 describes the *positioning space* that contains the observations of the four-dimensional continuous field in space over time. Observations are measurement values at every point in space and time, based on some measurement scale, which may be quantitative or qualitative. Besides, observations are always marked by some degree of uncertainty, which depends on the type of sensors being used for collecting the location and movement information of mobile entities, such as X, Y, Z coordinates, speed and time. They can be navigation sensors (e.g. GPS, INS, MEMS sensors, digital compasses, etc.), remote sensing sensors (e.g. frame-based cameras, thermal cameras, laser scanners, etc.) and wireless technologies.

In this tier, the movement metaphors can be used to infer some empirical knowledge from the discovered patterns, such as density clusters of points in space. It is important to point out that we have only a set of observations of a finite sequence of time-referenced locations, represented by points where the movement of an entity starts at t_0 and ends at t_{end} in space. We do not have a trajectory representation of these points yet. However, it is possible to distinguish the observations according to four point representations. We distinguish among the following:

- Stop: A cluster of points that represent stops with a very short duration of some minutes due to traffic light or stop signs.
- Stop over: A cluster of points that represent a change of speed. For example, a road accident.
- Short stay: A cluster of points that represent stays with a short duration of some hours due to an activity such as working, shopping or leisure.
- Long stay: As cluster of points that represent stays with a long duration of several hours that will correspond to the sensor device being switch off or being at home.

The reasoning task consists of allowing one to infer a consequence from a set of point patterns. The consequences are drawn from these point patterns down to a specific metaphor such as, for example, the *movement-as-urban forms* metaphor. Human behaviour is constrained by urban forms, such as urban morphology, transportation network, availability of facilities and density of a city and the relative location of neighbourhoods. The *movement-as-urban forms* metaphor is needed to provide the premises the knowledge about the point patterns generated at this tier, for example, the shape characteristics of urban morphologies such as radial, linear, concentric and grid. Figure 2.4 illustrates some examples of these urban forms and their associated shape characteristics.

(1) Ring network in a concentric city: concentric pattern
(2) Radial network in a lob city: radial pattern
(3) Linear poly-nuclear city: linear concentric pattern
(4) Concentric poly-nuclear city: circular concentric pattern
(5) Linear network in a linear city: linear pattern
(6) Grid city: square concentric pattern

In the Real-Time Graz experiment in Austria, observations of cellphone usage have been collected through the city based on a location system where the movement of the cellphones was recorded and tracked with the agreement of the customers [42]. Figure 2.5 illustrates the visual density clusters found after a 24 h experiment in the city of Graz. It is already possible to realise the important role of a metaphor such as movement-as-urban forms in order to infer some knowledge about the clusters found in this experiment. In this example, it is possible to visually identify the circular concentric patterns representing possibly a concentric poly-nuclear city,

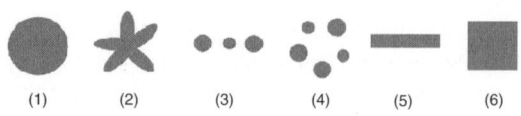

Fig. 2.4 Examples of possible urban forms as a movement metaphor

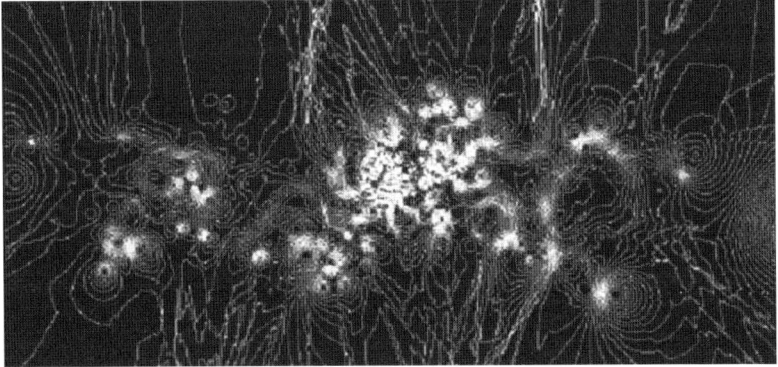

Fig. 2.5 Plan view of cellphone usage in the city of Graz – http://senseable.mit.edu/graz/

with some linear poly-nuclear patterns as well. It is also important to point out that in this example, the trajectory representation does not exist yet.

In terms of privacy, the main issue is to make sure that the granularity of data collection is in accordance with the privacy requirements of the sensor carriers. The domain experts can also make sure that the granularity of positioning data set is appropriate for the needs of the application domain and it complies with the privacy requirements of the different sensor carriers.

2.5.2.1 Tier 2: The Geographic Space

Tier 2 represents the 'geographic space', where our cognitive system from an array of properties values, which is capable of forming trajectories and reasoning about them. In geometrical terms, the movement of an entity is termed as a trajectory (we will use 'movement' and 'trajectory' interchangeably as already proposed by Pfoser and Jensen [37]. From the movement metaphors and representations already defined in the previous tier 1, a trajectory is therefore defined in this tier as any polyline between stops, stop overs, short stays and/or long stays. Moreover, in this tier is where the privacy requirements of the sensor carriers need to be implemented using security constraints. If there are sensor carriers who want their trajectories to be unobservable or who want to remain anonymous, then the necessary steps need to be taken here by applying different trajectory privacy-preserving methods such as cloaking and mixes.

The deductive reasoning task is characterised by inferring descriptive knowledge such as the trajectory characteristics (e.g. space and time), the geographic environment where the trajectory occurs (i.e. landscape), the topological relations between the trajectories and the association between the trajectories and specific features of a landscape. Such information can be a set of properties that are associated to

Fig. 2.6 Examples of possible types of local streets in an urban area

individual trajectories themselves, or a pre-defined group of trajectories. The overall goal is to help the experts to deduce the consequences for the existence of linear patterns of the movement of the trajectories. A set of movement metaphors is necessary to be defined by using some kind of classification scheme, set of association rules or clustering. For example, the aim might be to discover linear clusters that may be understood by which of these categories, allowing the experts to generate one or more internal theories to explain the discovered linear clusters of trajectories.

Currently, the main metaphor being used at this ontological level is *accessibility*, which can be obtained by the calculation of trajectory distances (lengths). This is carried out by using the average distances between zone centroides (regional scale) or using the distance between the origin and destination zone centroides (city scale). Depending on the size of zones, the actual trajectory distance may be significantly different to the distance calculated using average centroid distances [4, 50, 53]. The calculations also do not account for the configuration of the transport network in order to establish the actual route distances. In fact, they are only based on straight line distances between origin and destination zones [51].

However, the GKDD process is entirely propositional and different movement metaphors need to be taken under consideration at this ontological level. For example, the metaphors of *movement-as-urban form* and *movement-as-accessibility* can be used to deduce the consequences of the linear patterns from the trajectories based on the generalisation of the trajectories using the transportation networks such as the types of streets in an urban area (Fig. 2.6). Accessibility is constrained by urban forms such as the transportation network. For example, the tree street patterns usually impede the movement of people on reaching a destination; meanwhile, grid street patterns facilitate the fast reaching to a particular destination. A GKDD process might provide, for example, the data mining query mechanism necessary to discover an anomaly in the trajectories corresponding to different types of local streets, the similarities and dissimilarities among the trajectories according to the different characteristics of street types and finally discover point clusters of non-movement among trajectories and their association with the type of local street.

Previous initiatives on gathering information about the trajectories of people at the street level include the being currently realised under the OpenStreetMap Project.[1] This experiment is generating data about the movement behaviour of people in several cities in Europe (Fig. 2.7).

---

[1] More information available at http://wiki.openstreetmap.org/.

Fig. 2.7 This map was generated by the OpenStreetMap contributor using GPS data, and licensed as creative commons

2.5.3 Tier 3: The Social Space

The tier 3 encompasses the model that underlies our daily trajectories and their fundamental relations with human activities. Traditional spatial planning theory usually considers the geographic environment as a space where human activities take place and represent the geographic environment according to the goal of a spatial planning. For example, if the national government develops its spatial policies for a country, it requires a representation of the geographic environment that contains, for example, the cities, main infrastructure, population densities, nature areas, etc., flows of people and goods. Municipalities, on the other hand, developing their detail spatial policy for their cities use less abstract representations of the geographic environment. They need detailed, high level information about the individual functions of the buildings, detail infrastructure and social compositions of different neighbourhoods. The information is usually described as in terms of transportation modalities (e.g. car or public transport), commuting time or distance, spatial distribution of jobs and housing locations, total vehicle miles travelled, average trip lengths and congestion on links and intersections. Finally, from socio-economic statistics, information can be obtained about the geographic environment, such as income and education of a neighbourhood.

The above examples show that the planners use various metaphors for the geographic environment, depending on the context of the spatial planning. These metaphors are, however, currently based on mostly static models of activities. Relations between representations and activities are based on assumptions, spatial-analysis or activity-based models. However, the same geographic environment should also be considered as the result of movement patterns of people represented by their invisible footprints of trajectories on the landscape. Pulselli [39] has already pointed out that although positioning data sets of mobile entities are becoming increasingly available, surprisingly enough, they have not been used to describe the social and spatial systems.

A social system consists of individuals, groups and organisations that maintain relations through intentional (cooperative) activities based upon a more or less common set of rules, norms and values, and acts within the boundaries of the institutions that are derived from it [13]. The spatial system is composed of biotic and abiotic components, processes that alter these components and relations between them. An important difference between social and spatial systems is that the latter has been mostly described in geographic terms, while the first has not. One exception is found in the work of Hagerstrand [21], where three types of constraints have been formalised to represent the location of trajectories according to the human activities in both spatial and social systems. They are capability constraints (limit of activities of individuals due to their physical capabilities and/or resources), coupling constraints (constrain where, when and for how long and individual can join others to produce, transact and consume) and authority constraints (impose certain conditions of an individual's accessibility).

Therefore, it is important to realise that a trajectory takes place in a *social-spatial system*. The main metaphor used at this tier level is *movement-as-activity*. The socio-spatial organisation concept defines social-activities in a spatial perspective and can be used to analyse the interactions between social developments and the spatial system [60]. The spatial system and the social system are strongly intertwined and should not be analysed separately. There is a structural coupling between the geographic environment and the social system that acts upon it. Processes in the social system such as the economic, political, or cultural subsystems have spatial consequences and vice versa.

In this tier, it requires that some sort of target must already have been identified, and the task becomes one of uncovering 'what if' scenarios to explain the trajectory patterns within a social context. For example, instead of finding the best location for a supermarket based on the proximity of objects on the landscape, the problem becomes about finding the best location based on the patterns of the trajectories of people, which in turn, suggest that the human activity on a landscape is potentially more complex and probabilistic. Consequently, induction is most commonly applied at the tier 3 ontological level in order to determine a model that can best 'fit' the trajectories. Some examples are given below:

1. Discover patterns that explain the occurrence of a certain activity (e.g. shopping, recreation)
2. Discover the dependencies between different characteristics of activities
3. Discover the activities subsets and time periods with the corresponding patterns
4. Detect the occurrence of an unexpected activity

The tier 3 is where the 'classification anonymity' or the 'categorisation anonymity' requirements of the sensor carriers need to be guaranteed. The data miner and the domain expert need to make sure that exact rules that contain the complete data population do not breach any of the 'group privacy' requirements. Further, there may be privacy requirements in the following form:

- If the size of an identified group is less than 10% of the complete population, then a sensor carrier wants to be unobservable, or it should not be inferred that the sensor carrier belong to this group.

or

- If an unexpected activity (example number 4 above) is detected, and this contains information about clearly identifiable locations, small set of trajectories or small groups of people, then this information should either not be released or only accessible to trusted parties.

In tier 3, the experts need to be aware of sensor carrier's privacy requirements about inferred information and the context in which this information may be used. Inferences about the movement on trucks on highways and city centres may be meaningful for traffic balancing, but may be a threat to companies whose weaknesses on product distribution may then become inferable by other companies.

2.5.4 Tier 4: The Cognitive Space

The tier 4 represents the 'Cognitive Space' where the goal of a geographic knowledge discovery process is to gain knowledge through abductive reasoning that can function in the absence of pre-determined hypotheses, training examples or rules. Abduction is flexible because it is not restricted to using existing knowledge of pre-defined patterns (deduction) or else learns from examples that are presented or selected (induction), but instead free to create new structures that help to explain the patterns of a data mining process. Cognitive tacit knowledge is a non-linguistic non-numerical form of knowledge that is highly personal and context specific, and deeply rooted in individual experiences, ideas, values and emotions. It refers to ingrained schema, beliefs and mental models that are taken for granted [34].

Therefore, it is important to point out the difference between tacit and implicit knowledge. Implicit knowledge is something experts might know, but not wish to express, while tacit knowledge is something that experts know but cannot express, it is personal, difficult to convey, and which does not easily express itself in the formality of language. Searle [46] argues that cognitive tacit knowledge is not a form of knowledge (such as beliefs, theories and empirical hypothesis) but rather the preconditions of forming an individual's background knowledge. This raises the possibility that at least some metaphors of background knowledge can be confronted with the ones of cognitive tacit knowledge, which implicates that the features of the world are not independent of the mind.

2.6 Future Application Domains for a Privacy-Aware GKDD Process

This section relates the geographic knowledge discovery process as described in the previous section with the prospect of potential application domains. Three application domains have been selected to illustrate the expected innovations on applications in transport management, spatial planning and marketing.

2.6.1 Transport Management: The Integration of Multimodal Choices

Transport or transportation refers to the movement of people and goods from one place to another. The term is derived from the Latin trans (across) and portare (carry). Transport management is aimed at solving the problems between infrastructure (e.g. transport networks) and operations (e.g. road traffic control). Modalities are a combination of infrastructure and operations. In this scheme, both private and public transport modalities are managed by a planning authority having control of some decision variables: road pricing, transit ticket prices and the service characteristics of transit. The multimodal transport system is subject to some constraints: physical and environmental capacity constraints, and budget constraints. For example, in some cases an upper bound is imposed on the ticket price, in order to help people who are captive of transit.

The overall goal of the integration of multimodal transportation is to develop innovative solutions to fundamental problems, working in multidisciplinary teams that explore a range of expertise including statistical analysis, operational research, psychology, engineering, marketing, visual culture and aesthetics, and IT. One of the problematic issues of multimodal transportation is the routing of people towards and through inner cities. Cities tend to expand at their boundaries. As a consequence the city centre needs to handle increasing traffic flows of different modalities coming into the centre. As result of the growth of many cities the intensity of the traffic in many centres exceed by far the capacity of the centre's infrastructure as it was designed originally. Congestion, conflicts between various modes of transportation, parking problems and the exploitation of alternative routes through residence areas causing nuisance to the residence are the result. Traditionally, the problem of congestion of inner cities is dealt with by imposing all kind of parking restrictions, one-way street policies and road toll.

2.6.2 Tier 0: The Reality Space

The prospects of providing travellers and planners with multimodal information about their trajectory behaviour will, for the first time, integrate people's behaviour information with transportation information (infrastructure and operations). If such integration can be achieved based on the patterns observed from the trajectories of people and their relations between space, time and activities, we will have the possibility of presenting to travellers, and in particular drivers, with comparable information on travel options across modalities. An integrated multimodal information service will have a great potential to inform and influence travel choices, as well as identify the requirements from, and potential benefits of knowing about the patterns of people's trajectories. Geographic knowledge about the dynamics of movement, the distribution and composition of modalities might improve the knowledge about accessibility of inner cities as well.

2.6.3 Tier 1: The Positioning Space

The travel data is usually described as in terms of individual modalities (e.g. car or public transport), commuting time or distance, spatial distribution of jobs and housing locations, total vehicle miles travelled, mode of average trip lengths and congestion on links and intersections. However, evidence from the National Travel Survey data for Great Britain suggests that measures such as driving speed at different times of the day do not show large variations [51]. Moreover, Crane [11] firmly states that there is evidence to suggest that the travel diaries systematically overstate household travel, and as a result, short journeys may be under recorded.

Most common metaphors used at this tier are related to the *movement-as-accessibility* metaphor that defines how people move from one location to another as pedestrians, or taking cars, bikes or public transportation. For a geographic knowledge discovery process, this might imply the deductive search as one of the following examples:

1. Discover how a set of point patterns evolves from time t_1 to time t_2, in terms of a specific mode of transportation
2. Discover an observation window (spatial and temporal extends) where point patterns reveal a change of mode of transportation
3. Discover the rules that explain a spatial distribution of a set of point patterns at a given time in terms of a specific mode of transportation

The domain experts need to consider if any of these discovered patterns are in conflict with the privacy or security requirements of any of the sensor carriers. Privacy conflicts may arise at different granularity for different stakeholders. Especially in areas where the data are sparse, it may be possible to identify individuals or organisations. Once an individual is identified, his/her trajectories may be highlighted and many inferences that breach his/her privacy may be possible. In the case of an organisation, for example a business in a remote area, the traffic in and out of the building may be identifiable. If this information is made public, it may allow competing businesses to analyse the traffic of the remotely located business, allowing them to make inferences about their activities. A related example actually is in the US, where Bill of Lading and Ship Manifests collected in a system called 'US Customs Automated Manifest Systems' that include information about the trajectories of ships, the origin and the target organisations are made public in adherence to the 'Freedom of Information Act.' As a result, competing companies in the US are able to profile the import/export activities of many European companies with the Americas [45].

2.6.4 Tier 2: The Geographic Space

The geographic knowledge discovery process of patterns of moving people and their respective trajectories will have an impact on understanding the relationship between modality choices and trajectory patterns. For example, an increase of trajectory patterns at a local scale may result in a reduction on *accessibility*, which is defined as the ease of reaching a particular destination at this tier. Moreover, the knowledge of such trajectory patterns will play an important role in studying the route conditions from effects such as hazards, noise, traffic jams and visual pollution.

Metaphors are needed to allow the choice of the scale where a type of trajectory patterns can be found. One example is given by information about *accessibility*, which is usually known by gathering data from models and empirical surveys about how various individuals organise their daily activities in time and space, as well as the travel involved. Simulations, questionnaires as well as statistical data (when available) are some examples of sources used to infer knowledge about how people commute during the day, month or year. Most common socio-economical statistics used are population growth (e.g. annual growth rate), social statistics (e.g. levels of education, sex, age), economic growth (e.g. GNP, household incomes), employment structure (e.g. commercial, financial and service sectors), land use policy and regulations, land use patterns and built up areas (e.g. urban growth), transport statistics, commuter's demographic and household characteristics.

This kind of information would allow the experts to check if information can be inferred from the trajectory representation that may breach the privacy requirements of any of the sensor carriers. If so, the experts may consider using one or a number of privacy-preserving methods to protect these stakeholders.

2.6.5 Tier 3: The Social Space

Trajectory patterns are the outcome of a highly complex interplay between personal and household characteristics and features of the urban/rural environment. The development of a geographic knowledge discovery process will be essential to be able to gain knowledge on how some assumed spatiotemporal relationships between urban/rural forms and trajectories behaviour will result in activity patterns such as direction of commuting (e.g. within inner, within outer, or cross commuting in a metropolitan area), modality (distance, type of transport – private, public, non-motorised) and time (e.g. time spent by commuters who are travelling to work). Some examples are given below:

1. Discover accessibility patterns that explain the occurrence of shopping activity with its corresponding transportation modality
2. Discover the dependencies between working and leisure activities according to a specific transportation modality
3. Detect the occurrence of an unexpected activity

Here, the data miners and domain experts need to observe the classifications that they infer and check to see whether the 'category anonymisation' requirements of the different sensor carriers are breached.

2.6.6 Spatial Planning: The Adaptation of Space to Human Behaviour

Spatial planning is aimed to change the organisation of a geographic environment to meet the demands of society. Demands of society continuously change as the result of change in the society and also due to change in the geographic environment itself. Demands result into claims upon existing spatial functions. As space becomes a limited resource the geographic environment is expected to fulfil multiple functions [54]. People compete for the same resources. Especially rural areas are under increasing pressure and need to fulfil multiple functions [7]. They have to be attractive for recreation but also productive in terms of agriculture while they provide the space necessary to meet the claims of expanding urban areas. There is a clear shift of rural areas having primarily a production function towards rural areas that are regarded as differentiated residence area. At the same time, planning shifts from planning primarily based upon hierarchical principles towards more actor oriented and participatory types of planning [7, 17, 61].

In spatial planning, location-allocation representations and methods have been developed when positioning data sets were scarce and difficult to obtain, and models were deterministic or entirely predictable. In principle, mobile technologies made it possible to gather very large data sets containing movement information from mobile devices over time. This has opened the opportunity to deal with the location-allocation problems from a people's perspective in spatial planning.

2.6.6.1 Tier 0: The Reality Space

Currently, decision making in spatial planning often takes into account the land uses of a spatial environment. Currently, land use describes the *activity* the landscape is used for. At the national level, current types of land use are important to decide about new development scenarios for a region in terms of defining recreational areas, nature and urban areas. At local levels, land use is an important parameter for deciding about locations for living, industrialisation and leisure.

The geographic knowledge about the patterns of the movement of people on a spatial environment might refine the concept of land use in spatial planning. Currently land use is mostly based on the knowledge obtained from a spatiotemporal classification of features on a landscape. These classifications are static as they do not include human behaviour. Knowledge about movement of people and masses might add additional insight beyond that of the traditional land use concept, mainly in terms of understanding the effects of the landscape on the movement of people and vice versa. As a result, land use could become the *activity* metaphor based on the movement of people, rather than the location of its features.

A second metaphor is related to the general concept of functionality of space. Functional spaces are spaces that are designed to fulfill a specific task. Typical examples focussed on in this research are shopping areas, airports, areas for large scale events and parks. Knowledge of the movement of people in these types of areas is required for the situating of shops, shop-types and checkpoints. More over the dimensioning of pathways, gateways and emergency evacuation routes might benefit from additional knowledge about the spatial temporal dynamics of moving crowds.

2.6.6.2 Tier 1: The Positioning Space

The above mentioned examples of applications in spatial planning would benefit from the gathering of positioning data of moving people on a landscape into a trajectory data warehouse. This type of data neither is commonly used in the process of designing spaces nor is it commonly available. Many of the design decision are traditionally based on estimations, extrapolation from known cases and simulation models.

2.6.6.3 Tier 2: The Geographic Space

Very little is known about the growth, shift or even decrease of movement patterns of people in urban and rural areas. There is no universally accepted standard classification of human *activities*, and the association among activities that generate patterns of trajectories at individual, organisational and urban/rural form levels is not well understood. The basic assumption is that individuals and households try to meet their basic needs and preferences by participating in activities, while the environment (urban/rural forms) they live in offers them the opportunities and constraints to do so. The geographic discovery process needs to be essentially targeted to finding linear patterns of trajectories that can be understood by a planner and designer. This implies that attributes of trajectories depend on the type, the scale and the goal of the planning and the function the space needs to fulfil. Little or no knowledge exist about these issues yet.

The main metaphor at this tier is *movement-as-urban form*. Mobility is the trajectory of individuals that is dependent on urban forms such as transportation networks and land use. Land uses support human activities. Those activities are spatially separated. People need transport to go from one place to another (home → work → shop → home, for instance). Transport is a 'derived demand,' in that transport is unnecessary but for the activities pursued at the ends of trips. Therefore, a certain land use type might enable common activities to occur close to a specific place (e.g. housing and food shopping), as well as places with higher density development closer to transportation lines and hubs. Poor land use concentrates activities (such as jobs) far from other destinations (such as housing and shopping).

2.6.6.4 Tier 3: The Social Space

Multifunctional land use can be defined as combining various socio-economic *activities* in the same area. The basic idea is to save scarce space [55]. An important aspect in the planning of multifunctional land use is to try to integrate activities as harmonised as possible and when possible strive to a synergy between two (or more) activities. In many European countries, there is a constant friction between the aim of keeping particular environments as undisturbed as possible to facilitate the habitat functions for flora and fauna or allowing these environments to function as leisure areas for citizens to divert from their demanding daily lives. The concept of multifunctional land use tries to integrate both activities of a geographical environment. The challenging factor is to provide ample opportunities for recreational activities while preserving and developing nature. Traditional instruments, which are part of the 'toolbox' of the planner, are zoning, routing and temporary closing. Mountain bikes are for example barred from highly sensitive area or only allowed on special tracks. Some breeding areas are closed for traffic during certain hours a day or during the mating and breeding seasons.

Most of these decisions are made in absence of knowledge about the effects of the dynamics of recreational activities on the quality of the nature in that area. No knowledge about the current behaviour of the various leisure seekers is present. Information about intensity and followed routes are mostly estimated or based on incidental counting. Knowledge about the patterns of movement of visitors of nature areas and the type of activities might improve the harmonisation of multifunctional use of nature areas. Decisions can be made based on measured patterns of movement and related to observed effects on the environment. This might lead to more precise or flexible zoning, routing or access policies.

Additionally geographic knowledge discovery might be aimed at finding patterns that represent conflicting behaviour. In relatively small areas (like the ones found in the Netherlands) often there are irritations between, for example, walkers and mountain bikers or horseman. Insight in the periods and locations of potential conflicting trajectories might allow managers to improve the way they handle the various types of visitors leading to fewer conflicts. The research challenge for these applications is to discriminate amongst the different type of movement patterns or trajectories and assign the properties that are relevant to these trajectories. The question of how to analyse, the differences between, for example, the patterns of trajectories of a hiker and a mountain-biker is not a trivial one. No methods are known yet that can deal with recognition and classification of these subtleties. At the same time, the possibility of identifying such subtleties opens up a number of privacy-related questions. Individuals may not want their leisure activities to be so clearly identifiable. In fear of possible negative outcomes for their activities, institutions or organisations may also argue against the collection of data of such high granularity.

2.6.7 Marketing: The Shift Towards Movement-Aware Marketing

Currently marketing is mostly done based on customers' profile, which, normally, is statically defined. Such profiles are based on characteristics like gender, age, income, family situation and purchase history. On the basis of the customer's characteristics, strategies can be developed to determine a need for purchasing certain products amongst potential or existing customers. Traditionally, the success of marketing depends on what is called the marketing mix of four P's: product, price, promotion and placement. This rather traditional view on marketing has been criticised, since its main focus is on a company or marketer rather than the consumer. Furthermore, the traditional marketing mix model does not serve the marketing of services very well. In the last decades, suggestions for change in the various elements of the marketing mix have been articulated. For example, placement needs to be converted to convenience, and promotion to communication. More elaborate knowledge about potential customers also needs to be improved in marketing mix models.

Current computer and database technology enable the storage, processing and analysis of much more variables of human behaviour that are relevant to marketing. An important development is that of geo-marketing. Geo-marketing implies the use of GIS to add location information into the marketing mix. Based on spatial analysis, additional knowledge and insight might be gained about, for example, the spatial distribution of income and demographic composition of districts.

The main metaphor at this tier is that of *movement-as-personalisation*. Using movement data, marketers and service organisations can better target their information and services to specific users depending on their activities, relations and locations. The 'scary' outlook of many people to be spammed by location-based advertisements, generated by relative dumb LBS, can be alleviated by providing more intelligent information based on movement behaviour.

2.6.7.1 Tier 0: The Reality Space

Recently, LBS have been added to the geo-marketing sector as a new marketing tool. Using LBS, marketers can pinpoint their marketing mix and enhance their communication with potential customers based on their exact location and time. Most obvious examples are push marketing based on SMS messages sent when a customer passes a shop he or she might be interested in (mobile advertising).

As a next step, LBS might develop further towards movement-based services (MBS). One of the differences with 'traditional' LBS is that it will take into account the history, behaviour and relation with other movements. LBS only provide the context from the users, and the environment (*who* is *where* at time *t*). MBS have the potential to add to this, knowledge about *what* he/she did, *how* he/she did it and with *whom*.

2.6.7.2 Tier 1: The Positioning Space

The data used for LBS usually describe only the location in space, a caller id and a time stamp. For MBS, information about the followed tracks, the movement characteristics (speed, acceleration, periodicity in movements, etc.) need to be added and stored. This is a substantial shift in how to deal with the data. Currently LBS do not need the analysis and storage of locations per se. The majority of LBS are user or event based. The discrete event approach of LBS may only requires the data for the moment the data is requested. MBS typically require the maintenance and storage of data to be able to infer patterns out of it. One of the concerns at this stage is privacy. As data are required and requesting of the movements of individual people preserving privacy is an important requirement. The ubiquitous nature of the collection of movement data makes privacy an even more pressing concern. The main metaphors in this tier are related to construction of the public/private divide and freedom from intrusion [20]. The control about what and when data about movement activities should be private or public need to be clear to and perhaps in control of the person carrying a sensor. The right to be let alone is a basic right of humans. Especially in marketing-based application, the control of the right should be part of the decision making about what part of the reality space should be sampled and registered.

2.6.7.3 Tier 2: The Geographic Space

Little is known about the use of movement patterns for marketing purposes. There is barely research after the marketing-related behaviour and movement behaviour. In principle there are two models that make use of movement data: the first, the consent model is based on informing or assisting users with information or services based on authorisation given by them. This means that people decide what type of services or information they are interested in. On the basis of their movement behaviour, these services/information are provided to them when required or needed in an intelligent fashion. In the second model, the informed model, users receive targeted information based on their movement behaviour, location, time and the behaviour of others, i.e. the behavioural pattern they are part of. The challenge of the informed model is to couple information about movement behaviour with other sources of (behaviour) information like shopping history and non-behaviour information. For both models, the information should, in principle, be able to infer the basic knowledge on the following:

- What someone is doing
- How someone is moving and with whom
- Who else is moving in a similar manner (coinciding patterns)

Therefore, the definition of privacy in such an application depends on which of the above location information the sensor carrier is unwilling to have analysed. Further, any collection of information about clusters of people around an individual is also information about those in the cluster. This may result in a privacy conflict, between those in the cluster, who do not want their participation in the cluster to be known, and the sensor carrier. Both, the collection and analysis of data, as well as the selection of privacy-preserving knowledge discovery methods depend on the specification of privacy goals and resolution of privacy conflicts among stakeholders. Privacy-aware services can be developed, which are currently hard to realise, such as presenting services of information based on the type of movement. If you are, for example, driving a car on a crowded highway you probably would like to have information presented differently than when walking around in a city. So movement-based behavioural information might also facilitate the means and methods by which information is presented to customers.

2.6.7.4 Tier 3: The Social Space

On the basis of tier 2, the inferential space tries to discover the causes and consequences of movement behaviour. The discovery of geographic knowledge related to privacy aware marketing is mainly targeted to finding groups that show similar behaviour and to determine if this behaviour is interesting, given a certain marketing goal. Examples of typical knowledge are the following:

- Discover the general patterns that explain the behaviour of certain groups of people given a marketing perspective. Using the characteristics of these groups, marketing can be targeted and personalised.
- Discover the dependencies between movement behaviour and the effects of personalised movement-aware marketing. Can movements of people be influenced by certain marketing actions or are the effects of marketing dependent of movement behaviour?
- Discover the type of information appreciated by people when they are moving at a certain time, modality and location.

Currently the above types of knowledge discovery cannot be carried out or only limited based on marketing research.

2.7 Conclusions

This chapter introduces a geographic knowledge discovery process, in which the primary goal of identifying, associating and understanding patterns is used to infer the location, identity and relationships among mobile entities, and their respective trajectories in a spatial environment. In this case, the different types of inferences play a different role according to what a domain expert wants to infer, i.e. the location, changes, properties, identity or relationship among the appropriate metaphors. It is the metaphor, and only after it makes sense that an unknown set of patterns can be interpreted and understood by a domain expert. Basically, three modes of reasoning are presented using a multi-tier ontological framework. They are deductive, inductive and abductive modes of reasoning.

In the deductive mode of reasoning, the geographic knowledge discovery process involves the search for common attributes among a set of mobile trajectories, and then the arrangement of these trajectories into classes, clusters or patterns according to a meaningful metaphor. The focus is on applying statistical approaches (probability distributions, hypothesis generation, model estimation and scoring) for exploring classes, clusters or patterns from a data set.

In the inductive mode of reasoning, the geographic knowledge discovery process is based on learning due to the reduction of uncertainty in knowledge. Several techniques have been developed, such as rule induction, neural networks, genetic algorithms, case-based learning and analytical learning (theorem proving). Many techniques partition the target data set into as many regions as there are classes by using a function, for example, a posterior probability or linear discriminate functions. These techniques provide a data fit, in the sense that the main goal is to generate derived knowledge describing the data, often called concept hierarchies.

In the abductive mode of reasoning, the importance of cognitive tacit knowledge needs to be considered. Will the information have the same meaning and weight (in terms of privacy) if the patterns are used in contexts other than it was meant to be? In this case, the value of the discovered knowledge is judged and the decision is taken on its role in making decisions for the application domains such as transport management, spatial planning and geomarketing. It might turn out that final decisions made are not in line with patterns suggested by the knowledge discovery process. The political, economic or social realities of the decision-making process are sometimes prevalent above the rational knowledge inferred from a geographic knowledge discovery process. Questions like why do people choose certain modality of transportations at certain times of the day, and why are certain transportation modalities more present in area a than in area b are some examples where new metaphors could explain the relations between certain movement behaviour and the characteristics of a geographic environment.

The inevitable challenge facing the research community at the moment is directed towards a more complete integration of these modes of reasoning and their association to movement metaphors within a geographic knowledge discovery process. It is in this context of attempting to build bridges between them that three application domains are identified and explained using the proposed multi-tier ontological framework on transport management, spatial planning and marketing.

This chapter has also shown our first attempt at integrating privacy requirements into a multi-tier ontological framework of a geographic knowledge process. Definitions of privacy, kinds of possible privacy threats and the complexity of the different requirements for privacy required by stakeholders have been discussed within a geographic knowledge discovery process.

References

1. G.D. Abowd, C.G. Atkeson, J. Hong, S. Long, R. Kooper, and M. Pinkerton. Cyberguide: A mobile context-aware tour guide. *Wireless Network*, 3(5):421–433, 1997.
2. C. Aldridge. A theoretical foundation for geographic knowledge discovery in databases. In *Proceedings First International Conference on Geographic Information Science (GI-Science'00)*, 2000.
3. K.W. Axhausen and T. Gärling. Activity-based approaches to travel analysis: Conceptual frameworks, models and research problems. *Transport Reviews*, 12:324–341, 1992.
4. D. Banister, S. Watson, and C. Wood. Sustainable cities, transport, energy, and urban form. *Environment and Planning B: Planning and Design*, 24(1):125–143, 1997.
5. B. Brunk and B. Davis. Sdat enterprise: Application of geospatial network services for collaborative airspace analysis. In *Proceedings of the CADD/GIS Symposium*, 2002.
6. Y. Cai, N. Cercone, and J. Han. Attribute-oriented induction in relational databases. In G. Piatesky-Shapiro and W. Frawley (eds.), In *Proceedings of Knowledge Discovery in Databases*, pp. 213–218, 1991.
7. H.V.D. Cammen and M.A.D. Lange. Ontwikkelingen in wetenschap en technologie: Sturingstheorieen en landelijke gebieden. In *Nationale Raad voor Landbouwkundig Onderzoek*, 1998.
8. S. Card, J. Mackinlay, and B. Shneiderman. Information visualization. In *Readings in Information Visualization*, pp. 1–34. Morgan-Kaufmann, 1998.
9. G. Chen and D. Kotz. A survey of context-aware mobile computing research. Technical Report TR2000-381, Department of Computer Science, Dartmouth College, November 2000.
10. D. Cook, A. Buja, J. Cabrera, and H. Hurley. Grand tour and projection pursuit. *Journal of Computational and Graphical Statistics*, 2:225–250, 1995.
11. R. Crane. The influence of urban form on travel: An interpretive review. *Journal of Planning Literature*, 15(1):3–23, 2000.
12. P. Deadman. Modelling individual behaviour and group performance in an intelligent agent-based simulation of the tragedy of the commons. *Journal of Environment Management*, 56:159–172, 1999.
13. F. Kleefmann. *Planning als zoekinstrument*. VUGA, 's Gravenhage, 1984.
14. U. Fayyad, G. Piatetsky-Shapiro, and P. Smyth. From data mining to knowledge discovery: An overview. In *Advances in Knowledge Discovery and Data Mining*. AAAI/MIT Press, 1996.
15. A.U. Frank. Ontology for spatio-temporal databases, pp. 9–78, 2003.
16. M. Gahegan, M. Wachowicz, M. Harrower, and T. Rhyne. The integration of geographic visualization with knowledge discovery in databases and geocomputation. *Cartography and Geographic Information Science (special issue on research challenges in geovisualization)*, 28(1):29–44, 2001.
17. S. Geertman and J. Stillwell. *Planning Support Systems in Practice*. Springer, 1996.
18. E. Gendlin. Crossing and dipping: Some terms for approaching the interface between natural understanding and logical formulation. *Journal Minds and Machines*, 5:547–560, 1995.

19. T.M. Georges, J. Harlan, L. Meyer, and R. Peer. Tracking hurricane claudette with us air force one over the horizon radar. *Journal of Atmospheric and Oceanic Technology*, 10:441–451, 1993.
20. S. Gürses, B. Berendt, and T. Santen. Multilateral security requirements analysis for preserving privacy in ubiquitous environments. In *Proceedings of Workshop on Ubiquitous Knowledge Discovery for Users (UKDU'06)*.
21. T. Hägerstrand. What about people in regional science? *Papers of the Regional Science Association*, (24):7–21, 1970.
22. J. Han and Y. Fu. Discovery of multiple-level association rules from large databases. In *Proceedings of 21st International Conference on very Large Data Bases (VLDB'95)*, pp. 420–431, 1995.
23. J. Han, K. Koperski, and N. Stefanovic. GeoMiner: A system prototype for spatial data mining. In *Proceedings of 1997 ACM-SIGMOD International Conference Management of Data (SIGMOD'97)*, pp. 553–556, 1997.
24. M. Harvey and J. Han. Geographic data mining and knowledge discovery: An overview. In *Geographic Data Mining and Knowledge Discovery*, pp. 3–32. Taylor and Francis, 2001.
25. C. Hsu and C.A. Knoblock. Using inductive learning to generate rules for semantic query optimization. In *Advances in Knowledge Discovery and Data Mining*, pp. 425–445. MIT Press, 1996.
26. A. Hunter, N. El-Sheimy, and G. Stengouse. Close and grizzly gps/camera collar captures bear doings. *GPS World*, February Issue: 24–31, 2005.
27. J. Huttenlocher, L. Hedges, and S. Duncan. Categories and particulars: Prototype effects in estimating spatial location. *Psychological Review*, 98:352–376, 1991.
28. M. Johnson. Metaphorical reasoning. *Southern Journal of Philosophy*, 21(3):371–389, 1983.
29. G. Lakoff. *Women, Fire, and Dangerous Things*. University of Chicago Press, 1987.
30. A. Ligtenberg, M. Wachowicz, A.K. Bregt, A.J.M. Beulens, and D.L. Kettenis. A design and application of a multi-agent system for simulation of multi-actor spatial planning. *Journal of Environment Management*, 72:43–55, 2004.
31. A. MacEachren, M. Wachowicz, R. Edsall, D. Haug, and R. Masters. Constructing knowledge from multivariate spatiotemporal data: Integrating geographic visualization (gvis) with knowledge discovery in databases (kdd). *International Journal of Geographic Information Science*, 13(4):311–334, 1999.
32. T.P. McNamara, J.K. Hardy, and S.C. Hirtle. Subjective hierarchies in spatial memory. *Journal of Experimental Psychology: Learning, Memory, and Cognition*, 15:211–227, 1989.
33. G. Navratil and A. Frank. Data quality for spatial planning – An ontological view. In *Proceedings on Competence Center of Urban and Regional Planning (CORP'06), (Geomultimedia'05)*, 2006.
34. I. Nonaka and H. Takeuchi. *The Knowledge Creating Company*. Oxford University Press, 1995.
35. R. Oppermann and M. Specht. A context-sensitive nomadic exhibition guide. In *Proceeding of Second International Symposium on Handheld and Ubiquitous Computing (HUC'00)*, pp. 127–142, 2000.
36. D. Peuquet. *Representations of Space and Time*. The Guilford Press, 2002.
37. D. Pfoser and C.S. Jensen. Querying the trajectories of on-line mobile objects. In *Proc. 2nd ACM Intl Workshop on Data Engineering for Wireless and Mobile Access*, pp. 66–73, 2001.
38. B. Poore and N. Chrisman. Order from noise: Toward a social theory of geographic information. *Annals of the Association of American Geographers*, 96(3):508–523.
39. R.M. Pulselli, F.M. Pulselli, C. Ratti, and E. Tizzi. Dissipative structures for understanding cities: Resource flows and mobility patterns. In *Proceedings of the First International Conference on Built Environment Complexity (BECON'05)*, pp. 271–279, 2005.
40. J.R. Quinlan. Learning logical definitions from relations. *Machine Learning*, 5(3):239–266, 1990.
41. N. Ramakrishnan and A.Y. Grama. Data mining-guest editors' introduction: From serendipity to science. *Computer*, 32(8):34–37, 1999.

42. C. Ratti. Space syntax: Some inconsistencies. *Environmental and Planning B: Planning and Design*, 31:487–499, 2004.
43. I. Reginster and G. Edwards. The concept and implementation of perceptual regions as hierarchical spatial units for evaluating environmental sensitivity. *Journal of Urban and Regional Information Systems Association*, 13(1):5–16, 2001.
44. J. Richmond. Simplicity and complexity in design for transportation systems and urban forms. *Journal of Planning Education and Research*, 17:220–230, 1998.
45. C. Schulzki-Haddouti. U.S. unternehmen veröffentlicht sensiblen daten aus seefrachtverträgen. *http://www.heise.de/newsticker/meldung/76598*, 2006.
46. J. Searle. *The Construction of Social Reality*. Free Press, 1995.
47. S. Shekhar, B. Hamidzadeh, A. Kohli, and M. Coyle. Learning transformation rules for semantic query optimization: A data-driven approach. In *Special Issue on Learning and Discovery in Knowledge-Based Databases*, Vol. 5(6), pp. 950–964. Institute of Electrical and Electronics Engineers, Washington, 1993.
48. M. Siegel. Automatic rule derivation for semantic query optimiser. In *Proceedings of the International Conference on Databases and Expert Systems (DEXA'88)*, pp. 371–385, 1988.
49. C. Smyth. Mining mobile trajectories. In *Geographic Data Mining and Knowledge Discovery*, pp. 337–361. Taylor and Francis, 2001.
50. N. Spence and M. Frost. Work travel responses to changing workplaces and changing residences. In *Cities in Competition: Productive and Sustainable Cities for the 21st Century*, pp. 359–381. Longman Australia Pty Ltd., 1995.
51. D. Stead and S. Marshall. The relationships between urban form and travel patterns: An international review and evaluation. *European Journal of Transport and Infrastructure Research*, 1(2):113–141, 2001.
52. E. Tiezzi. *Beauty and Science*. WIT Press, 2004.
53. P. Troy. Let's look at that again. *Urban Policy and Research*, 10(1):41–49, 1992.
54. A. van der Valk. The dutch planning experience. *Landscape and Urban Planning*, 58:201–210, 2001.
55. R. Vreeker, H.D. Groot, and E. Verhoef. Urban multifunctional land use: Theoretical and empirical insights on economies of scale, scope and diversity. *Built Environment*, 20(4):289–307, 2004.
56. M. Wachowicz. *Object-Oriented Design for Temporal GIS*. Taylor and Francis, 1999.
57. M. Wachowicz. Geoinsight: An approach for developing a knowledge construction process based on the integration of gvis and kdd methods. In *Geographic Data Mining and Knowledge Discovery*, pp. 239–259. Taylor and Francis, 2001.
58. C. Wallace. Classification by minimum-message-length encoding. In *Proceedings of the International Conference on Advances in Computing and Information*, pp. 72–81. Springer, 1990.
59. W. Wang, J. Yang, and R. Muntz. Stinga: Statistical information grid approach to spatial data mining. In *Proceedings of 23th International Conference on very Large Data Bases (VLDB'97)*, pp. 186–196, 1997.
60. J. Wisserhof. Landelijk gebied in onderzoek: Ontwikkeling en toepassing van een interdisciplinair conceptueel kader. In *KU Nijmegen*, 1996.
61. C.V. Woerkum. Communicatie en interactieve beleidsvorming. In *Sam-som, Alphen aan den Rijn*, 2000.

Chapter 3
Wireless Network Data Sources: Tracking and Synthesizing Trajectories

C. Renso, S. Puntoni, E. Frentzos, A. Mazzoni, B. Moelans, N. Pelekis, and F. Pini

3.1 Introduction

Due to inexpensive modern sensing technologies and extensive use of wireless communication, location information about moving objects is increasing rapidly. Some positioning technologies are based on GPS-equipped devices, while others utilize the infrastructure of the underlying communication network. This opens new opportunities for offering, monitoring, and decision-making novel applications in a variety of fields. To name a few, we have location-based services (LBS), fleet management and traffic control applications, emergency, navigation, and geocoding services. These compose a subset of existing applications where such kind of data comprise the core of the underlying business.

Nevertheless, a new class of applications will take advantage from GeoPKDD approach, where the core information is the *movement of people*, i.e., sequences of positions of users over time.

Therefore, starting from the analysis of people's movements, a novel class of services, denoted movement-based services (MBS), can be enabled. LBS can be rephrased as *Give me some service depending on where I am now*, whereas MBS can be rephrased as *Give me some service depending on where I and other people have been in the past*.

Movement data to be analyzed can be *real* or *synthetic*. Indeed, real movements come from collecting trajectories of people; these can be represented as synthetic trajectories that simulate specific kinds of movements. Having synthetic data sets is extremely useful for correct development, verification, and testing of data analysis algorithms such as data mining [10]. Indeed, having a predictable data set allows developers to test algorithms on extreme situations and to verify the correctness of results.

C. Renso
KDD Laboratory, ISTI-CNR, Pisa, Italy, e-mail: chiara.renso@isti.cnr.it

Sequences of user positions have to be collected to draw the (possibly approximated) trajectory. Wireless network positioning technology allows one to locate a device inside the network with different levels of accuracy depending on the specific method used. However, methods to regularly collect a number of user positions need to be activated for user tracking.

This chapter presents the sources of user movement data, both real and synthetic. Indeed, the first part of the chapter is devoted to a survey of positioning technologies available with today's network, cellular, satellite vehicles, and more recent technologies such as Wi-Fi and Bluetooth. The second part of the chapter focuses on surveying tools capable to produce synthetic movement data sets, simulating objects either moving on free space or constrained on the network.

Structure of the chapter follows. Sect. 3.2 presents positioning technologies categorized from the point of view of the supporting technology: global system for mobile communications (GSM), satellite, and other wireless technologies. Section 3.3 introduces the concept of mobile location service with a brief overview of available technology. Section 3.4 discusses some approaches to obtain user trajectories collecting positioning information. Then, in Sect. 3.5, we present some approaches for synthetically generating trajectories. Approaches are categorized by the kind of movement generated, free or constrained by a network. Eventually, Sect. 3.6 draws the conclusions and a road map for wireless tracking and trajectory synthesizing in GeoPKDD.

3.2 Categorization of Positioning Technologies

In this section, we survey some techniques to locate users moving in a wireless network. Positioning means finding where a user is, at a given time instant, inside the network.

The main quality measure of positioning techniques is *accuracy*. Assuming that we can measure time precisely, measuring accurate location data depends on several factors. Among them, we recall the inevitable introduction of inaccuracy in the real world (e.g., distortion of radio waves due to atmosphere irregularities and natural obstacles) and the level of accuracy provided by the corresponding positioning technology in use. Whatever the positioning technology is, industry has introduced the concept of quality of position (QoP) [4], signifying the fact that reporting the location of a mobile object should also be accompanied with some indication of the incorporated inaccuracy. More specifically, QoP is usually prescribed by the following three dimensions:

- The accuracy of the location information, i.e., how closely we can determine the position of a mobile object in a worst-case scenario
- The age of the location information, i.e., how long ago the location information was collected
- The confidence in the accuracy information, i.e., the probability of the error (e.g., the accuracy is 50 m with 67% probability)

Accuracy	Example applications
Regional up to 200 km	weather reports, localized weather warnings, traffic information (pre-trip)
District (up to 20 km)	local news, traffic reports
Up to 1 km	vehicle asset management, targeted congestion avoidance advice
500 m to 1 km	rural and suburban emergency services, manpower planning, information services (where are ...?)
100 m (67%)–300 m (95%)	U.S. FCC mandate (99-245) for wireless emergency calls using network based positioning methods
75 m–125 m	urban SOS, localized advertising, home zone pricing, network maintenance, network demand monitoring, asset tracking, information services (where is the nearest?)
50 m (67%)–150 m (95%)	U.S. FCC mandate (99-245) for wireless emergency calls using handset based positioning methods
10 m–50 m	asset location, route guidance, navigation

Fig. 3.1 Accuracy in LBS examples (with respect to range) [26]

Naturally, different applications require different qualities of position. In Fig. 3.1, we present a representative set of applications with respect to the requirements in location accuracy.

Positioning techniques are categorized in the next paragraphs as cellular based [GSM/universal mobile telecommunications system (UMTS)] and satellite based (GPS, Galileo, and others), depending on the supporting communication network.

3.2.1 Cellular-Based Positioning Technologies

In this section, we shortly discuss about GSM/UMTS positioning technologies nowadays used by cellular telecommunication companies to enable LBS for customers and third-party companies.

3.2.1.1 Technology

Two of the most widely used telecommunication systems for mobile telephones are GSM and UMTS protocols [8]. These protocols are based on a network architecture called *cellular*. In general, in a cellular architecture, a geographical area is covered by a number of *antennas* (or base transceiver stations, BTS) emitting a signal to be received by mobile devices. Each antenna covers an area called *cell*. In this way, the covered area is partitioned in a number of, possibly overlapping, cells, uniquely identified by the antenna. In urban areas, cells are close to each other and small in area (even micro- or picocells), the diameter can be from 100 m to a kilometer. In rural areas, the radius of a cell can reach a maximum of 30 km. The presence of a device in a cell is detected by the system periodically to maintain correctness and validity of the location information subsystem. Position data are maintained in GSM/UMTS location databases (called VLR/HLR – visitor/home location register) [7]. These registers maintain location information within substructure of the network such as location areas, a group of antennas covering a quite extended area (100–300 km^2). The actual location of a device in GSM network is registered at this level. However, since these registers are specialized in routing rather than positioning, with the advent of LBS in last few years, another class of location register has been introduced in network vendors. These registers are specialized in retrieving and calculating the position of subscribers. Furthermore, the introduction of the serving mobile location center (SMLC) server allows to calculate the location information and an estimation of the achieved accuracy.

3.2.1.2 Data Format

We have different data format to represent location information, depending on which method is used to calculate the position on the network. For example, antenna-based location information is represented by the spatial information as the position of the antenna plus some additional information, such as its orientation, coverage area, and signal power. In some methods, it is possible to use location information derived by two, three, or more antennas, to obtain a more precise and reliable information. In general, data format of the location information should indicate:

- The time of detection that can be represented by a tuple indicating the detection date and time, e.g., dd–mm–aaaa and hh:mm:ss GMT + hh
- A position information represented by a precise location with some geographic reference system (WGS84, UTM32, etc.)
- The identifier of the detected mobile user such as its *international mobile subscriber* (IMSI) and telephone number (MSISDN)

3.2.1.3 Methods

Various methods can be used to calculate an approximate position of a mobile device. Methods are *device independent* when no further requirements are needed on the handset to be localized. Similarly, they are *network independent* when they do not require additional technology to be implemented in the network infrastructure.

Figure 3.2 shows an overview of all accuracy levels of various methods. We can note that the most accurate is GPS method and its accuracy level decreases in urban areas due to high buildings and a number of other obstacles.

In the following, we briefly discuss the most used cellular positioning techniques.

3.2.1.4 Cell Identity

In the cell identity (CI) positioning method, the location of a device is identified by the cell where it is connected. This information is available at the network as well as at the handset. The antenna identifier is converted to a geographic position by means of the existing knowledge residing in the coverage database SMLC. Accuracy

3 Wireless Network Data Sources

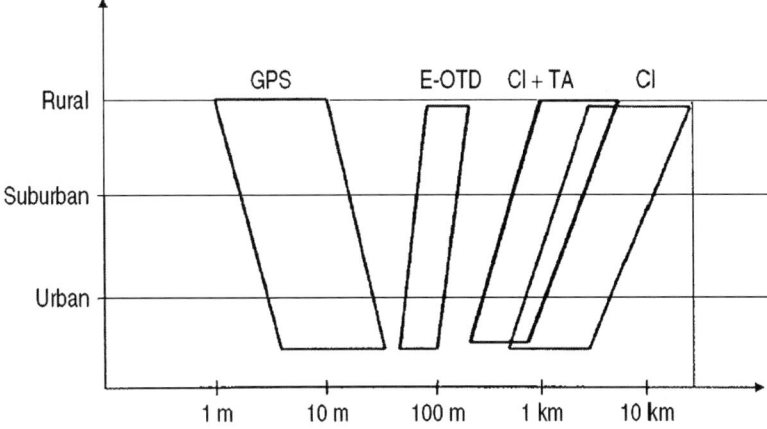

Fig. 3.2 Position information accuracy (with respect to range of cells)

depends on the cell size and the antenna type (circular or sectorial) and can vary from 100 m to few kilometers. This method can be improved with *timing advance*, a measure of the distance between the antenna and the device.

3.2.1.5 Cell Identity and Timing Advance

This method improves CI using measurement reports that contain power level at the handset from the serving cell and cells on the neighbor list. The power level at the handset can be used to estimate distances from device to antenna using simple wave propagation models. Accuracy is slightly better than CI, and depends on the location of antennas and environmental conditions that can affect the signal strength.

3.2.1.6 Enhanced Observed Time Differences

In *enhanced observed time differences* (E-OTD) positioning method, the handset measures the arrival time of signals transmitted from three or more antennas. Two specific methods can be implemented depending on the available underlying technology: E-OTD MS assisted and E-OTD MS based. In MS assisted technology, measurements are made by the handset and then transferred to the SMLC that calculates, by triangulation, the position of the device. In the MS-based E-OTD, the position calculation function resides at the handset and the calculated position is returned to the SMLC. Accuracy can vary from 50 to 100 m. It is worth noting that, in terms of resources, E-OTD is a very expensive method since it needs some additional and specific equipments to be added to the network. For this reason, it is not widely used by LBS vendors that prefer some cheaper and simpler solution, such as *assisted GPS* (A-GPS).

3.2.1.7 Assisted GPS

In A-GPS, the handset measures the arrival time of signals transmitted from three or more satellites (satellite-based methods are described in detail in Sect. 3.2.2). This technology has a quite low impact on the network because it requires only the support at the SMLC level. Positioning performances are better in rural space and poor in urban space where buildings and other obstacles disturb signals from satellites, thus accuracy can vary from 2 m in rural areas to 20 m in urban areas. This method is quite efficient, in terms of quality/cost ratio, and reliable in terms of quality of given information. At the other side, it has a technological dependence since the handset needs to be GPS compliant to receive signals from satellite.

3.2.2 Satellite Vehicles-Based Positioning Technology

Modern localization satellite vehicles (SVs) techniques (or terrestrial equipments based) for positioning are based on electromagnetic impulses traveling time between the transmitters and receivers.

SVs positioning techniques are usually categorized in two main classes:

1. Mobile terminated (MT)
2. Mobile originated (MO)

Most of LBS known to mass market customers belong to the first family (GPS, Galileo, Glonass) in which the transmitters are installed on board of SVs and the receivers are held by users. With this technology solution, the localization measure is available to customer equipments. This is often used to automatically route to a destination. Frequently, localization receivers are integrated with transmitters (GSM, satellite communicators, classic VHF radio link) to send the information to service centers.

In *mobile-originated* techniques, the transmitters are installed on board of the user equipment. The services that use this kind of localization include *search and rescue services*, oriented to nautical, aeronautical, or other specialized applications. They are called *SARSAT satellite services*, where the user equipment transmits signal to the satellites. From the SVs, signals are immediately forwarded to safety centers. With this technology solution, customer equipments do not allow the users to know the position measure. The localization information is obtainable by a surveillance center of monitored vehicles (e.g., boats and airplanes). To increase the security and safety effects of people and vehicles, the current transmitters are equipped with a GPS receiver and with a radio frequency voice channel. In case of distress, therefore, the customer equipment transmits, besides the identification of the equipment that is strictly associated to the vehicle, also the GPS position. In addition, the first aid personnel are equipped with special receivers to better detect the exact distress point and to provide voice assistance. Therefore, GPS-equipped transmitter allows to automatically route to a destination.

In both cases, the position is computed defining a mathematical model of the media between the transmitters and receivers. However, the random positions of SVs and receivers make it difficult to compute the path between them. A highly precise localization can be obtained by using a sophisticated and computational expensive mathematical model of propagation paths. The challenging difficulty is due to the fact that, during the paths, the waves pass through a media changing their propagation characteristics as nonlinear multidimensional continuous function. The propagation parameters are function of the height above sea level, day time, seasons, pollution, and meteorological conditions.

3.2.2.1 GPS

The GPS project [16], funded by USA Department of Defense, is based on 24 SVs moving on six orbital plans, tilted 55° respect of equator on an altitude of about 20,000 km. Actually 29 SVs are operative of which three are IIR-M class (the last generation). As already anticipated, GPS project implements a mobile-terminated solution, where transmitters are on board of SVs and therefore all the information (positioning, speed, and timing) are available only at receiver level. GPS works on two different frequencies: $L_1 = 1,575.42$ MHz and $L_2 = 1,227.60$ MHz. L_1 frequency is both for mass market and military (or special) applications whereas L_2 one is devoted only to military (or special) applications.

Over L_1 frequency, two signals are coded: C/A for mass market and P(Y) for military (or special) applications. Over L_2, only the P(Y) signal for military (or special) applications is coded. In the near future, over L_2 frequency, a new civil signal, called *L2CS*, will be broadcasted. This signal will allow higher precision and better availability.

The GPS receivers allow to measure, besides the location, two other important values: speed and time. It means that GPS can be used as tachometer and as a high-precise time reference instrument. This second feature is particularly useful in communication networks where data synchronization is essential.

The accuracy performances of the GPS system, for civil applications (low profile), are basically:

- Positioning: 32 m
- Speed: 0.1 m s^{-1}
- Timing: 1 µs
- Time to first fix (`ttff`): 1 min

The `ttff` parameter is the time elapsed from the receiver turning on to the position data availability. This value is particularly meaningful in all services where the position information must be available immediately turning on the receiver, such as services integrated with mobile phone.

The error sources can be SVs clock errors, SVs position errors, bad propagation model in the ionosphere, bad propagation model in the troposphere, and multipath effects. All these effects cause an error (e) of about 16 m per SV. Since four SVs are necessary to get the position, in the worst case, the error is 64 m. Since the four errors are statistically independent, the resulting error is 32 m with 67% probability.

When the sea level is not considered, latitude and longitude can be computed by three SVs and therefore the positioning performance increases, reaching 28 m. The main source of error here is the bad ionosphere modeling that causes 10 m error. Modern commercial GPS receivers can compute the positioning using more than four SVs, until 32, if viewable, achieving a better guess of measures [15].

To increase the GPS performances, several solutions have been designed. The most well known are European geostationary navigation overlay service (EGNOS), A-GPS (already described in Sect. 3.2.1.7), and the new satellites constellations.

3.2.2.2 European Geostationary Navigation Overlay Service

EGNOS is a project designed by Europe Space Agency (ESA) and by the main aerospace industries. It is based on the concept that the propagation characteristics of the atmosphere are quite stable in a wide area. This means that the error of the measures is prevalently the same in the entire area. The project architecture is based on a network of reference stations installed in Europe, in high precision georeferenced sites. The reference stations are composed by a sophisticated GPS (called *reference GPS*) and a computer to measure and send, in real time, the GPS system error to the main center. The GPS system error can be measured as the difference between the known exact position of the reference station and the position measured by the reference GPS. All errors are sent to the main station, whereas they are packed and sent to the delivering infrastructure. The delivery infrastructure uses two types of media: three geostationary SVs (different from GPS constellation) and a Web server. The geostationary SVs transmit the information using L_1 frequency, thus allowing commercial GPS devices to receive the error information by just upgrading the firmware. This solution can work very well in every part of the world, except city centers of old towns, since urban canyon makes it difficult to set up a link to geostationary SVs. The second solution uses a radio link or GSM (GPRS/UMTS) connection to the server hosting the error information. EGNOS project increases the precision performance up to 2 m. Other solution analogous to EGNOS are the American WAAS and the Japanese MSAS. In addition to that, other private and public solutions based on this principle (differential GPS – DGPS) are achievable.

3.2.2.3 New Satellites Constellations

By the end of February 2008, a new generation of SVs will be launched. It will provide a deep evolution of GPS services. They will transmit with higher power and will introduce L_2 frequency for civil user (L2CS). The L_2 civil services will allow the advantages of increasing the GPS performances in terms of precision, availability, and reducing of `ttff`. Under military point of view, new encryption will allow to benefit of a better protection antijamming. In the far future (2011), an additional frequency labeled L5 will be set up to further increase GPS performances.

3.2.2.4 Galileo

Galileo will be the European answer to American GPS. It will be operative by 2010; it uses three frequency bands at around 1,500 MHz and will use 30 SVs to offer five services:

1. *Open service* (OS). It is targeted to general purpose customer similar to current GPS users.
2. *Safety of life service* (SoL). It is targeted to people involved in safety activities. It is a double frequency service.
3. *Commercial service* (CS). It is targeted to service as parking, auto route payments, or other commercial services. It is a double frequency service.
4. *Public-regulated service* (PRS). It is targeted to government applications: military, police, and similar customers. This is a high availability service, it is also protected by voluntary disturbs.
5. *Search and rescue service* (SAR). This service will enhance the quality of service actually available in search and rescue area. That is because it will integrate the Galileo receiver with current SARSAT transmitters that guarantee a higher response rapidity and a higher precision.

3.2.2.5 Glonass

Glonass is a satellite-based localization system delivered by ex-USSR. At the moment, the Glonass localization service is accessible only to less than 50% of the world, that is mainly over 50° North and under 50° South latitude. The full world wide coverage will be available by the end of 2010. Glonass localization performances are similar to GPS ones. Glonass receivers are not targeted to consumer market, since they are used for professional applications only. Most of Glonass receivers are therefore equipped also with GPS receivers, it means that the availability and the precision of the Glonass localization improve with respect to GPS only receivers. The copresence of two receivers (Glonass and GPS) in the same localization set guarantees also a faster localization acquisition. This performance is really appreciated in all professional applications where high precision measures are required. Both GPS and Glonass use two frequency bands called L_1 and L_2. The main difference between the two localization technologies is that GPS uses one channel of 20 MHz in each frequency, while Glonass uses 25 channels of 562.5 kHz in each frequency.

3.2.3 Nonconventional Positioning Technologies

Apart from cellular- and satellite-based network, new, nonconventional, positioning technologies are becoming of widespread use. These technologies include the indoor global positioning system (indoor GPS), the Bluetooth, and the Wi-Fi positioning. The former utilizes a number of pseudosatellite devices simulating the GPS system under indoor conditions, while Bluetooth and Wi-Fi positioning techniques are based on the same principles, utilizing trilateration between a mobile device which is connected through Bluetooth or Wi-Fi, and at least three Bluetooth receivers (or Wi-Fi access points) with known positions.

3.2.3.1 Indoor Global Positioning System

Generally speaking, GPS positioning techniques are not able to operate indoors due to the fact that the satellite signal strength is too low to penetrate the infrastructure of a building. As such, the recently developed indoor GPS focuses on exploiting the advantages of GPS for developing a location-sensing system for indoor environments. The indoor GPS operates by transmitting the navigation signal by a number of pseudosatellite devices that generate a GPS-like navigation signal called *pseudolites*. The signal is designed to be similar to the GPS signal to allow pseudolite-compatible receivers to be built with minimal modifications to existing GPS receivers. As in the GPS system, at least four pseudolites have to be visible for navigation, unless additional means, such as altitude aiding, are used. Indoor GPS solutions can be applicable to wide space areas where no significant barriers exist, such as airport terminal stations, conference centers, etc. Moreover, indoor GPS takes into account the low power consumption and small size requirements of wireless access devices, such as mobile phones and handheld computers [38].

3.2.3.2 Bluetooth Positioning

Tracking and positioning using Bluetooth are a relatively easy and low-cost task. To track the position and accordingly the movement of a Bluetooth device within a large area, one has to use trilateration. In fact, using trilateration, one can determine the location of a mobile phone based on the distances of a mobile device from three Bluetooth receivers installed in known positions. It is possible to calculate the distance of a Bluetooth-enabled phone from a given receiver by using techniques involving signal levels and other analysis, as shown in [18], while the accuracy obtained by the trilateration process is about 1.7 m, which is a rather satisfying accuracy.

In general, when at least three Bluetooth receivers are installed in known locations, using the trilateration technique it is possible to locate a Bluetooth device and track the device's movement with a good accuracy, while as the number of receivers involved in the positioning process grow, the method achieves even greater accuracy.

A limitation of the method is seen when dealing with large areas, since the maximum range of a standard Bluetooth dongle today has ranged up to only 100 m. Therefore, to achieve a full coverage of the area of interest, it would be necessary to install a receiver at least every 100 m (typically in smaller distances). Subsequently, it would be a great challenge and require thousands of Bluetooth receivers to cover large outdoor areas. On the other hand, it seems that the Bluetooth positioning has many advantages when dealing with indoor conditions.

3.2.3.3 Wi-Fi Positioning

Although Wi-Fi positioning systems (WPS) were initially developed for indoor purposes, recently they have been employed also for outdoor purposes. In particular, since GPS systems do not function well inside urban areas due to the height of the buildings and to the large glassy surfaces usually found in large buildings, we need to focus on different directions to improve our positioning technologies. One of these new approaches includes the usage of the large number of Wi-Fi access points (which is also growing rapidly), by measuring the distance between at least three access points and the mobile device of interest, and then, apply basic trilateration to determine its position. Recent developments [37] can provide accurate positioning up to 6 m. Moreover, WPS needs less than 1 s to determine the device position. WPS is a very promising positioning technology, being however depended on the number of neighboring access points, leading thus to its employment on suburban, urban, and indoor areas, while its availability in rural areas is rather low.

3.3 Mobile Location Systems

In the recent years, we have been witnessing the explosion of emerging nontraditional applications, such as mobile services. Someone could cite a series of such applications focusing on user requirements parallel with the lines of research of the GeoPKDD project. Such a hot application could target on traveling service providers (from taxi drivers to courier boys or on-demand mobile medical desks) rambling and waiting for customers [30]. For instance, a user (tourist, businessman, consumer) moving around a city equipped with a user friendly next-generation mobile terminal (e.g., 3G cell phone or PDA enhanced by the presence of a GPS receiver and a digital compass), receiving hints of information, commercial spots, etc. Besides mobile user management, other application areas include fleet management or navigational systems. As the number of mobile commerce, or in general, mobile services, increases rapidly every day, the need for effective systems about location data (*mobile location systems* – MLS) is vital. Given the above fact, a prospective definition of an MLS as presented in [28] is the following one: MLS is a location system, including applications that determine the geographic position of mobile subscribers and provide them with relevant information and services. Analyzing this definition, an MLS is associated with one or more positioning technologies (any of the satellite-based or telecommunication network-based technologies described in Sect. 3.2), while it further supports information exchange between the system and the end user. In the subsequent paragraphs, we briefly present the design principles and a representative set of existing MLS.

3.3.1 Architectural and Operational Aspects of MLS

The big picture of each MLS shown in Fig. 3.3 involves four main components: Fig. 3.3a – the *mobile network* providing the technological framework for the positioning of the mobile entities, the location data which are disseminated with the help of Fig. 3.3b – a *location center* which interacts with Fig. 3.3c – the *application/ database* level which in its turn provides Fig. 3.3d – the user *mobile services*. Focusing on the communication between the location center and the mobile application components, there are standards and protocols to facilitate the interchange of data. Toward this direction, the OMA *location working group* (WG) [29] has been created to develop specifications to ensure interoperability of mobile location services on an end-to-end basis. The location WG adopts relevant specifications developed by the former *location interoperability forum* (LIF) and the former WAP Forum, and converges other relevant industry initiatives as needed.

More specifically, the *mobile location protocol* (MLP) is an application-level protocol for getting the position of mobile stations (mobile phones, wireless personal digital assistants, etc.) independent of underlying network technology. The MLP serves as the interface between a location server and a *location service* (LCS) client. This specification defines the core set of operations that a location server should be able to perform. Possible realizations of a location server are the GMLC, which is the location server defined in GSM and UMTS, and the MPC, which is defined in ANSI standards. Since the location server should be seen as a logical entity, other implementations are possible. In most scenarios, an LCS client initiates the dialogue

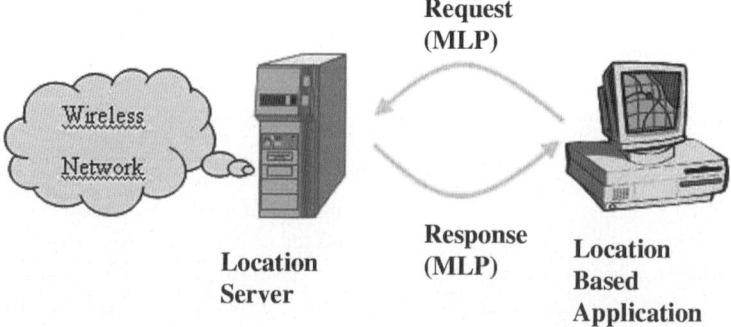

Fig. 3.3 The big picture of MLS

3 Wireless Network Data Sources

by sending a query to the location server and the server responds to the query. This specification has been prepared by LIF to provide a simple and secure application programming interface (API) to the location server, but that also could be used for other kinds of location servers and entities in the wireless network. The API is based on existing and well-known Internet technologies as HTTP, SSL/TLS, and XML, to facilitate the development of location-based applications. The above-discussed specification protocols support a series of functions applicable to mobile location data. Suggestively, we present the subsequent operations composing a representative set of supported services:

- Tracing: Example services contain position finding of a stolen car or locating persons in an emergency situation (e.g., 911 or 112 calls).
- Simple or reverse (or de-)geocoding: Simple geocoding includes validation and conversion of human friendly address formats to lat/lon geographic coordinates. Reverse geocoding performs the mirrored task.
- Mapping: This function generates location-dependent maps and provides scale, zoom, etc., operations.
- Routing: It calculates the optimal (in terms of network distance or travel time) route taking transport means into consideration. Variations of this function include:
  – Travel planning: specification of travel destination and intermediate waypoints
  – Route guidance: this service determines deviation from the route and sends a message to the user
- Spatial querying: This function retrieves location-dependent information from a database. Examples include:
  – Window search queries: location-based yellow pages (what-is-around services)
  – Nearest-neighbor search: position finding of the nearest point of interest (POI)

3.3.2 Commercial MLS

3.3.2.1 Ericsson's Mobile Location Solution

Mobile location solution (MLS) is Ericsson's name for a location system, including applications, which determines the geographic position of mobile subscribers and provides them with relational information and services [28] depicted in Fig. 3.4. MLS does not require any kind of transformation to GSM mobile devices and comprises a server-based solution that allows the application of location services in any GSM network consisting of Ericsson switching systems. Ericsson provides specialized HLR, MSC, and BSC software as well as additional required hardware. The heart of MLS is the *mobile positioning server* (MPS) that allows the applications to access the location information of the mobile devices. There is also an API that

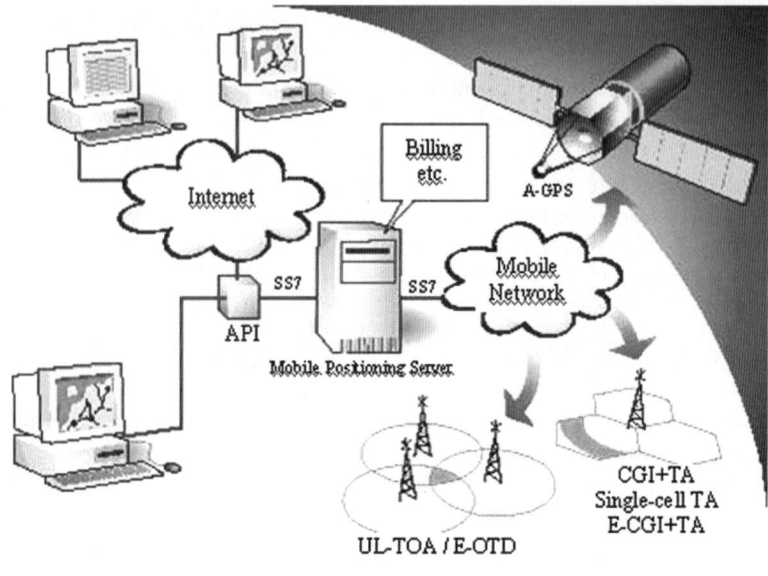

Fig. 3.4 Ericsson's mobile location solution [28]

permits the development of independent applications. The MPS is also responsible for ensuring the privacy of the subscribed members and provides the users with the ability to choose whether they want to be traced or not. Except from emergency call services, network routing of fleets and tracing of stolen cars, there are a variety of mobile services supported by the MLS commercial examples which include: weather and traffic reports, localized advertising the yellow pages, etc. To conclude, MLS supports most of the positioning methods, both network based and mobile assisted.

3.3.2.2 ESRI's LBS Solution

ESRI's LBS solution (http://www.esri.com) (Fig. 3.5) provides application developers with geospatial server software and a Web services platform (ArcWeb Services) for developers to integrate mapping geographic information system (GIS) content and capabilities into applications or ArcGIS. The available functionality consists of a variety of choices from spatial database gateways to geocoding and map rendering operations. Furthermore, ESRI has introduced Tracking Server as its solution product used to collect and send real-time data from many data sources and formats to Web and desktop clients. Tracking Server is an enterprise-level technology that is integrated with ESRI's other server and service products. With these servers and services, one can build new Web and wireless applications or enhance existing ones with location, addresses, points of interest, dynamic maps, and routing directions.

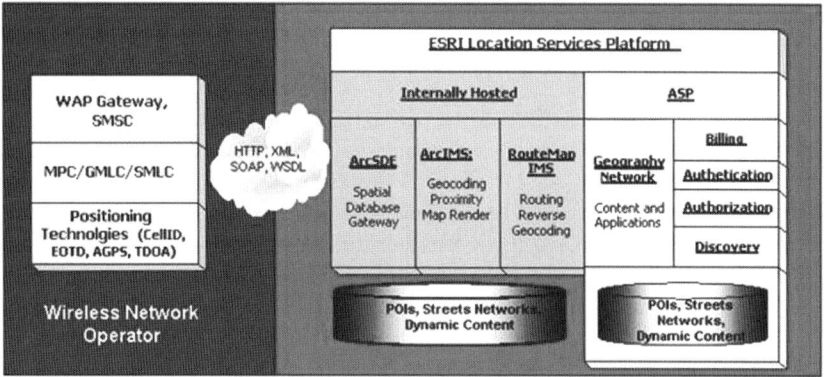

Fig. 3.5 ESRI's LBS architecture

3.3.2.3 Nokia mPosition

The Nokia mPosition (http://www.nokia.com) solution is a complete end-to-end solution providing operators with LBS for mobile networks. More specifically, mPosition offers a wide selection of location-based applications (such as traffic and weather reports, restaurant, theater or movie ticket bookings, emergency services), middleware and integration services. Nokia mPosition can be implemented within an operator's network regardless of the network vendor. It has established an open developer community, while it uses open standards and common industry forum developments for application interfaces. To conclude, one of the services supported by Nokia mPosition is mCatch. Nokia mCatch is an LBS system that supports GSM networks with an upgrade path to GPRS, EDGE, and 3G networks. It is designed to support the basic positioning methods of current standard GSM phones and in the future, it will support high accuracy positioning technologies as E-OTD, A-GPS, and IPDL-OTDOA.

3.3.2.4 CellPoint

The CellPoint system (http://www.cellpt.com), which is already in commercial use, requires no expensive overlays or modifications to the cellular networks, while it can be controlled by a remote location. The CellPoint technology conforms with GSM-compatible terminals as well as with cell phones supporting the WAP protocol and provides GSM operators with competitive advantages in offering value-added mobile location services. Basically, it is a software-based solution that uses the Sim Toolkit of the cell phone and an Internet connection. The positioning methodology is based mainly on the CGI, TA, and the network measurement result (NMR), while provides future support for A-GPS, E-OTD, and TDOA. The technology permits the indoor positioning of cell-phones even in the case where the cell-phone is inside a pocket or a briefcase, cases where the GPS does not function. Finally, CellPoint gives the capability to third-party software development companies to build their own applications via the MLP.

3.3.2.5 SnapTrack

SnapTrack (http://www.snaptrack.com) provides high-quality location applications by commercializing A-GPS technology. In detail, SnapTrack develops the elements necessary to implement an end-to-end location functionality on a wireless network. This includes a location server, a wide area reference network, client technology, the underlying system architecture, and location protocols. SnapTrack was recently acquired by QUALCOMM and became a subsidiary of QUALCOMM. The two most prominent products are the A-GPS location server technology and the A-GPS client technology. The SnapTrack location server technology has been implemented in QPoint software product. QPoint is widely deployed in the world today and is made available though system integration partners of SnapTrack, who provide a broader, commercially packaged solution containing QPoint software and other elements to wireless operators and location service providers for in-network use or as a hosted service. QPoint software is also licensed to certain test equipment manufacturers and deployment service houses. The SnapTrack A-GPS client technology has been licensed and implemented by several major semiconductor manufacturers which serve the wireless industry.

3.3.2.6 Cambridge Positioning System Cursor

Cambridge positioning system (CPS) (http://www.cursor-system.com) has introduced matrix technology that offers a powerful combination of sub-100 m accuracy, all-area coverage, and rapid time-to-fix. Matrix has been adopted by a number of operators, local service providers, and industry leading partners. With a seamless and standards-compliant evolution from GSM to *wideband code division multiple access* (WCDMA), matrix has minimal impact on the operator network enabling services to be quickly launched to market. Matrix needs no network hardware or satellites to deliver high accuracy location. Instead, it utilizes everyday network synchronization signaling to determine a location. In detail, cellular base stations emit periodic synchronization signals, which are listened by mobile devices (e.g., to enable handover from cell to cell). A matrix-enabled device measures the time these synchronization signals arrive at the device and sends this information to a matrix server in the network. The server contains precise information of each base stations location, meaning that a flight time between base station and mobile device can be calculated. This time can be turned into a distance and with three timings from three separate base stations, a calculation can be performed to effectively triangulate an x/y location for the device. This calculation takes less than 3 s ensuring a rapid location fix. This procedure of requesting a location to a device and its returning timing measurements can take place over standard SMS, or more cost effectively, using GPRS.

3.3.2.7 TruePosition Wireless Location System

The TruePosition wireless location system (http://www.trueposition.com) enables wireless carriers and public safety organizations to determine the geographic position, direction of travel, and velocity of mobile transmitters. TruePosition lies in the development of advanced location systems which include handset, network, and hybrid location solutions. At the heart of the TruePosition MLS offering is the TrueNorth location system, which combines the widest variety of positioning technologies, including network- and handset-based location solutions such as U-TDOA, Cell-ID, enhanced Cell-ID, A-GPS, and future hybrid solutions.

3.3.2.8 SignalSoft Corporation

The SignalSoft (http://www.signalsoftcorp.com) software house has experience in GISs, radio location, and online transaction processing (OLTP). The company offers a complete package of professional LBS and multiple deployment options. Its main benefit is the location server (middleware) that supports all the proposed positioning technologies as well as a few more (i.e., TOA/TA, TDOA, AOA, A-GPS, E-OTD, Cell-ID/TA, Cell-ID/NMTC, MAP ATI). The location manager is the technological core of SignalSoft, meaning the calculation module of a subscriber's location. The available information is combined in an appropriate means to achieve the greatest possible accuracy in the estimation of the longitude/altitude of the cell phone's location. What is more location studio is software that runs in the network and it is responsible for the secure interface between the subscriber and an LBS application, as well as for the privacy- and authentication-related issues.

3.4 From Positioning to Tracking: Collecting User Movements

As we have shown in previous sections, positioning is an atomic real-time operation that gives the position of a device at a given time instant. Collecting a series of positions over time means *tracing* the movements of a user. Positioning can be done regularly, e.g., at each fixed time interval, or based on events, e.g., a telephone call. A temporal sequence of positions gives the trajectory (or trace) of a user. The trajectory approximates the movements of the user with an error that depends both on the accuracy of the positioning technology and on the frequency of positioning operations (the sampling rate).

User tracking methods currently available depend on the supporting network infrastructure. For example, in cellular-based networks, user tracking can be performed by a mobile trace procedure defined by the GSM protocol [6]. Here, a subscriber can be traced by the network operator by collecting all the communication signals transmitted between the device and the infrastructure. This kind of tracing is the most accurate available at the moment in GSM-based networks. However, this procedure has two main drawbacks. The first one is that it tends to overload the network structure, thus usually only a few (about 10–50) traces may be collected simultaneously. Another disadvantage is that it is possible to select a set of users to trace, whereas it is not possible to select a specific geographical area to cover. Indeed, users can move in very wide areas traveling between regions, whereas movement-aware applications tend to analyze movements in a given, possibly restricted, area. Therefore the choice of the sample of users to track may become critical for some applications.

Another source of user tracing data in GSM networks is the billing information. Indeed, all out-coming calls from a device are stored by the network operator in registers where we have User ID, data and time of the call, duration of the call, the cell where the call began, and the cell where the call finished. This kind of data set is global for all the network and so it is quite easy to select calls that occur in a given predefined area. However, the main drawback is that the accuracy of trajectories is very low, since only the starting and ending cells are sampled. The use of geographic background knowledge along with duration of the call may help in approximating a trajectory.

Other types of data the network operator usually collects on user activity in the network are statistical data about *cell density*, i.e., the number of users that are active inside a given cell at a given time interval. These data are already aggregated, therefore it is not possible to derive the trajectory of a single user. However, it can be useful to integrate and compare other types of movement data.

In satellite-based positioning methods, accuracy is very high compared with cellular-based ones. However, in mobile-terminated technologies – the most common in today's user applications, data are collected at the handset, making necessary a further explicit step where a central server collects data from user receivers. Similarly, in mobile-originated technologies is the service provider that may collect all user trajectories.

From this brief survey of positioning and tracking technologies, it comes out that obtaining collections of user trajectories is far from being a simple and obvious task. Instead, it usually demands for network technology upgrade, as well as high costs in bandwidth use and store requirements. Furthermore, from the great interest that LBS shown in the market in the last few years, and since they just need instant position, no technology investments have been done by communication providers to collect such data.

In Sect. 3.5, we survey some tools, available in the literature, that aim at building synthetic trajectories data sets. As already pointed out in previous sections, having synthetic trajectories allows one to test and validate analysis algorithms, even when real user trajectories are not available.

3.5 Synthetic Trajectory Generators

In this section, we briefly present spatiotemporal data generators recently proposed in the literature. These tools aim at producing trajectories of *moving objects* that evolve following user parameters. The spatial movement can be of two kinds: free or network constrained. Free-movement objects mean that objects can move freely in two-dimensional space, whereas network-constrained objects can move only inside a network infrastructure. Network-constrained objects are usually used to simulate the movement of trains, metro, buses, trucks, and so on. Following this distinction, in Sect. 3.5.1, we present some spatiotemporal data generators.

3.5.1 Free-Movement Generators

In Sect. 3.5.1.1, we present the generate spatiotemporal data (GSTD) generator, one of the most well-known spatiotemporal synthetic data generator in the literature, followed by its extension, cellular network trajectory reconstruction environment (CENTRE), which is implicitly designed to generate random movements of users in a wireless network. Our survey on free-movement generators is concluded by presenting *G-TERD*, which allows one to generate random moving areas, and *Oporto*, an application-specific data generator which mimics the fishing at sea scenario.

3.5.1.1 The Generate Spatio-Temporal Data Generator

The GSTD generator [34] was initially built upon a few basic yet general principles discussed in [33]. As a result GSTD currently supports the generation of both points and minimum bounding rectangles (MBRs). The generated data sets are transaction time oriented and rather amnesic since future events do not depend on past states. Furthermore, the cardinality of the data set is assumed to be constant throughout the data generation process. The data generation process is controlled by the following three parameters that allow the generation of a wide variety of scenarios:

1. The duration of an object, i.e., the elapsed time between two timestamps in which a change of position occurs
2. The shift of an object, i.e., the traveled distance along each space dimension between two subsequent object position updates, which also implicitly controls its speed
3. The resizing of an object (applicable only to objects of type MBR), i.e., the shrinking/enlargement of objects between two subsequent object updates

For each one of those parameters, the user can choose one of the three supported statistical distributions to be followed: Uniform, Gaussian, and Skewed (Zipfian). Moreover, the values of the above three parameters (i.e., duration, shift, and resizing) can be bounded by determining their maximum and minimum values. Finally,

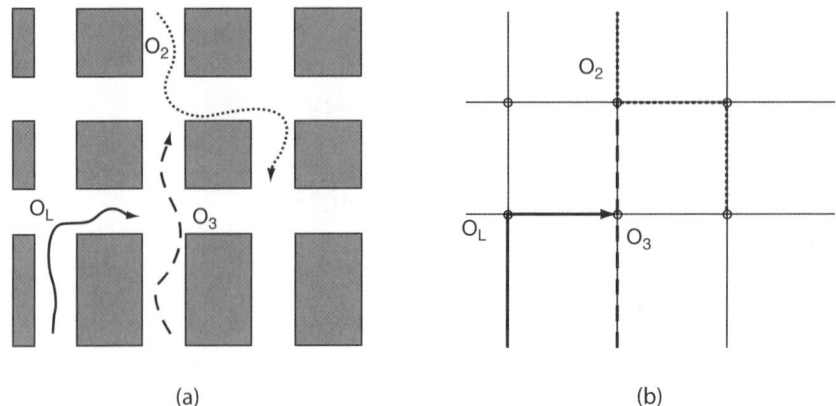

Fig. 3.6 (**a**) Movement of objects restricted by the infrastructure in urban area and (**b**) movement of objects on fixed network being an abstract description of the same urban area

GSTD provides three different ways for handling the case of points leaving the unit space (1) in the radar approach, objects may leave the unit space and while not displayed (and not reported) are still considered since they can eventually return (and be redisplayed), (2) objects can "bounce off" the space coordinates in the adjustment approach, and (3) in the toroid approach, as the name suggests, the dataspace is assumed to be toroidal, hence objects never leave it. Some enhancements over the original GSTD algorithm were introduced in [24]. First, the idea of nervousness is introduced, i.e., varying the object's shift. In GSTD's initial design, the changes in the objects' shift were to take effect during the whole simulation lifetime. The introduction of this parameter allowed the generated objects to change their behavior in a systematic way, following again a statistical distribution. A second modification was the introduction of the notion of infrastructure, i.e., spatial objects which obstruct movement. Infrastructure can be composed of real objects or synthetically generated MBRs. In the latter case, MBRs could change their shape/size and move as well. This approach as also discussed in [5, 22] can be used to simulate the restrictions posed against moving objects by existing urban infrastructure, instead of using the abstract representation of network edges (Fig. 3.6).

Initially developed as a stand-alone application, GSTD was improved and reimplemented as a Web-based application, available in [11, 29, 31]; both sites also provide source code for the data generator, so that it can be locally run [32]. Its current version allows one to generate and to store on the Web server several data sets in each run. One or more of those data sets can be visualized (in an animated manner) at the same time. The user can download the data set (in XML format) for future use and/or distribution. To illustrate some of the above GSTD features, Fig. 3.7 shows a snapshot of a generated data set which exhibits points moving freely (adjustment approach) from a central cluster (Gaussian) following a random movement distribution leading thus the moving objects to spread all around the dataspace. The GSTD data generator is very widely used and is considered to be the standard data

3 Wireless Network Data Sources

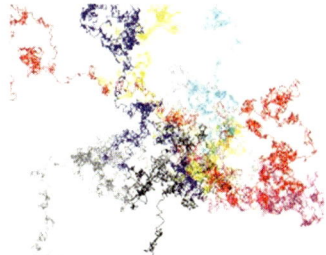

Fig. 3.7 A set of trajectories generated using GSTD

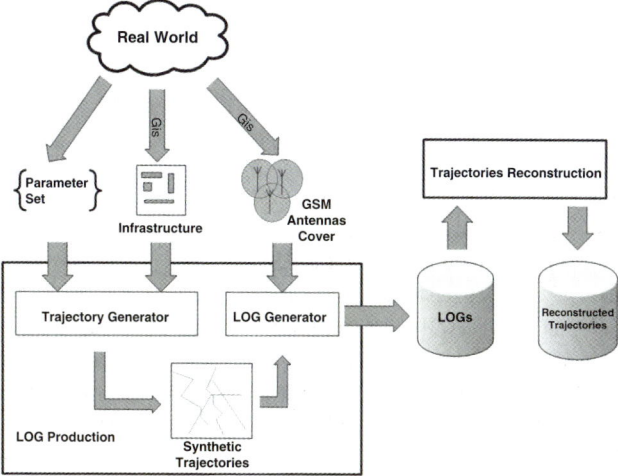

Fig. 3.8 CENTRE: general architecture

generator employed in the majority of the research papers regarding trajectories of objects moving in an unconstrained two-dimensional space. Among others, it has been used to evaluate trajectory indexes [3, 23, 39], query processing techniques over trajectories [9, 22], spatial data warehouses [21], and to influence the research in other fields, such as spatial access methods under frequent updates [14]. There are two major factors which drive the broad approval and adoption of GSTD by the research community: the first is the lack of real spatiotemporal trajectory data, while the second is the idea that, as long as GSTD's users publish the values of the GSTD parameters they used, anyone can reproduce (and use) exactly the same data set – this is also the chief goal of GSTD, namely, to remove the ad hoc nature of evaluating and comparing different systems.

3.5.1.2 Cellular Network Trajectory Reconstruction Environment

CENTRE has been proposed recently in [10] as a system for randomly generating movement data of users through cellular network. The system (Fig. 3.8) has three components (1) the *synthetic trajectories generation*, able to generate possible object behaviors on a specific space trajectories, (2) the *Log generation*, which is designed to take into account the various network technological requirements, and (3) the *approximated trajectories*.

The synthetic trajectories generation module is based on an extension of GSTD algorithm presented in Sect. 3.5 [24, 35]. The extension is mainly concerned with the capability to drive the generation process with different group behavior defined by the user. Each group defines typical aspects of a moving object, such as velocity, direction, and agility, that can be obtained setting a number of probability distributions. It is possible to define, for each group, an obstacles infrastructure, that is a collection of rectangles, that must be avoided by group objects. Another feature is the possibility for an object to change group during the evolution, simulating the movement behavior of a person that may change during the day (taking car, then walking, then taking the train, and so on).

Each group is characterized by a parameter setting representing typical aspects of a moving behavior, such as speed, direction, and agility. These settings correspond to probability distributions that are then combined to generate typical trajectories. Moreover, the groups themselves may be fatherly combined to change the object behavior.

The Log generation module simulates the network positioning task. Indeed, it takes a set of synthetic trajectories represented as a collection of points and a topology of a network of antennas, named *antenna cover map*, and returns a set of antenna detection Logs, thus simulating the CGI method. In fact, each point of the trajectory is transformed according to the intersection with the detection area of one or more antennas. The sampling rate is interval based. The user may define his/her own antenna cover map through a GIS so that he/she may generate different simulation scenarios.

Once the Logs of mobile data are available, either synthetically generated or as real data provided by network operator, data have to be prepared for the analysis algorithms. The *trajectory reconstruction* module takes as input the set of Logs, an antenna cover map, and possibly background knowledge on the region (geography, road infrastructures, urban infrastructures, etc.) and tries to reconstruct the original (synthetic or not) trajectories. Reconstructed trajectories are, typically, an approximation of the original ones, due to the information loss of the Log generation step (that comes out as a combination of the accuracy of positioning technology and sampling rate of the tracking). This additional functionality provides a minimal kernel of the system where data mining experiments may really be carried on.

A more recent version of CENTRE 2.0 has been developed to deal with the special requirements of data mining algorithms. In particular, the new engine of CENTRE 2.0 is now based on an attraction principle, where objects evolve in space toward specific areas from which they are attracted. This method allows to build

3 Wireless Network Data Sources

data sets particularly suited for testing density clustering algorithms [17]. The definition of attraction areas allows the user to define dense clusters to be found by the algorithm.

3.5.1.3 G-TERD

Another generator related to GSTD [12] is the generator for time-evolving regional data (G-TERD), originally presented in [36]. G-TERD differs from GSTD in that it generates sequences of raster images, though it still is a synthesizer of data which evolve with time. In particular, G-TERD is specifically designed for applications stemming from the field of time-evolving regional data. The basic parameters controlling the objects generated by G-TERD are their structure as two-dimensional regional objects, their maximum speed, the zoom and rotation angle per update, the influence of other moving or static objects, the position and movement of the scene-observer, the statistical distribution of each changing factor, and finally time. Moreover, G-TERD also controls the object's color.

Obviously, G-TERD allows the user to set more parameters than GSTD does. It supports the statistical distributions supported by GSTD and a few additional ones. While GSTD generates moving points and MRRs, G-TERD is able to generate regions of more general shapes, which may, e.g., rotate, enlarge, or shrink again in a systematic way. The coloring of regions is also supported. Like GSTD, G-TERD allows for the specification of obstacles to movement. GSTD's radar approach allows objects to leave the dataspace; the viewable area in GSTD is fixed and cannot be changed. In G-TERD, the dataspace is typically larger than what the user sees, and a so-called scene-observer capability allows the user to change point of view, e.g., follow a particular object's path in time or "fly" over the dataspace. Whereas GSTD is Web based, G-TERD is an MS Windows-based application; its source code for the (stand-alone) data generator is publicly available through the Web. The generated data can be visualized (although not animated as for GSTD) using an accompanying application.

From the above discussion, it is clear that G-TERD controls a variety of parameters in a more sophisticated way than GSTD does. However, its use is very limited through the research community, mainly due to the fact that the research community's focus is on the management of moving points (trajectory) data, rather than on moving regions, which are not considered applicable to real-world applications.

3.5.1.4 Oporto

The Oporto generator [20, 27] was not designed to be as general as GSTD or G-TERD; instead, it is designed to mimic the very specific scenario of fishing at sea. In particular, Oporto models fishing ships, which leave harbors following shoals of fish while at the same time avoiding storm areas. The shoals of fish themselves are attracted by plankton areas. Harbors are static objects, while ships, storms, and plankton areas, so-called bad and good spots, are dynamic ones. Ships and harbors are represented by moving and static points, respectively, while spots are MBRs, which can vary in shape and size, but do not move. In addition, they always grow and subsequently shrink. Shoals of fish, on the other hand, can change size, shape, and position over time. The user can model a shore line along with the location of harbors on it.

Unlike GSTD and G-TERD which follow the amnesic approach between two subsequent object's position updates, the underlying model of the Oporto generator is based on the notion of attraction and repulsion. That is, ships (fish) are attracted by fish (plankton), whereas storm areas repel the ships. Therefore, data sets generated by the Oporto generator are more likely to be closest to real-world scenarios. Oporto allows the user to generate and visualize animated data sets through its Web interface, while it is also available as a stand-alone application.

Though Saglio and Moreira [27] argue that Oporto is capable of generating data sets representing several scenarios, it seems to be quite limited when compared to GSTD and G-TERD. For example, it has limited capability of generating data according to different distributions since it only supports the uniform one. Nevertheless, it is still one of the few generators based on a well-known real application, generating thus data sets being close to real ones. Oporto has limited success in the research community compared to GSTD, though it is more popular than G-TERD is, since it generates realistic trajectories of moving points.

3.5.2 Network-Based Generators

In network-based data generators, objects may move only on a prefixed network. Usually these kind of data are useful to simulate vehicle traffic, such as trains and buses. The most well-known tool available in the literature is the Brinkhoff generator, described in Sect. 3.5.2.1. After that, we introduce a new tool (*macroscopic traffic synthesizer*) explicitly designed to simulate car traffic data.

3.5.2.1 Brinkhoff Network-Based Data Generator

Brinkhoff [1, 2] proposed one of the most cited work on network-based generators of moving objects. Indeed, previous approaches for generating spatiotemporal data did not consider that moving objects often follow a given network, such as trains, metro, and trams. Therefore, moving object benchmarks require data sets consisting of such network-based moving objects. Essential aspects of Brinkhoff generator are the maximum speed and the maximum capacity of connections, the influence of other moving objects on the speed and the route of an object, the adequate determination of the start and destination of an object, the influence of external events, and the time-scheduled traffic. This generator combines real data (the network) with

3 Wireless Network Data Sources

user-defined properties of the resulting data set, and the generation process requires the following steps:

1. Loading the network from simple binary files. A tool that allows converting TIGER/Line files exists.
2. The definition of the required user-defined functions and parameters.
3. The computation of the objects and their moves.
4. The report of the generated data into user-defined text files. To simplify the definition of the user-defined functions and parameters, the generator supports an ad hoc visualization of the generated data.

Brinkhoff built a Java-based framework for performing these four steps, and performance tests have shown that the generator computes large data sets within a reasonable time using a Java interpreter on a standard personal computer. An interactive demo is also available on Brinkhoff's generator Web site (http://fh-oow.de/institute/iapg/personen/brinkhoff/generator/) with some example scenarios, documentation, and useful tools.

3.5.2.2 Macroscopic Traffic Synthesizer

The macroscopic traffic synthesizer (MTS) was originally developed at the Theoretical Computer Science Group of Hasselt University to test data mining techniques on traffic data. This tool takes two inputs:

1. *Road network.* A directed graph, where the edges represent a road segment containing data like maximum speed and number of lanes
2. *Source–destination matrix.* Telling how many cars are leaving a certain starting point (source) at a specific time to a certain destination

It then generates as output a sequence of four-tuples (ID, x, y, t), stating that the car with ID as identifier is at time t at a position with coordinates (x, y).

The main feature of the algorithm is that the output is not generated by just using a standard shortest path algorithm for route planning (e.g., A^* [19]). Such algorithm constructs the trajectory between the source and destination point writing out, with a certain time interval, where each car should be located at time t in x, y coordinates. This would give unrealistic data because such a standard route planning algorithm does not take into account traffic jams. Therefore, it would be possible to have, e.g., 100 cars on one specific place at the same time.

To avoid such anomalies, MTS uses three important characteristics concerning traffic systematics which are explained in the highway capacity manual [13]. This manual describes relations between speed, density, and flow of traffic. In this way, MTS simulates not only traffic jams, but also the intelligence of car drivers that will use, when possible, an alternative route when a traffic jam is on the planned route.

MTS has been extended to generate as output a sequence of three-tuples (Car-ID, Cell-ID, t), telling that, at time t, the car with identifier Car-ID is in the cell with identifier Cell-ID. The case, where cells have an overlap, is solved by giving the intersection of cells a unique Cell-ID. Of course, MTS needs to know the position of the antenna that can be obtained in a preprocessing step, so that each road segment of the road network, on top of data such as maximum speed and number of lanes, also contains a Cell-ID. In case a road segment crosses more than one cell, this road segment is subdivided.

3.6 Conclusions and Open Issues

Due to the great interest that LBS are attracting in today's wireless applications, positioning technologies are becoming of primary importance in wireless networks, having increasing technological support and improvements. On the other hand, tracking technologies have very little support by wireless networks. Indeed, only few and ad hoc methods may allow to collect user movements. However, we believe that the great potential of new generation movement-aware applications may push vendors to design and implement new tracking procedure as well as improving current ones.

From the point of view of the accuracy level (in space and time), the availability of new generation satellite positioning systems and the use of the associated receivers are becoming of a widespread use in everyday life for an ever-increasing number of mobile users. As already pointed out in Sect. 3.2, the accuracy level is going to decrease down to a few meters, so we can expect, in the near future, to have a great amount of highly accurate user trajectories.

As far as synthetic trajectory generators are concerned, many direction can be followed to make these tools more *GeoPKDD oriented*. For example, synthetic trajectory generators can be extended, to support more realistic movements, by allowing the user to configure the movement of real user movements. For example, the memory-less approach employed by GSTD (which is also used in other generators) is a rather artificial methodology hardly found in real world's conditions, since the majority of the real spatiotemporal objects is moving from a particular origin to a prespecified destination. Furthermore, the movement of real moving objects is determined by other parameters, such as speed and direction, which cannot be fully controlled by the existing GSTD's interface. In addition, there are certain types of query processing algorithms and indexing techniques that require the management of other parameters influencing the performance of the algorithms; as such, algorithms exploiting object's speed and direction would have to be tested against moving objects with known speed or direction distributions. Another direction that can to be followed is the extension of tools toward data mining needs, thus combining realistic behavior with more specific algorithm requirements, following the preliminary ideas of CENTRE 2.0.

Furthermore, another development direction for "GeoPKDD data synthesizers" is to produce an integration of the different tools. The level of the integration can vary from loose to tight. A loose integration means to design a set of interface specifications to make these tools producing a common output format (standard trajectories format). A tight integration, on the other hand, means to integrate the tools in a unique software architecture, a unique language, and a unique user interface. This means to design an architecture where interfaces are defined to produce a common output format and where a set of guidelines is drawn to direct the user to the suitable tool based on his/her requirements.

References

1. T. Brinkhoff. Generating network-based moving objects. In *Proceedings of the 12th International Conference on Scientific and Statistical Database Management (SSDBM'00)*, p. 253. IEEE Computer Society, Silver Spring, MD, 2000.
2. T. Brinkhoff. A framework for generating network-based moving objects. *Geoinformatica*, 6(2):153–180, 2002.
3. V.P. Chakka, A. Everspaugh, and J.M. Patel. Indexing large trajectory data sets with SETI. In *Proceedings of the Conference on Innovative Data Systems Research (CIDR'03)*, 2003.
4. J. Hjelm. *Creating Location Services for the Wireless Web*. Wiley, London, 2002.
5. E. Frentzos. Indexing objects moving on fixed networks. In *Proceedings of 8th International Symposium on Advances in Spatial and Temporal Databases (SSTD'03)*, pp. 289–305, 2003.
6. ETSI/GSM. Digital cellular telecommunications system (phase2+); subscriber and equipment trace. GSM12.08, version 5.1.1, Release 1996.
7. ETSI/GSM. Home location register/visitor location register – report 11.31–32 [8].
8. ETSI/GSM. Technical reports list. http://webapp.etsi.org/key/key.asp?full_list=y.
9. E. Frentzos, K. Gratsias, N. Pelekis, and Y. Theodoridis. Nearest neighbor search on moving object trajectories. In *Proceedings of 9th International Symposium on Advances in Spatial and Temporal Databases (SSTD'05)*, pp. 328–345, 2005.
10. F. Giannotti, A. Mazzoni, S. Puntoni, and C. Renso. Synthetic generation of cellular network positioning data. In *Proceedings of the 13th Annual ACM International Workshop on Geographic Information Systems (GIS'05)*, pp. 12–20, 2005.
11. GSTD, 2006. http://db.cs.ualberta.ca:8080/gstd.
12. G-TERD, 2006. http://delab.csd.auth.gr/stdbs/g-terd.html.
13. *Highway Capacity Manual*. Transportation Research Board, Washington, DC, 2000.
14. M.L. Lee, W. Hsu, C.S. Jensen, B. Cui, and K.L. Teo. Supporting frequent updates in r-trees: A bottom-up approach. In *Proceeding on 29th International Conference on Very Large Data Bases (VLDB'03)*, pp. 608–619, 2003.
15. R. Lojacono, F. Pini, S. Angelucci, and J.L.G. Marin. Gps: quanto sono precisi? *Wireless, Tecnologie per il management dei processi aziendali*, p. 65, 2007.
16. P. Misra and P. Enge. *Global Positioning System – Signals, Measurements and Performance*, 2nd edn. Ganga-Jamuna, Lincoln, MA, 2006.
17. M. Nanni and D. Pedreschi. Time-focused density-based clustering of trajectories of moving objects. *Journal of Intelligent Information Systems, Special Issue on Mining Spatio-Temporal Data*, 27(3):267–289, 2006.
18. M. Nilsson, J. Hallberg, and K.S. Positioning with bluetooth. In *Proceedings of 10th International Conference on Telecommunications (ICT'03)*, 2003.
19. N.J. Nilsson. *Artificial Intelligence: A New Synthesis*. Morgan Kaufmann, Los Altos, CA, 1998.
20. Oporto, 2006. http://www-inf.enst.fr/~saglio/etudes/oporto/.
21. D. Papadias, P. Kalnis, J. Zhang, and Y. Tao. Efficient olap operations in spatial data warehouses. In *Proceedings of 7th International Symposium on Advances in Spatial and Temporal Databases (SSTD'01)*, pp. 443–459, 2001.

22. D. Pfoser and C.S. Jensen. Querying the trajectories of on-line mobile objects. In *Proceedings of the 2nd ACM International Workshop on Data Engineering for Wireless and Mobile Access*, pp. 66–73, 2001.
23. D. Pfoser, C.S. Jensen, and Y. Theodoridis. Novel approaches in query processing for moving object trajectories. In *Proceedings of 26th International Conference on Very Large Data Bases (VLDB'00)*, pp. 395–406, 2000.
24. D. Pfoser and Y. Theodoridis. Generating semantics-based trajectories of moving objects. In *International Workshop on Emerging Technologies for Geo-Based Applications*, 2000.
25. F. Pini. Machine to machine, un segmento in espansione nelle comunicazioni. *Wireless, Tecnologie per il management dei processi aziendali*, p. 58, 2006.
26. T.T.G.P. Project. The third generation partnership project, 2006. http://www.3gpp.org.
27. J. Saglio and J. Moreira. Oporto: A realistic scenario generator for moving objects. *GeoInformatica*, 5(1):71–93, 2001.
28. G. Swedberg. Ericssons mobile location solution, 1999. http://www.ericsson.com/ericsson/corpinfo/publications/review/1999_04/93.shtml.
29. Y. Theodoridis. The r-tree-portal, 2003. http://www.rtreeportal.org.
30. Y. Theodoridis. Ten benchmark database queries for location-based services. *The Computer Journal*, 46(6):713–725, 2003.
31. Y. Theodoridis. R tree portal, 2006. http://www.rtreeportal.org.
32. Y. Theodoridis and M.A. Nascimento. Generating spatiotemporal datasets on the www. *SIG Management of Data Record*, 29(3):39–43, 2000.
33. Y. Theodoridis, T.K. Sellis, A. Papadopoulos, and Y. Manolopoulos. Specifications for efficient indexing in spatiotemporal databases. In *Proceedings of 10th International Conference on Scientific and Statistical Database Management (SSDBM'98)*, pp. 123–132, 1998.
34. Y. Theodoridis, J.R.O. Silva, and M.A. Nascimento. On the generation of spatiotemporal datasets. In *Proceedings of 6th International Symposium on Advances in Spatial Databases (SSD'99)*, pp. 147–164, 1999.
35. Y. Theodoridis, J.R.O. Silva, and M.A. Nascimento. On the generation of spatiotemporal datasets. *Lecture Notes in Computer Science*, 1651, 1999.
36. T. Tzouramanis, M. Vassilakopoulos, and Y. Manolopoulos. On the generation of time-evolving regional data. *GeoInformatica*, 6(3):207–231, 2002.
37. S. Wireless. Skyhook wireless, 2006. http://www.skyhookwireless.com.
38. V. Zeimpekis, G.M. Giaglis, and G. Lekakos. A taxonomy of indoor and outdoor positioning techniques for mobile location services. *SIGecom Exchanges*, 3(4):19–27, 2003.
39. H. Zhu, J. Su, and O.H. Ibarra. Trajectory queries and octagons in moving object databases. In *Proceedings of International Conference on Information and Knowledge Management (CIKM'02)*, pp. 413–421, 2002.

Chapter 4
Privacy Protection: Regulations and Technologies, Opportunities and Threats

D. Pedreschi, F. Bonchi, F. Turini, V.S. Verykios, M. Atzori, B. Malin, B. Moelans, and Y. Saygin

4.1 Introduction

Information and communication technologies (ICTs) touch many aspects of our lives. The integration of ICTs is enhanced by the advent of mobile, wireless, and ubiquitous technologies. ICTs are increasingly embedded in common services, such as mobile and wireless communication, Internet browsing, credit card e-transactions, and electronic health records. As ICT-based services become ubiquitous, our everyday actions leave behind increasingly detailed digital *traces* in the information systems of ICT-based service providers. For example, consumers of mobile-phone technologies leave behind traces of geographic position to cellular provider records, Internet users leave behind traces of the Web pages and packet requests of their computers in the access logs of domain and network administrators, and credit card transactions reveal the locations and times where purchases were completed. Traces are an artifact of the design of services, such that their collection and storage are difficult to avoid. To dispatch calls, for instance, the current design of wireless networks requires knowledge of each mobile user's geographic position. Analogously, DNS servers for the Internet need to know IP addresses to dispatch requests from source to destination computers.

What happens to traces of information? In certain instances, the traces that we generate are *discarded* once they are no longer required for service delivery. However, increasingly, our traces are *stored* by service providers for record keeping, quality assurance, and by legal mandates. The ability to capture and store large quantities of data is supported by decreasing costs in digital storage technology. Yet, while traditional information systems manage *business-oriented* information – such as sales, customers, and billing-related records – the traces generated by ICT-based services reveal finer-grained *process-oriented* information about what individuals do and the functions of complex organizations. As such, the collected information

---

D. Pedreschi
KDD Laboratory, Dipartimento di Informatica, Università di Pisa, Italy, e-mail: pedre@di.unipi.it

carries a potential wealth of *knowledge* regarding the processes that govern the life of complex economical and social systems.

As an illustration of traces in geographic systems, consider the information that wireless phone networks gather on their consumers. Network providers capture highly detailed traces about an individual's activities through the combination of two factors: (1) the pervasiveness of mobile technologies and (2) the accuracy of positioning technologies. To situate the pervasive nature of mobile technologies in context, the number of mobile-phone users worldwide was recently estimated at 1.5 billion in 2005. In certain regions, such as Italy, the percentage of the population that uses mobile phones is close to maximum capacity. In other regions, especially developing countries, the percentage of the population using mobile phones is rapidly increasing. Moreover, as explained in Chap. 3 of this book, the suite of location-based technologies currently used by wireless carrier operators can provide accurate estimates of a user's location. The integration of various positioning technologies – such as GPS-equipped mobile devices, Wi-Max, Wi-Fi, and Bluetooth for indoor positioning – is expected to enable even more detailed localization capability in the future.

As a result of the knowledge that may be discovered in the traces left behind by mobile users, the information systems of wireless networks pose potential opportunities for enhancing services, but threats abound. It should be noted that data itself, are neither "good" nor "bad." Rather, it is how the data are processed and applied, i.e., the purpose that leads to a distinction between seemingly acceptable and unacceptable uses. Thus, it is necessary for service providers to

- Specify the knowledge that can be collected and searched from digital traces
- Define the purposes for which traces can be stored, analyzed, and shared
- Indicate who has the right to inspect stored traces of personal information

To characterize how traces are used, and by whom, we introduce the following metaphors:

The Spy. The goal of this individual is to discover knowledge about the behavior of an individual, or group of related individuals, for investigative or surveillance purposes. Often, but not always, this individual is motivated by malicious intent.

The Historian. The goal of this individual is to perform archeological investigations. More specifically, this individual strives to characterize the behavior of communities for the purpose of analysis – understanding the dynamics of these communities and the way they live. In contrast to the Spy, the Historian's intentions are often benevolent, but not always.

Before the ICT revolution, there was little chance of overlap in the time of criminal investigation and the time of historical investigation: the Spy looks for traces in the present, while the Historian looks for traces in the past. Today, we are faced with the possibility of pursuing an *archeology of the present*: discovering, in real time, precise representations of what we do, how and where we do it through the digital traces of our mobile activity.

The analysis of information systems provides opportunities to produce usable, timely, and reliable knowledge. In principle, it is now possible to learn from recent history to design more efficient and novel services, thereby enhancing the way we live today. This knowledge will be extracted from traces left as recently as a few moments ago, perhaps the previous day. More concretely, we can use the traces captured from location-based technologies to reconstruct our mobile behavior. In essence, we can represent the way we move at community level. This enables us to improve decision making in mobility-related issues, including the way we:

- Plan traffic and public mobility systems in metropolitan areas;
- Plan physical communication networks, such as new roads or railways;
- Localize new services in our towns;
- Forecast traffic-related phenomena;
- Organize postal and logistics systems;
- Avoid repeating mistakes that emerge from the freshly analyzed movement behavior;
- Timely detect changes that occur in the movement behavior.

Simply put, the collection and extraction of knowledge from mobility data endow us with the ability to construct services that address important public interests.

Yet, by enabling the analysis of mobility data, we create potential threats to personal privacy. The mobility data in question are detailed and each record corresponds to a specific individual. If such data are made available for the aforementioned purposes, we put at risk our right to secrecy in our movements and meetings. By revealing mobility information, we also reveal the places we visit, the places at which we live or work, and the people we meet. The more information an individual reveals, the more (s)he paints a picture of her/his life. Personal mobility data, as gathered by the wireless networks, can consist of highly sensitive information. The disclosure of such information can potentially violate the privacy rights, as specified in an increasing number of laws and regulations defined at the national and international level (see Sect. 4.2).

Here, we introduce a third member to our metaphor:

> *The Scientist.* It is the goal of this individual to prevent the Spy from violating an individual's privacy and anonymity in mobility data without harming the Historian's ability to perform studies.

A Scientist may observe that, for the Spy to achieve his surveillance goal, he needs to know the movements of individuals. However, for analytical purposes, the Historian does not need to know the identities of the individuals to whom the mobility data correspond. Rather, anonymized trajectories are sufficient to construct aggregate movement behavior that pertains to groups of people.

Is this reasoning correct? Can we conclude that the Historian's analysis for public interest poses no risks? Is it possible that the Historian could inadvertently jeopardize the privacy of the individuals?

Unfortunately, hiding identities, or "deidentifying" data, is not enough to guarantee privacy protection. In certain cases, it is possible to reconstruct the identities of individuals from released data, even when explicit identifiers – such as name, phone number, and address – have been removed and replaced with pseudonyms. Databases that are devoid of explicit identifiers may contain attributes that, in combination, can be linked to identified collections, thus leading to "reidentification" of the individuals. A combination of attributes that can uniquely represent individual entities is called a *quasi-identifier* [10]. A popular example of reidentification via quasi-identifiers is from a study conducted by Latanya Sweeney in 1990s. At the time, it was assumed that deidentified hospital discharge databases were protected from reidentification. Thus, the databases, where each record reports medical diagnosis and treatment information per an individual's hospital visit, were made available to the public, for a fee, by state and federal agencies. However, it was shown that the discharge databases contained a quasi-identifier on the following patient-specific attributes: (1) date of birth, (2) gender, and (3) residential five-digit ZIP code. Alone the combination of these attributes does not breach privacy of the patients. However, using data from 1990 US census, Sweeney demonstrated that the aforementioned quasi-identifier is unique in approximately 87% of the US population, or 216 million individuals at the time of the study [24]. Thus, if there exist publicly accessible resources that contain the quasi-identifier, then a significant portion of the population could be reidentified.

In fact, Sweeney demonstrated that such a reidentification was possible with data from the state of Massachusetts. In a more specific experiment, Sweeney reidentified discharge databases from the state of Massachusetts. In the investigation, the public identifiable database consisted of the poll list of voters of the county of Cambridge, MA (available for sale for political campaign purposes). This list contained person-specific attributes including residential ZIP code, date of birth, sex, name, address, party affiliation, and mode. By joining the discharge and voter data on the quasi-identifiers, it was possible to uniquely link sensitive medical information to specific individuals. In fact, Sweeney showed that the records for the governor were reidentifiable as well: six people had his date of birth, only three of those were men, and only one had his ZIP code. As a consequence, the medical records of the governor were uniquely identified from publicly available resources.

Clearly, the removal of identity is not a fail-safe solution to identity protection, especially when quasi-identifiers are used as pseudonyms. This vulnerability motivates the study of techniques, briefly discussed in Sect. 4.3, that attempt to mitigate the danger of quasi-identifiers while preserving the usefulness of data for analysis. Yet, even a brute force solution, such as the replacement of identifiers with unintelligible codes, may not be sufficient when the data to be disclosed correspond to mobile and geographic information, such as personal trajectories. An example of such insufficiency is discussed in [6, 18]: consider a trajectory, devoid of identifiers, that occurs periodically every working day from location A in the suburbs to location B downtown during the morning and in the reverse direction (from B to A) in the evening. This trajectory can be linked to the people who live in A and work in B. Therefore, when locations A and B are known at a detailed granularity, it may be possible to reidentify specific persons to their daily routes. This example typifies

4 Privacy Protection

how, in mobility data, the consideration of geographic space and time can function as a powerful quasi-identifier.

Another interesting model of privacy concerns is given with respect to location-based reidentification [19–21]. In these works the term *reidentification* refers to correctly relating seemingly anonymous data to explicitly identifying information (such as the name or address) of the person who is the subject of those data. While reidentification has usually been associated with data released from a single data holder, they show how an individual could be related to a "trail" of seemingly anonymous and homogeneous data left across different locations. Successful reidentifications are reported for DNA sequences left by hospital patients and for IP addresses left by online consumers.

Of course, for identity protection, alternatives to deidentification may be applied, such as through the addition of noise to the data in such a way that exact positioning is obfuscated without preventing the extraction of useful aggregated knowledge [14, 17].

The Scientist might argue that, in the end, it is unnecessary to disclose the individual records that have been collected. Rather, the Historian may be given limited access to the data, such as by functioning as a trusted civil servant, to produce the knowledge (e.g., mobility patterns or geographic models) that is disclosed for the public utility. In this sense, only aggregated information is divulged beyond the confines of the initial collection. The source data are kept secret while an aggregated version is shared. In terms of Chap. 2, only synoptic, as opposed to elementary, questions are allowed to be executed on the secret source data. Thus, questions such as "Who was near the leaning tower yesterday between 3 and 4 a.m.?" cannot be answered; questions such as "How many people were near the leaning tower yesterday between 3 and 4 a.m.?" can be answered. This layer of protection can be extended to prevent queries with small number results from being answered.

Since aggregated information concerns large groups of individuals, it is tempting to conclude that its disclosure is safe. Once again, however, this reasoning is flawed. As explained in [5], rules with high support (i.e., concerning many individuals) can sometimes be used to deduce new rules with very limited support that can precisely identify one, or few, individuals with their sensitive attributes. As an example, assume that the following rule can be mined from the source data:

$$\text{Age} = 27 \ \wedge \ \text{ZIP} = 45254 \ \wedge \ \text{Diagnosis} = \text{HIV} \rightarrow \text{NativeCountry} = \text{USA}$$
$$[\text{sup} = 758, \text{conf} = 99.8\%].$$

This rule informs us that 99.8% of 27 years old in a specified geographic area that has been diagnosed an HIV infection are born in the US. From this rule, we can infer that 0.2% of 758 persons are 27 years old, live in the given area, have contracted HIV, and were not born in the US. Furthermore, a simple calculation reveals that there is only one person with this combination of characteristics. Now, the combination of age, ZIP code, and native country is a quasi-identifier, and it is possible that we can find, via a register with county demographics, that there is only one 27-year-old resident in the given area who was not born in the US. This revelation would constitute a privacy violation, as such a person will be uniquely identified.

The moral of this story is that protecting privacy when disclosing information is nontrivial. Anonymization and aggregation do not necessarily prevent attacks to privacy, and this difficulty makes the problem socially relevant: for the very same reason, the problem is scientifically attractive. As often happens in science, the problem is to find an optimum in the tradeoff between two conflicting goals: utility and privacy. With respect to utility, we need to provide the Historian with fine-grained knowledge of an individual's mobility data for analytical purposes. Yet, simultaneously we must uphold and we must shield to prevent the Spy from violating the privacy of, or extracting sensitive information from, an individual's mobility. It is interesting that the same conflict – essentially between opportunities and risks – can be read as either a mathematical problem or a social (or ethical, or legal) challenge. Indeed, the privacy issues related to ICTs are unlikely to be solved by exclusively technological means: paraphrasing Rakesh Agrawal, one of the first researchers to address privacy issues in data management, any real solution to privacy problems can only be achieved through an alliance of technology, legal regulations, and social norms [2]. It is exactly with this observation in mind that we created an observatory on privacy regulations within the GeoPKDD project, whose aims are described in Sect. 4.4. Before delving into the observatory and project, we briefly provide an overview of the current international situation from the (1) legal front: in terms of laws and regulations about protection of personal data (Sect. 4.2) and (2) the technical front: in terms of privacy-preserving technologies in data management and data mining (Sect. 4.3). A more detailed discussion on the latter theme can be found in Chaps. 8 and 11.

4.2 Privacy Regulations

Of all human rights, privacy is perhaps the most difficult to define because it varies widely according to context, culture, and environment. Nonetheless, privacy is a fundamental human right that underpins dignity, freedom of speech, and association. In many countries, the right to privacy is recognized and the concept has been fused with data protection, which interprets privacy in terms of management of personal information. From an international perspective, privacy rights are protected in the Universal Declaration of Human Rights, the International Covenant on Civil and Political Rights, and in many other international human rights treaties.

Privacy can be interpreted as the individual right to choose freely what to do with their one's personal information. Nearly every country in the world includes the right of individuals' freedom in its constitution, and this should generally imply the right to privacy.

At a minimum, these provisions include rights of inviolability of the home and secrecy of communications. Most recently written constitutions include specific rights to access and control one's personal information. In many countries,

4 Privacy Protection

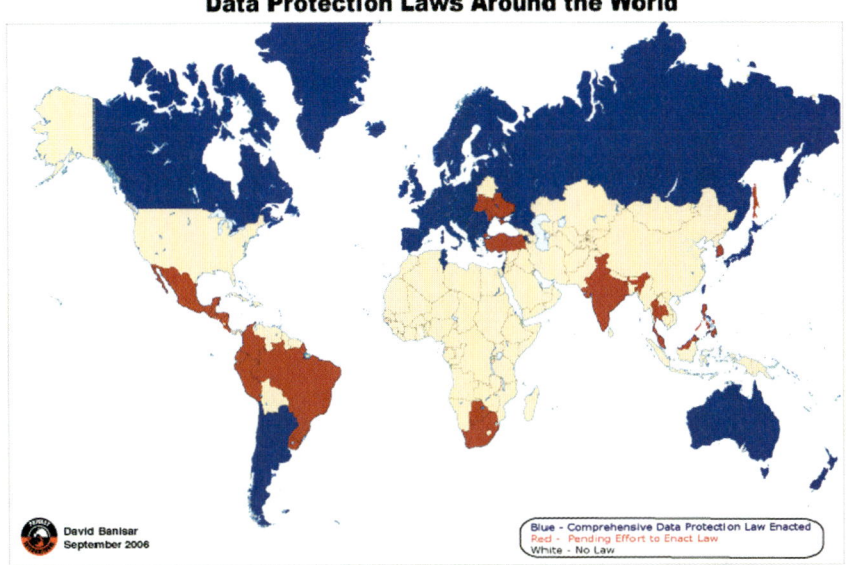

Fig. 4.1 Global map of data protection up to September 2006 (source: *Privacy International*)

international agreements that recognize privacy rights have been adopted into law,[1] as shown in the global map of data protection (Fig. 4.1).

4.2.1 Aspects and Models of Privacy Protection

Among the various aspects by which privacy can be considered, there are two aspects that are crucial to our study (1) information privacy and (2) privacy of communications:

Information privacy. It involves the establishment of rules governing the collection and handling of personal data, such as credit, medical, and government records. It is also known as "data protection."

Privacy of communications. It covers the security and privacy of mail, telephones, e-mail, and other forms of communication.

There are also four major models for privacy protection, which can be applied either independently or simultaneously. In the countries that protect privacy most effectively, all of the models are used together to ensure privacy protection. The models are defined as follows:

Comprehensive laws. These are general laws that govern the collection, use and dissemination of personal information by both the public and private sectors.

---

[1] Part of the material presented in this section is borrowed from the "Overview of Privacy" of the 2005 edition of the Privacy and Human Rights Report (www.privacyinternational.org).

There is an oversight body that ensures that data collectors comply with the enacted laws and regulations. This is the preferred model of most countries that have adopted data protection laws. Furthermore, this model was adopted by the European Union to ensure compliance with its data protection regime. A variation of these laws, which is described as a "coregulatory model," was adopted in Canada and Australia. Under this approach, industry, as opposed to government, develops rules for the protection of privacy. The rules are defined by industry members and are overseen by a privacy agency.

Sectoral laws. Some countries, such as the United States, have avoided the enactment of broad data protection rules in favor of specific sectoral laws. For example, different laws and regulations govern the confidentiality of video rental records [28], financial information [27], and medical records [26]. In such cases, enforcement is achieved through a range of mechanisms. A major drawback with this approach, however, is that it requires new legislation to be introduced with each new technology. As a consequence, technology protections frequently lag behind policy specification. The lack of legal protections for an individual's privacy on the Internet in the United States is a striking example of its limitations. In addition, there is a lack of a centralized oversight agency for monitoring and addressing privacy violations. In many countries, sectoral laws are used to complement comprehensive legislation by providing more detailed protections for certain categories of information, such as telecommunications, police files, or consumer credit records.

Self-regulation. Data protection can also be achieved, at least in theory, through various forms of self-regulation, in which company and industry bodies establish codes of practice, or codes of conduct and engage in self-policing. However, in many countries, there is little evidence that the aims of the codes are regularly fulfilled. Major problems associated with this approach is lack of adequacy and enforcement. In addition, industry codes in many countries tend to provide weak protections and lack enforcement.

Technologies of privacy. With the recent development of commercially available technology-based systems, various aspects of privacy protection have moved into the hands of individual users. For instance, users of the Internet, and of physical applications, can employ a range of programs and systems that provide varying degrees of privacy and security of communications. These include encryption, anonymous remailers, proxy servers, and digital cash. In this model, there is no government or industry oversight. Rather, the protection of one's personal data falls squarely on the shoulders of consumers.

4.2.2 The Evolution of Data Protection

Interest in the right of privacy increased in the 1960s and 1970s with the advent of information technology. The surveillance potential of powerful computer systems prompted demands for specific rules governing the collection and handling of personal information. The genesis of modern legislation in this area can be traced back to the first data protection law in the world, which was enacted in the Land of Hesse in Germany in 1970. This was followed by national laws, enacted in Sweden (1973), the United States (1974), Germany (1977), and France (1978).

Two crucial international instruments are evolved from these laws (1) The Council of Europe's 1981 Convention for the Protection of Individuals with regard to the Automatic Processing of Personal Data[2] and (2) The Organization for Economic Cooperation and Development (OECD) Guidelines Governing the Protection of Privacy and Transborder Data Flows of Personal Data.[3] These policies set out specific rules covering the handling of electronic data and rules that describe personal information as data that are afforded protection at every step from collection to storage and dissemination.

4.2.2.1 The European Union Data Protection Directives

In 1995, the European Union enacted the Data Protection Directive[4] to harmonize member states' laws. The goal of the directive is to provide consistent levels of protections across Europe for citizens to ensure the free flow of personal data within the European Union. The directive sets a common baseline level of privacy that not only reinforces current data protection law but also establishes a range of new privacy rights. The directive applies to the processing of personal information in electronic, as well as paper, files.

A key concept in the European data protection model is "enforceability." Data subjects are endowed with rights that are established in explicit rules. Every European Union country must have a data protection commissioner, or agency, that enforces the rules. Moreover, it is expected that the countries with which Europe does business must provide a similar level of oversight.

The directive established several basic principles for European citizens. These principles include the following rights:

- The right to know where the data originated
- The right to have inaccurate data rectified
- The right of recourse in the event of unlawful processing
- The right to withhold permission to use data in some circumstances

For example, individuals have the right to opt-out, free of charge, from direct marketing. The directive contains strengthened protections over the use of sensitive personal data relating, for example, to health, sexual orientation and endeavors, religious preference, and philosophical beliefs.

The 1995 directive imposes an obligation on member states to ensure that the personal information relating to European citizens has the same level of protection when it is exported to, and processed in, countries outside of the European Union.

---

[2] http://conventions.coe.int/Treaty/en/Treaties/Html/108.htm
[3] http://www.oecd.org/document/18/0,2340,en_2649_34255_1815186_1_1_1_1,00.html
[4] http://www.cdt.org/privacy/eudirective/EU_Directive.html

This requirement has resulted in growing pressure outside of Europe for the passage of more strict, as well as internationally governed, privacy laws. Countries that refuse to adopt adequate privacy laws may find themselves unable to continue certain types of information flows with Europe, particularly if they involve sensitive data.

In 1997, the European Union supplemented the 1995 directive through the introduction of the Telecommunications Privacy Directive.[5] The 1997 directive established specific protections for emerging technologies covering telephone, digital television, mobile networks, and other telecommunication systems. The 1997 directive imposed wide-ranging obligations on carriers and service providers to ensure the privacy of users' communications, including Internet-related activities.

In July 2000, the European Commission issued a proposal for a new directive on privacy that would apply to the electronic communication sector. The proposed amendments were to strengthen privacy rights for individuals through the extension of protections that were already in existence. During the process, however, the Council of Ministers pushed for the inclusion of data retention provisions that would require Internet service providers and telecommunication operators to store logs of all telephone calls, e-mails, faxes, and Internet activity for law enforcement purposes for up to 2 years. The goal of data retention was to assist in the prevention of terrorism and organized crime. However, the proposal for the inclusion of data retention provisions was met with varying degrees of opposition, in the fear that such collection and storage put an individual's control over their information, and their privacy, at greater risk.

Following the events of September 11 in the United States, the political climate changed and the European Parliament was under increasing pressure from member states to adopt the Council's proposal for data retention. It finally reached a deal in favor of the Council's position and on 25 June 2002 the European Union Council adopted the new privacy and electronic communication directive as voted in the Parliament. Under the terms of the new directive, member states may now pass laws mandating the retention of the traffic and location data of all communications taking place over mobile phones, SMS, landline telephones, faxes, e-mails, chatrooms, the Internet, or any other electronic communication device. Similar data retention regulation proposals are currently under heated debate in the United States Congress.

4.2.2.2 The APEC Privacy Initiative

In 2003, the Asia–Pacific Economic Cooperation (APEC), which consists of 21 countries, commenced on the development of an Asia–Pacific privacy standard. This is one of the most significant international privacy initiatives since establishment of the European Union's Data Protection Directive in the mid-1990s. In February 2003, Australia submitted a proposal for the development of APEC privacy principles, and

---

[5] http://www.dataprotection.ie/viewdoc.asp?m=&fn=/documents/legal/6aiii.htm

recommended the use of the 20-year-old OECD Guidelines on the Protection of Privacy and Transborder Flows of Personal Data (1980) as a starting model. A privacy subcommittee, composed of Australia, Canada, China, Hong Kong, Japan, Korea, Malaysia, New Zealand, Thailand, and the United States, was established to handle the composition of the principles. In March 2004, Version 9 of the APEC privacy principles was released as a public consultation draft.[6]

The APEC privacy initiative is notable in that it has the potential to encourage the development of stronger privacy laws in APEC countries. Currently, the APEC members provide little in the way of standardized privacy protection. The development of a common directive will help in providing a regional balance between the protection of privacy and the economic benefits of trade involving personal data. Yet, the development of a directive has the potential negative consequences. Specifically, the adoption of privacy principles is dangerous to long-term regional privacy protection if it becomes a means by which the APEC economies accept a second-rate standard. At the present time, criticisms of the APEC principles emphasize that they do not satisfy, let alone strengthen, the 20-year-old OECD standards, which are now too weak in the face of the information age.

4.2.2.3 Data Havens and the Safe Harbor Arrangement

The ease with which electronic data flow transnational borders is caused for a concern that data protection laws could be circumvented through the transfer of personal information to third countries where the law of the country of origin does not apply. By doing so, the data could be processed in the receiving country, sort of "data paradises" to avoid compliance with strict privacy laws[7] without legal limitations. For this reason, most data protection laws include restrictions on the transfer of information to other countries unless information protection in the receiving country is considered acceptable by the originating country. This requirement has resulted in growing pressure outside of Europe for the passage of strong international data protection laws. Determination of a data haven's system for protecting privacy is made by the European Commission based on the principle that the level of protection in the data haven must be "adequate" rather than "equivalent." An alternative model of protection is to allow the originating country to rely on a private contract that contains standard data protection contractual clauses. This type of contract would bind the data processor in the data haven to respect the fair information practices such as the right to notice, consent, and access. This model would permit data processors to define "adequate protection" in a context-specific manner. At the same time, however, a limitation to such a model is that data protection standards in a data haven would not be standardized, which could cause conflicting levels of privacy for transferred data.

---

[6] http://www.bakercyberlawcentre.org/appcc/

[7] Note that "data havens" are instead software and computer networks (e.g., *Freenet*) aimed at protecting privacy in countries where no privacy protection laws exist.

The European Commission never issued a formal opinion on the adequacy of privacy protection in the United States, although serious doubts were put forward regarding whether the United States' sectoral and self-regulatory approaches to privacy protection provide an adequate standard as specified in the directive. The European Union commissioned two prominent United States' law professors to investigate this matter. The result was a detailed report on the state of United States privacy protections and pointed out the many gaps in United States protection.[8] Despite concerns, the United States government strongly lobbied the European Union, and its member countries, to rule that the United States model of data protection was adequate. In 1998, the United States initiated the negotiation of a "Safe Harbor" agreement with the European Union to ensure the continued transnational flow of data.

The main premise of the Safe Harbor clause is that organizations in the United States will voluntarily self-certify to adhere to a set of privacy principles specified by the United States Department of Commerce and the Internal Market Directorate of the European Commission. The organizations in the United States would certify the adequacy of their safeguards, with respect to the principles, and thus would continue to receive personal data from organizations in the European Union. On 26 July 2000, the Commission approved the agreement, but promised to reopen negotiations on the arrangement if the remedies available to European citizens proved to be inadequate. Privacy advocates and consumer groups both in the United States and Europe are highly critical of the European Commission's decision to approve the Safe Harbor clause. Many believe it will fail to provide European citizens with adequate protection for their personal data. The agreement rests on a self-regulatory system whereby organizations promise not to violate their declared privacy practices. Under the current model, there is little enforcement, or systematic review, of compliance. Furthermore, European citizens are not granted the opportunity to appeal data transfer, nor are they granted the right to compensation at the time of self-certification.

4.2.3 Privacy Constraints in the GeoPKDD Context

In the context of the European GeoPKDD project, the standard for privacy regulations is the directive 95/46/EC of the European Parliament and Council we already mentioned, that was approved on 24 October 1995. The directive provides a number of definitions that are applicable to privacy-preserving data publishing and mining. In addition, the directive poses a number of open questions that are left to be addressed in legislation of national countries. As a result, there are many opportunities for contributions of the GeoPKDD project at the international, as well as national, level. In this section, we present some of the directive's definitions and questions, and relate them to the goals of the GeoPKDD project.

---

[8] Paul M. Schwartz and Joel R. Reidenberg. Data Privacy Law. Michie, 1996.

There are several basic definitions that were established by the directive that are important for knowledge discovery. First, definition (a) states that:

> "Personal data" shall mean any information relating an identified or identifiable natural person ("data subject"); an identifiable person is one who can be identified, directly or indirectly, in particular by reference to an identification number or to one or more factors specific to his physical, physiological, mental, economic, cultural or social identity.

Second, definition (b) states that:

> "Processing of personal data" ("processing") shall mean any operation or set of operations which is performed upon personal data, whether or not by automatic means, such as collection, recording, organization, storage, adaptation or alteration, retrieval, consultation, use, disclosure by transmission, dissemination or otherwise making available, alignment or combination, blocking, erasure or destruction.

With respect to GeoPKDD, it is important to recognize the following general statement in the directive, known as Premise 2:

> Data-processing systems are designed to serve man; they must, whatever the nationality or residence of natural persons, respect their fundamental rights and freedoms, notably the right to privacy, and contribute to economic and social progress, trade expansion and the well-being of individuals.

This general statement establishes two important constraints that are applicable to knowledge discovery techniques (1) the respect of an individual's freedom in the derivation of services and (2) the "economic and social progress, trade expansion, and the well-being of individuals."

The previous establishes a precedent for protection of personal information; however, it does not specify how data should be protected. This issue is addressed in Premise 26 of the directive, which establishes a general concept of *identifiable* and *anonymous*. Premise 26 reads as follows:

> The principle of protection must apply to any information concerning an identified or identifiable person; whereas, to determine whether a person is identifiable, account should be taken of all means likely reasonably to be used either by the controller or by any other person to identify the said person; whereas the principles of protection shall not apply to data rendered anonymous in such a way that the data subject is no longer identifiable; whereas codes of conduct within the meaning of article 27 may be a useful instrument for providing guidance as to the ways in which data may be rendered anonymous and retained in a form in which identification of the data subject is no longer possible.

The concepts of identifiability and anonymity are central to knowledge discovery, and received already attention by the research community, as discussed in Chaps. 8 and 11 of this book. Nonetheless, further research is needed to provide a firm technical ground for these concepts in the context of mobility data, addressed in the GeoPKDD project, which are bound to become a major source of privacy threats and analysis opportunities.

The last part of Premise 26 specifies that the directive may be integrated by member states as "codes of conduct." For example, the Italian Data Protection Authority has issued a code of conduct for journalists with respect to the collection and dissemination of personal information.[9] One aim for the GeoPKDD project is the

---

[9] http://www.garanteprivacy.it/garante/doc.jsp?ID=487496

suggestion of definitions for privacy-preserving data mining that can be incorporated into a code of conduct for mobile data analysts.

It is worth noting that the directive shows a clear awareness to favor the use of data analysis for the public interest, which implicitly includes data mining. In fact, Premise 29 of the directive states:

> The further processing of personal data for historical, statistical or scientific purposes is not generally to be considered incompatible with the purposes for which the data was previously collected provided that member states furnish suitable safeguards; whereas these safeguards must in particular rule out the use of the data in support of measures or decisions regarding any particular individual.

It is interesting to note that the directive implicitly refers to a concept that resembles a formal computational privacy model, such as k-map [25]. Informally, this model states that a disclosed piece of data is k-mapped, when there are no less than k individuals in a population that are represented by the data.[10] In the context of the directive, this notion is codified in Premise 40:

> It is not necessary to impose this obligation[11] [...] if would involve disproportionate efforts, which could be the case where processing is for historical, statistic or scientific purposes; whereas in this regard the number of data subjects, the age of the data, and any compensatory measures adopted may be taken into consideration.

4.3 Privacy-Preserving Data Analysis

In a routine day, organizations collect the details of an individual's interactions with mobile and location-based services. Individuals are often willing to provide personal information in exchange for a perceived benefit, such as a location-based service. For instance, a GPS service provider can record an individual's movements by monitoring how people use GPS-based navigation systems to travel from one location to another, such as driving an automobile from London to Paris. The information that organizations collect is complex and large in size, but data mining methodologies present an opportunity to extract patterns and discover knowledge regarding how individuals use location-based systems, as well as about the individuals themselves. Data mining is the development and application of computational methodologies to organize and discover knowledge embedded in large quantities of data. Data mining technologies are intent neutral; they harbor neither benevolent nor malevolent intent with respect to the individuals to whom collected data correspond. Nonetheless, it is often the case that individuals relinquish their data for a service without understanding what their data reveal or how they can be used by the data collector.

Local, national, and international laws, as well as various regulatory directives, have been enacted to prevent the misuse of personal information collected by public and private organization. Legislative actions, such as those described earlier in this

---

[10] Note, k-map is a particular type of privacy model. Alternative models exist, and use varying measures of protection, such as information theory [23].

[11] of communicating to the owner of data the use of it.

chapter, mandate that data stewardship facilities respect the private nature of personal information and refrain from applying this data for purposes other than those for which the data were collected. In addition, regulations often require that data stewards inform data subject about the different usages of their information. Bear in mind, however, a common exception to this rule is that personal information can be disclosed without a data subject permission upon a court order.

To adhere to the existing legislation, scientists are investigating new methodologies for personal and corporate data deidentification in such a way that data mining will not impose any threat to the privacy of subjects participating in a study, or to the confidentiality of corporate secrets of competitors. Therefore, *privacy-preserving data mining*, i.e., the study of data mining side effects on privacy, captures growing attention from researchers and administrators across a large number of application domains [4, 8, 29]. This is made evident by the fact that major companies – including IBM, Microsoft, and Yahoo! – are allocating significant resources to study this problem. For example, IBM has sponsored a Privacy Institute[12] and developed data privacy-related products, such as "Hippocratic Databases"[13] (e.g., [3]). Recent techniques that have been proposed to serve this purpose include the masking of raw data by adding noise (e.g., [14, 17]), the swapping of values (e.g., [11– 13, 16]), the aggregation of neighboring values, cryptographic techniques for the secure sharing of private data in a collaborative environment (e.g., [9, 22]), and the hiding of sensitive knowledge in shared data. A more recent trend is the investigation of techniques to achieve privacy-preserving data integration.

As personal data collection becomes more ubiquitous, the power to build database systems that link an individual's information across disparate data repositories becomes easier. As a consequence, the privacy provisions integrated into each data collection, as well as the privacy regulations that safeguard personal data, are weakened. Recent research in privacy-preserving data integration attempts to mitigate these threats and specifically addresses issues created by the collection of location, mobile, and trajectory data.

Many privacy-preserving data mining methodologies have been developed; yet there remain many open issues that require further investigation. One of today's critical challenges is that – despite an increasing interest in privacy from academic, corporate, and government agencies – there remains a lack of technology transfer in privacy-preserving data mining technologies [7]. This problem stems from a combination of technical, political, and economic challenges. First, privacy concerns and data mining endeavors vary across application domains. It is unclear how technical solutions from specific applications can be generalized into principles that can be reused in other application domains. Second, there exists a communication gap between the scientists that develop theories and technical solutions and the lawyers that define the regulations regarding privacy issues for data analysis, collection, and dissemination. Third, there is no well-defined incentive structure for organizations to protect the privacy of their data subjects [15]. Similarly, there is a lack of incentive

---

[12] http://www.research.ibm.com/privacy/
[13] http://www.zurich.ibm.com/pri/projects/hippocratic.html

for individuals to recognize the sensitivity, and protect the privacy, of their personal data. Thus, though privacy technologies exist and individuals can be informed of the potential risks to privacy if they exchange personal information for a service, most people choose the service over privacy [1].

The existing regulatory context can pose challenges and constraints to the development and adoption of novel technical solutions. Yet, we believe that real solutions to the challenges posed by the location-based applications, such as those studied within the framework of the GeoPKDD project, can be achieved through a combination of building technical tools, enacting legal regulations, and evolving social norms. The passing of new regulations with explicit privacy protections can help transition of the current norm from a closed to an open society that penalizes the misuse of personal information. New solutions need to provide feedback and opportunities for individuals, as well as organizations, to develop regulations and protections that are amenable to complex location-based technologies. Regulations need to support and not suppress the adoption of new technologies. It is necessary to achieve a more frequent cooperation and open dialogue between scientists and policy officials. Both communities need to communicate and inform in each other's endeavors.

Furthermore, given the current state of affairs, there are different technological issues that must be addressed for, and incorporated into, real world privacy solutions. First, we must characterize and standardize new forms of personal data, such as location-based information. Second, it is necessary to understand and model new forms of data collection processes, such as mobile commerce, GPS-based services, and ubiquitous environments with stream and sensor-derived data. Third, it is necessary to understand the inference capabilities and limitations of an increasing number of data mining techniques that are applied to extract knowledge from the various types of data collected. The goal is to support the collection, analysis, and sharing, of person-specific data without jeopardizing sensitive knowledge.

We have witnessed the mobilization of various initiatives in response to the aforementioned issues. These include the formation of data protection authorities and independent organizations that are focused on securing private data against misuse and misconduct. From a funding perspective, there are funding initiatives, such as GeoPKDD, to develop and disseminate state-of-the-art privacy solutions to a broader community.

4.4 The Role of the Observatory

The GeoPKDD project investigates and develops technical advances that are needed to embed privacy into data mining tools. Yet, we recognize that technology must be disseminated, and be informed of the social context in which it resides. Thus, in addition to its technical endeavors, GeoPKDD has organized a *privacy regulation observatory* that brings together GeoPKDD technologists, representatives of the national and European privacy authorities, as well nongovernmental privacy-related associations. In summary, the aim of the GeoPKDD privacy observatory is to assist authorities as technical consultant in the field of privacy-preserving data mining.

More specifically, we believe that regulations and laws will be enacted as a response to existing and future privacy-preserving methods, including those developed within the GeoPKDD project. The goal of the observatory is to harmonize the resulting regulations with the activities of technologists and the GeoPKDD project. The activities of the observatory will include the creation, and maintenance, of relationships with the European Commission authority and the national authorities of the countries that are partners of the consortium. Such relationships are aimed to properly implement the resulting regulations into our methods and tools and to provide refinements of the technical regulations regarding privacy-preserving analysis methods.

A first step toward GeoPKDD's goal to establish relationships beyond technologists is the establishment of a relationship with the Italian Data Protection Commission (Garante per la Protezione dei Dati Personali[14]). Italy implemented the main European directive, Directive 95/46, in 1996 as law no. 675/96. Italian Data Protection Commission is endowed with the power to establish sanctions when they discover the violation of regulations in cases that are brought before them. The directives, both at the European level and at national Italian implementation, are subject to interpretation and thus cases of potential privacy infractions are addressed in a case-by-case manner.

Another important aspect of the GeoPKDD project is its potential to interact with, and inform, organizations that recognize the need for location privacy standards. For example, one such organization is Geopriv,[15] which is an Internet Engineering Task Force (IETF) working group that examines risks associated with location-based services. The IETF has proposed several requirements for location privacy, including limited identifiability and customizable rules for controlling data flows. A second example organization is Privacy International,[16] which is a human rights group formed in 1990 as a "watchdog" on surveillance projects that are run by governments and corporations. We anticipate that dissemination of GeoPKDD research results will include an annual GeoPKDD workshop devoted to the presentation of location-based privacy technology, as well as policy, achievements. The workshop will also serve as an international forum for spatiotemporal privacy-preserving data mining.

4.5 Conclusions

There is an increasing fear that the growing collection and dissemination of personal mobility data will weaken the privacy rights of individuals. In part, this is due to the fact that detective-like investigations have revealed that privacy threats

---

[14] http://www.garanteprivacy.it
[15] http://www.ietf.org/html.charters/geopriv-charter.html
[16] http://www.privacyinternational.org/

abound in the collection and dissemination of data derived from ICTs. Specifically, the dissemination of human mobility data, devoid of identifying information, such as pseudonymized traces, is not sufficient to prevent privacy breaches. Despite the threats to personal privacy, a service provider has the right to analyze data collected from mobility services to discover socially useful knowledge that benefits the individuals, community, and law enforcement. Nonetheless, journalists and regulators, in Europe and beyond, increasingly claim that the defense of an individual's right to privacy must come before the sharing of personal mobility data.

Despite the apparent opposition, the right to personal privacy does not necessarily preclude the right of a service provider to learn knowledge from collected mobility data and vice versa. Computer scientists have unearthed a fascinating array of problems related to privacy and mobility data. Research on these problems has led to the production of foundations, as well as basic applications, of privacy-preserving technologies. As research in this field progresses, the goal is to reach a win–win situation for privacy advocates and service providers: obtain the advantages of collective mobility knowledge without inadvertently divulging any individual mobility knowledge.

We believe that this research on privacy for mobile data collection and analysis must be tackled in a multidisciplinary way. The opportunities and risks are shared by technologists, social scientists, jurists, policy makers, and general citizens. Research will need to be informed by, as well as helps to inform, those that design laws and oversee jurisprudence. If this goal is achieved, the results will have an impact on the social acceptance, as well as the dissemination of, ubiquitous technologies.

References

1. A. Acquisti. Privacy in electronic commerce and the economics of immediate gratification. In *Proceedings of Electronic Commerce Conference (EC'04)*, pp. 21–29. ACM, New York, 2004.
2. R. Agrawal. Privacy and data mining. In *Proceedings of the 15th European Conference on Machine Learning and the 8th European Conference on Principles and Practice of Knowledge Discovery in Databases (ECML/PKDD'04)*, 2004. Invited Talk.
3. R. Agrawal and C. Johnson. Securing electronic health records without impeding the flow of information. *International Journal of Medical Informatics*, 76(5–6):471–479, 2007.
4. R. Agrawal and R. Srikant. Privacy-preserving data mining. In *Proceedings of the International Conference on Management of Data (SIGMOD'00)*, 2000.
5. M. Atzori, F. Bonchi, F. Giannotti, and D. Pedreschi. k-anonymous patterns. In *Proceedings of 9th European Conference on Principles and Practice of Knowledge Discovery in Databases (PKDD'05)*, 2005.
6. C. Bettini, X.S. Wang, and S. Jajodia. Protecting privacy against location-based personal identification. In *Proceedings of Second VLDB Workshop on Secure Data Management*, Vol. 3674. Lecture Notes in Computer Science. Springer, Berlin Heidelberg New York, 2005.
7. C. Clifton. What is privacy: critical steps for privacy preserving data mining. In *Proceedings of the IEEE Workshop on Privacy and Security of Aspects of Data Mining*, pp. 1–7, 2005.
8. C. Clifton, M. Kantarcioglu, and J. Vaidya. Defining privacy for data mining. In *Proceedings of National Science Foundation Workshop on Next Generation Data Mining*, pp. 126–133, 2002.

9. C. Clifton, M. Kantarcioglu, J. Vaidya, X. Lin, and M.Y. Zhu. Tools for privacy preserving distributed data mining. *SIGKDD Exploration Newsletter*, 4(2):28–34, 2002.
10. T. Dalenius. Finding a needle in a haystack – or identifying anonymous census records. *Journal of Official Statistics*, 2:329–336, 1986.
11. T. Dalenius and S. Reiss. Data-swapping: A technique for disclosure control (extended abstract). In *Proceedings of the Section on Survey Research Methods, American Statistical Association*, pp. 191–194, 1978.
12. T. Dalenius and S. Reiss. Data-swapping: A technique for disclosure control. *Journal of Statistical Planning and Inference*, 6:73–85, 1982.
13. V. Estivill-Castro and L. Brankovic. Data swapping: Balancing privacy against precision in mining for logic rules. In *Proceedings of the 1st International Conference on Data Warehousing and Knowledge Discovery (DaWaK'99)*, 1999.
14. A. Evfimievski. Randomization in privacy preserving data mining. *SIGKDD Exploration Newsletter*, 4(2):43–48, 2002.
15. J. Feigenbaum, M. Freedman, T. Sander, and A. Shostack. Economic barriers to the deployment of existing privacy technologies (position paper). In *Proceedings of the Workshop on Economics of Information Security*, 2002.
16. S. Fienberg and J. McIntyre. *Data Swapping: Variations on a Theme by Dalenius and Reiss*, Vol. 3050. *Lecture Notes in Computer Science*, pp. 14–29. Springer, Berlin Heidelberg New York, 2004.
17. J. Gouweleeuw, P. Kooiman, L. Willenborg, and P. de Wolf. Post randomisation for statistical disclosure control: Theory and implementation. *Journal of Official Statistics*, 14:463–478, 1998.
18. B. Hoh and M. Gruteser. Location privacy through path confusion. In *Proceedings of IEEE Conference on Security and Privacy for Emerging Areas in Communication Networks (SecurCOMM'05)*, 2005.
19. B. Malin. Betrayed by my shadow: learning data identity via trail matching. *Journal of Privacy Technology*, (20050609001), 2005.
20. B. Malin and E. Airoldi. The effects of location access behavior on re-identification risk in a distributed environment. In *Proceedings of 6th International Workshop on Privacy Enhancing Technologies*, Vol. 4258. *Lecture Notes in Computer Science*, pp. 413–429. Springer, Berlin Heidelberg New York, 2006.
21. B. Malin and L. Sweeney. How (not) to protect genomic data privacy in a distributed network: using trail re-identification to evaluate and design anonymity protection systems. *Journal of Biomedical Informatics*, 34:179–192, 2004.
22. B. Pinkas. Cryptographic techniques for privacy-preserving data mining. *SIGKDD Exploration Newsletter*, 4(2):12–19, 2002.
23. A. Serjantov and G. Danezis. Towards an information-theoretic metric for anonymity. In *Proceedings of the Second Workshop Privacy Enhancing Technologies*, Vol. 2482. *Lecture Notes in Computer Science*, pp. 41–53. Springer, Berlin Heidelberg New York, 2002.
24. L. Sweeney. Uniqueness of simple demographics in the U.S. population. Technical Report LIDAP-WP4, Laboratory for International Data Privacy, Carnegie Mellon University, Pittsburgh, PA, 2000.
25. L. Sweeney. *Computational Disclosure Control: Theory and Practice*. Ph.D. Thesis, Massachusetts Institute of Technology, Cambridge, MA, 2001.
26. U.S. Department of Health and Human Services. Standards for privacy of individually identifiable health information; Final Rule, *Federal Registrar* 45 CFR, Parts 160 and 164, 14 August 2002.
27. U.S. Federal Trade Commission. Privacy of consumer financial information; Final Rule, *Federal Registrar* 16 CFR, Part 313, 24 May 2000.
28. U.S. Video Privacy Protection Act, 1988. 18 USC 2710, PL 100618.
29. V.S. Verykios, E. Bertino, I.N. Fovino, L.P. Provenza, Y. Saygin, and Y. Theodoridis. State-of-the-art in privacy preserving data mining. *SIGMOD Record*, 33(1):50–57, 2004.

Part II
Managing Moving Object and Trajectory Data

Chapter 5
Trajectory Data Models

J. Macedo, C. Vangenot, W. Othman, N. Pelekis, E. Frentzos, B. Kuijpers,
I. Ntoutsi, S. Spaccapietra, and Y. Theodoridis

5.1 Introduction

Trajectory databases is an important research area that has received a lot of interest in the last decade. The objective of trajectory databases is to extend database technology to support the representation and querying of moving objects and their trajectory.

Moving objects are geometries, which may be points, lines, areas or volumes, changing over time. A trajectory consists in the description of the movement of those objects. A strict definition of 'movement' relates it to change in physical position. Physical movement implies an object and a reference system within which one can assess positions. Most frequently, the reference system is geographical space and we speak about objects moving in space, therefore, about trajectories of objects in space. As geographical space per se is continuous, physical movement is described by a continuous change of position, i.e. a function from time to geographical space. Movement also implies a temporal dimension as we can only perceive movement through comparison at two different instants. Therefore, a trajectory can be equivalently defined as the record of a time-varying spatial phenomenon.

Objects may move/change at specific instants in time, without any existence or any knowledge of their existence in between. A duck suddenly disappears from your perception and reappears somewhere nearby at a later moment. In these cases, movement is perceived as neither continuous nor stepwise, but a collection of separate instants or intervals. The question *what is a moving object?* can be answered tautologically as an object that moves. An object is an identifiable real-world element that may be perceived as having an existence dissociated from that of other objects. A person and a car are obvious examples of potential objects. An object that moves is an object that is not constrained to keep the same position during its

---

J. Macedo
Database Laboratory, École Polytechnique Fédérale de Lausanne, Switzerland,
e-mail: jose.macedo@epfl.ch

F. Giannotti and D. Pedreschi (eds.) *Mobility, Data Mining and Privacy.*
© Springer-Verlag Berlin Heidelberg 2008

whole existence. Objects that move become particularly interesting when we record their trajectory. Hence, hereinafter we restrict the term moving object to denote an object to which we can associate a trajectory.

Although a trajectory can be quite simply defined as a function from time to geographical space, its description, representation and manipulation happen to be more complex. Indeed, from an application point of view, a trajectory is the record of the movement of some object i.e. the record of the positions of the object at specific moments in time. Thus, although we naturally think of a nice curve representing the trajectory of the object, in reality the trajectory has to be built from a set of sample points, i.e. the positions of the object. And the nice curve is obtained by applying interpolation methods on the set of sample points. To find the more suitable curve connecting the sample points, various interpolation methods have been proposed. However, whichever interpolation method is used, the resulting curve will only be a guess of the probable trajectory. This guess is even worse when considering the possible measurement errors that inevitably happen when recording the original points. There is thus an inherent uncertainty associated to trajectories, which, depending on the cause, is either measurement or interpolation uncertainty. To model and manage adequately uncertainty, different interpolation methods and modelling concepts have been proposed. They are presented in Sect. 5.2.

Trajectory data modelling has received a lot of attention from the research community either from researchers applying existing spatiotemporal data models to trajectory data or from researchers proposing new models specifically dedicated to moving objects and their trajectories. Indeed simply considering trajectories as a function from time to geographical space, existing spatiotemporal models can be used to model trajectories. Those models, presented in more detail in Sect. 5.3.1, usually represent trajectories as time-varying geometry.

Another trend of research has considered constraint database models to represent trajectories. Indeed since trajectories can be seen as a collection of infinite points connecting a finite number of sample points, constraint database models can be specialized to represent moving objects and their trajectory (see Sect. 5.3.2).

Starting from this constatation that neither existing spatiotemporal models nor constraint database models were perfectly adapted for trajectory modelling, a parallel line of research focussing on modelling moving objects as well as supporting location-aware queries has emerged. Those works are presented in Sect. 5.3.3.

Even if quality work on moving objects exists, there are still many open issues regarding conceptual modelling of trajectories, uncertainty, multiple representation of trajectories, continuously acquired trajectories, and query capabilities. Those open issues are described in Sect. 5.4.

5.2 Basic Concepts: From Raw Data to Trajectory

For modelling concepts to represent trajectories in databases, basic concepts of trajectory data need to be presented. This is the objective of this section, where a more formal definition of a trajectory is first proposed. Then, through the description of different interpolation methods, the process of building a trajectory from the set of positions of real-world objects will be discussed. Finally, several methods to cope with uncertainty, an inherent component of trajectories, will conclude the section.

5.2.1 What Is a Trajectory?

In Sect. 5.1, we presented an intuitive definition of a trajectory as the description of the movement of some object. More formally, a trajectory T is the graph of a continuous mapping from $I \subseteq \mathbb{R}$ to $\mathbb{R}^2$ (the two-dimensional plane)

$$I \subseteq \mathbb{R} \to \mathbb{R}^2 : t \mapsto \alpha(t) = (\alpha_x(t), \alpha_y(t))$$

now $T = \{(\alpha_x(t), \alpha_y(t), t) \mid t \in I\} \subset \mathbb{R}^2 \times \mathbb{R}$

5.2.2 From Sample Points to Trajectories

The first and foremost restriction is of course that a trajectory connected to a data sample should contain the sample points, i.e. for all points (x_i, y_i, t_i) in the sample we have $(x_i, y_i, t_i) = (\alpha_x(t_i), \alpha_y(t_i), t_i)$. It is rather trivial to remark that if our sample points are ordered in time, i.e. if $i < j$ then $t_i < t_j$, then this order will be preserved along the trajectory.

Second, given a data sample, there is an infinite number of trajectories connected to that data sample. The trajectory is by no means unique. Finding a suitable curve connecting the 'dots', the sample points, is called interpolation.

5.2.2.1 Interpolating the Sample Points

Interpolation brings along its own problems. We wish it to be fast, easily manageable, flexible and accurate. Unfortunately, improving one property does not necessarily improve another. And as we will see, more often than not, these properties counteract each other.

Linear interpolation (Fig. 5.1) is the fastest and easiest of them all. The idea is to connect the sample points with straight lines, the linearity is expressed in the fact that equal jumps in time (between the same sample points) lead to equal jumps in space. For example, the segment between the points (x_i, y_i, t_i) and $(x_{i+1}, y_{i+1}, t_{i+1})$ is given by

$$(x, y, t) = (x_i, y_i, t_i) + \frac{t - t_i}{t_{i+1} - t_i} (x_{i+1} - x_i, y_{i+1} - y_i, t_{i+1} - t_i),$$

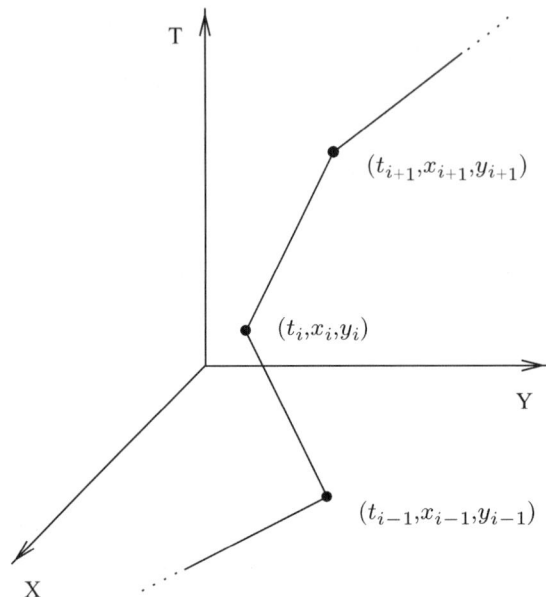

Fig. 5.1 Linear interpolation

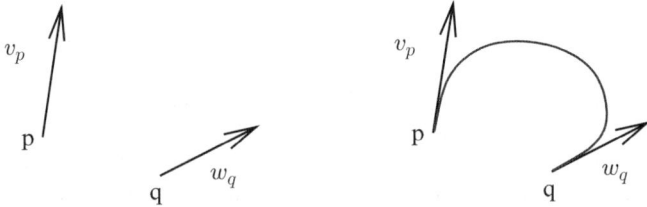

Fig. 5.2 Interpolation with Bezier curves

this is a straight line segment in $\mathbb{R}^2 \times \mathbb{R}$ parameterized by $t \in [t_i, t_{i+1}]$. The trajectory consists then of the concatenation of all these segments.

Interpolation in this manner is not so innocent, along the way some assumptions have been made. The first one is that the moving object has constant speed and direction between the sample points. Moreover, this speed is the minimal average speed needed to cover the distance between (x_i, y_i) and (x_{i+1}, y_{i+1}) in time $t_{i+1} - t_i$.

Second, changes in speed and direction at sample points are often abrupt and discontinuous, because of the sharp corners of the trajectory at the sample points. Note that the trajectory is continuous, but its speed and direction is not.

Third, it is fast. Fast to construct and to handle. Computing intersections, eliminating quantifiers (see the constraint model below) is 'easy' when we only consider objects described by linear equations (inequalities).

Interpolation with Bezier curves (Fig. 5.2) lends itself much better to create smoother curves. Given two sample points and a velocity vector, i.e. direction and

5 Trajectory Data Models

speed, in each sample point, a Bezier curve is a curve where each spatial coordinate is a third-degree polynomial of the time coordinate. The beginning and end point are exactly the respective sample points. The vectors in these sample points are precisely the velocity vectors to the curve in those points.

Bezier curves are fast to construct. Transitions over sample points are nice and smooth. The trajectory is everywhere differentiable. The downside is that it is a lot harder to handle. For example, computing distance along the trajectory and computing intersections with other trajectories become much less trivial task.

Plus, you need much more information to construct this trajectory. You need to know the object's speed and direction at each sample point. If those are unknown, one can still make educated guesses. Use the average direction taken from the direction from the previous to the current sample point and from the current to the next sample point. One can make similar guesses for the object's speed using the minimal average speed needed to get from the previous to the current sample point and from the current to the next sample point.

All these interpolation methods have one thing in common. The more sample points there are, i.e. the closer they are in time, the more accurate the trajectory will be. The two methods mentioned earlier will converge to the same trajectory when you increase the frequency of the sample points.

5.2.3 Uncertainty

Interpolation will only give you a guess of a probable trajectory and that guess leaves a certain amount of uncertainty about the chosen trajectory. This first kind of uncertainty will be referred to as *interpolation uncertainty*.

In the literature, uncertainty has been defined as the measure of the difference between the actual contents of a database, and the contents that the current user or application would have created by direct and perfectly accurate observation of reality [78]. Sources of uncertainty may be one of the following:

- Imperfect observation of the real world
- Incomplete representation language
- Ignorance, laziness or inefficiency

Interpolation uncertainty may be seen as a result of the two last points. Interpolation uncertainty can be managed with beads. The bead model works under the assumption that we know an upper bound for the object's speed in between sample points, and also that the position in the sample points is an exact position, although the latter can be tackled easily. As for the upper bound on the object's speed, the maximum speed limit of the area the object covers can be used for example.

Suppose the object's maximal speed is v, and that it travels from (x_i, y_i) at time t_i and arrives at (x_{i+1}, y_{i+1}) at time t_{i+1}. At any time $t \in [t_i, t_{i+1}]$, the distance of the object, at position (x, y), to (x_i, y_i) will be at most $v(t - t_i)$. This means that at any time $t \in [t_i, t_{i+1}]$, the object is somewhere in a disc with centre (x_i, y_i) and radius

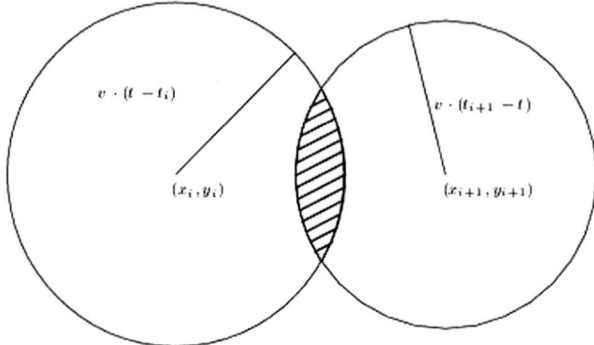

Fig. 5.3 The uncertainty area at time t

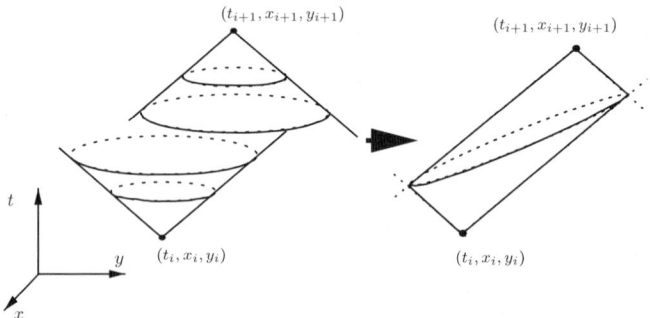

Fig. 5.4 An uncertainty bead

$v(t - t_i)$. Furthermore, in space–time this a cone, with its top in (x_i, y_i, t_i), an axis of symmetry parallel to the time axis and pointing backward in *time*.

At the same time $t \in [t_i, t_{i+1}]$, the object, at position (x, y), has to reach (x_{i+1}, y_{i+1}) in time $(t_{i+1} - t)$. That means its distance to (x_{i+1}, y_{i+1}) can be at most $v(t_{i+1} - t)$ and again the object is somewhere in a disc with centre (x_{i+1}, y_{i+1}) and radius $v(t_{i+1} - t)$. Similarly, this is a cone in space–time, but this time pointing forward in time.

So at any time $t \in [t_i, t_{i+1}]$ the object must be somewhere in the intersection of two discs as can be seen in Fig. 5.3. Or, more generally speaking, a point (x, y, t) *might* belong to a trajectory going from (x_i, y_i, t_i) to $(x_{i+1}, y_{i+1}, t_{i+1})$ if and only if it lies in the intersection of the cones as can be seen in Fig. 5.4.

The geometric object in space–time in Fig. 5.4 is called a (lifeline) bead [19]. Projecting this bead on the XY-plane yields to an ellipse with foci's in (x_i, y_i) and (x_{i+1}, y_{i+1}). This is easy to see, the distance between (x_i, y_i) and the object with coordinates (x, y) is *at most* $v(t - t_i)$, and that the distance to (x_{i+1}, y_{i+1}) is at most $v(t_{i+1} - t)$. Adding those two distances equals $v(t_{i+1} - t_i)$, which is constant and independent from t, for all $t \in [t_i, t_{i+1}]$. That means that the sum of the distances to (x_{i+1}, y_{i+1}) and (x_i, y_i) is at most a constant number $v(t_{i+1} - t_i)$ and that the geometric set of such points is, therefore, by definition the area bounded by an ellipse.

Beads are not easy to handle. To reduce the complexity of evaluating certain queries, minimal bounding circular and elliptic cylinders and even minimal bounding rectangles are used. Note that to manage uncertainty in this manner, we need a way to determine the maximal speed of the object.

Another kind of uncertainty needs to be considered, namely the kind introduced by measurement errors. Sensors introduce errors, e.g. the measurement error with GPS. We will refer to this kind of uncertainty as *measurement uncertainty*. A model that captures both kinds of uncertainty is described by [71].

In this model an *uncertainty threshold* is introduced. This threshold denotes the maximal distance of the object to the assumed location on the trajectory. After linear interpolation, this model assigns to each point on the trajectory a disc, parallel to the XY-plane, of radius equal to the threshold. Taking all those discs together in three-dimensional space–time results in a tube around the polyline connecting the sample points.

This threshold incorporates interpolation uncertainty and measurement errors all at once. It does not discriminate sample points from interpolated points, though we only need to take the measurement error into account when it comes to the sample points.

Now assume that the threshold is chosen so that the trajectory volume is a minimal bounding volume for the beads of the same sample points, then beads reduce uncertainty roughly by a factor 3, since a cone has one-third the volume of its minimal bounding circular cylinder.

However, considering the structure of the trajectory volumes, i.e. circles parallel to the XY-plane, these structures are much easier to handle computationally. For example, the alibi-query is child's play in this model, and the alibi-query determines whether two trajectories have a possible intersection. It is merely necessary to determine whether there exists a time instant in which the two trajectories are less than twice the threshold apart. Evaluating this query in the bead model is much less trivial [33], since it involves solving four quadratic equations.

In case of network-constrained movements, like cars in a highway or trains in railroads, the uncertainty between two consequent sampled positions could be further reduced by exploiting the network topology. Such an idea is depicted in [3] where authors provide equations that describe the geometry of the uncertainty area.

5.3 Modelling Approaches for Trajectories

Approaches for modelling trajectories fall in three categories: the first two categories, *spatiotemporal data models* and *constraint data models*, do not propose specific concepts for trajectories but can be used to represent trajectories. The last category, *moving objects data models* regroups attempts specifically developed for the modelling and querying of moving objects and thus the modelling and querying of trajectories.

5.3.1 Off-the-Shelf Spatiotemporal Data Models

Many spatiotemporal models have been proposed in the literature, stemming from either the entity-relationship approach (e.g. ST USM [40, 62], STER [72, 73]), the object-oriented approach (e.g. Perceptory [7, 8, 12, 43], Extended spatiotemporal UML [60,61], OMT-G [11] STOQL [34], spatiotemporal ODMG [13], Tripod [28]), or a logic-based approach based on constraints (e.g. [29, 42, 63]). A framework for characterizing spatiotemporal data models is given in [21, 50].

A common characteristic of these models is the use of data types as basic building blocks for developing spatiotemporal data management. The definition of standard two-dimensional spatial data types has reached a good level of consensus in the GIS community. Although temporal data types have been standardized in the GIS community [36], no such agreement exists in the database community: Proposed solutions [35, 65, 66] have not reached the acceptance status by the SQL committees [67] [16], and an alternative approach has been proposed in [17]. As for spatiotemporal data types, the work by [30,31] is foundational for building a general approach that is applicable to any modelling dimension.

In this section, we have chosen among the rich literature those models that can be considered as spatiotemporal data model and that are able to model moving objects and trajectories. Although these conceptual models do not attempt to describe the internal structure of trajectories, they may be used to describe time-varying geometry [1] that is useful for modeling trajectories.

5.3.1.1 ISO TC 211

The ISO TC 211 Geographic information/Geomatics is the ISO technical committee responsible for defining international standards related to geographic information. These standards specify methods, tools and services for acquiring, processing, analyzing, accessing, presenting and transferring geographic information between different users, systems and locations. In this section, we use two of the ISO TC 211 standards:

1. ISO 19107 Geographic information – Spatial schema [37] defining a set of spatial data types and operations for geometric and topological spaces. It only covers vector data.
2. ISO 19108 Geographic information – Temporal schema [36] defining a set of temporal data types and functions needed to describe spatial events that occur in time.

5.3.1.2 STER

The spatiotemporal ER (STER) model [72,73] is an extension of the entity- relationship model with constructs for modelling spatiotemporal information. The structural

---

[1] More specifically a time-varying point, which store the successive positions, (x,y) pairs, of the object over time.

5 Trajectory Data Models   131

concepts provided by STER are those of the basic entity-relationship model: entity type, relationship type, attributes, and is-a (generalization/specialization) link.

5.3.1.3 Perceptory

The approach proposed by [7, 8, 12, 43] is to define spatial, temporal, (and multimedia) plug-ins for visual languages (PVL) that can be added to any existing database design tool. This visual plug-ins consists in a set of elementary concepts with their graphical symbols and an associated grammar defining how the symbols can be combined to express more complex concepts. Perceptory provides for two-dimensional and three-dimensional spatial types but in the following we restrict ourselves to two-dimensional types. A temporal PVL has also been defined. It provides two symbols representing the basic types Instant and Period. As for the spatial PVL, combinations of these symbols allow to represent alternative temporalities or multiple temporalities.

5.3.1.4 MADS

MADS [79] is an object + relationship spatiotemporal conceptual data model. In this model, it is assumed that the real world of interest that is to be represented in the database is composed of complex objects and relationships between them; both are characterized by properties (attributes and methods) and both are possibly involved in a generalization hierarchy (is-a links). Spatiality and temporality may be associated at the different structural levels: object, attribute and relationship. The spatiality of an object conveys information about its location and its extent; its temporality describes its life cycle.

5.3.1.5 Comparison of the Models for Trajectory Modelling

In terms of data representation, trajectories may be modelled as a time-varying geometry. Except ISO TC 211, all the aforementioned spatiotemporal data models allow to represent time-varying geometries. For instance, in Fig. 5.5 is shown two alternative ways to model car trajectories using the MADS data model. The design (Fig. 5.5a) defines a car as a spatial object type having a time-varying geometry i.e. its spatiality is a set of pairs (point, instant). This is shown by the point icon at the top right corner followed by a function symbol including the instant icon. Alternatively, the time-varying spatial attribute may be kept in the trajectory attribute representing the movement of car objects as in Fig. 5.5b. An equivalent schema can be defined using the STER and the Perceptory models.

In case the model does not have the concept of time-varying spatial attribute, an approach for representing trajectories is to represent trajectories as objects on their own, independently of the object that generates them. Figure 5.5c illustrates a design of car trajectory using ISO TC 211 standard spatial and temporal hierarchies. In this

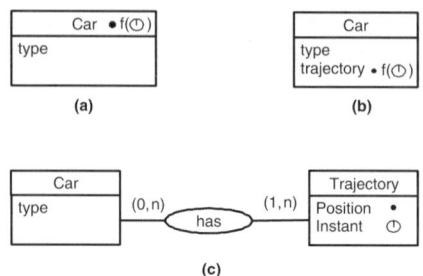

Fig. 5.5 Alternatives to trajectory modelling

design, an entity called trajectory is created with two attributes: position and time instant, which are ISO 19107 point (i.e. GM_Point class) and ISO 19108 instant (i.e. Instant class) type, respectively. For each entity that has a trajectory, we must associate it to the trajectory entity using a 1:N relationship.

To conclude we note that the representation of trajectories using the spatiotemporal data models presented above have the following limitations:

- The semantic of the above spatiotemporal concepts cannot express the exact semantic of what is a trajectory. For example, defining the car entity using a time-varying spatial attribute allows to represent the geometry of the trajectory adequately but does not encompass all the constraints and operations that are specific to trajectories. An example of such a constraint can state that a trajectory must contain at least two different instants, some specific operations can be the duration or the direction of the trajectory. A model for trajectory should provide for specific data structure for trajectories. The lack of specific structures for describing trajectories demands additional work to the designer in correcting the impedance between spatiotemporal constructor's semantics and correct trajectory representation.
- Although trajectories may be expressed as a spatiotemporal entity attribute, in some spatiotemporal data model, such as in STER, spatial attributes are inherited (or, obtained) from space, meaning that the entity attribute is defined whether or not the entity exists at specific position in space.
- None of the models propose any specific operations to analyse trajectories.
- None of the models propose any specific query language operators to query trajectories.

5.3.2 The Constraint Database Approach to Trajectories

During the past ten years, an acclaimed method for effectively representing infinite geometrical figures in databases is provided by the *constraint database model*. This model was introduced by Kanellakis, Kuper and Revesz in 1990 [39] and deeply studied during the second half of the 1990s (an overview of the area of constraint databases can be found in [49]). In the constraint model, a two-dimensional geometrical figure, for instance, is finitely represented by means of a Boolean combination of polynomial equalities and inequalities. These involve polynomials with two real variables that represent the spatial coordinates of a point in the plane. The set of points on the upper half of the unit circle, for instance, is in this context given by

$$\{(x,y) \in \mathbb{R}^2 \mid x^2 + y^2 = 1 \land y \geq 0\}$$

(in mathematical terminology, these figures are called *semi-algebraic sets* and for an overview of their properties we refer to [9]).

This way of representing fixed figures can easily be adapted to describe figures that change. Indeed, we can add a time dimension and consider geometrical objects in three-dimensional space–time that are described by polynomial equalities and inequalities that also have a time variable t. The set

$$\{(t,x,y) \in \mathbb{R}^{1+2} \mid x^2 + y^2 = (1-t)^2 \land y \geq 0 \land 0 \leq t \leq 1\}$$

models a shrinking half circle, whereas the set

$$\{(t,x,y) \in \mathbb{R}^{1+2} \mid (x-t)^2 + y^2 = 1 \land y \geq 0 \land 0 \leq t \leq 1\}$$

models a half circle that moves along the x-axis. In [41], an SQL-like query language was discussed for this kind of data that focusses on exploiting topological changes in moving or changing geometric figures.

Since, trajectories can be seen as a collection of infinite points connecting a finite number of sample points, the constraint model can be specialized to model moving points and trajectory data.

In this section, we discuss some specific attempts that can be classified under the constraint model, to deal with trajectories. In particular, we look at the *linear constraint model*, an approach based on *differential geometry*, and an approach using *equations of motion*.

5.3.2.1 The Linear Constraint Model

When the polynomials used to model data are restricted to be linear, we have the *linear constraint model*. Given a finite set of sample points, in space–time, the linear constraint model basically assumes that the moving object moves with constant speed along a straight line connecting two succeeding sample points. This speed is the minimal average speed needed to reach the destination. The graph of such a trajectory is a piecewise linear curve.

Suppose we have m time instants $t_1 < t_2 < \ldots < t_m$, and a function p that maps time t to a point

$$p(t) = (p_1(t), p_2(t), \ldots, p_n(t))$$

in $\mathbb{R}^n$, describing the trajectory of a moving object. In the linear constraint model each p_i is represented by the constraint $p_i(t) = b_{ij}t + c_{ij}$ for $t_j \leq t \leq t_{j+1}$ ($j = 1, \ldots, m-1$).

Obvious drawbacks are the discontinuities of the speed (and moving direction) of the moving object in the sample points. This makes the trajectory unsmooth and thus seemingly unnatural. On the other hand, a big argument for this approach is the ease of computation it allows.

5.3.2.2 The Differential Geometry Model

One approach by Su, Xu and Ibarra [69] is through the use of differential geometry. They use first- and second-order derivatives to describe direction, velocity and acceleration. On the other hand, they allow vector arithmetic on moving points to compute distances and speeds. A moving object is defined as a piecewise C^∞- curve in $\mathbb{R}^n$. Let $t_1 < t_2 < \ldots < t_m$ be m time instants and p a function that maps time t to a point $p(t) = (p_1(t), p_2(t), \ldots, p_n(t))$ in $\mathbb{R}^n$ where each p_i is a real continuous (for all t) function that is infinitely differentiable on $]-\infty, t_1[$, each $]t_{j-1}, t_j[$ and $]t_m, +\infty[$. The nature of the functions $p_i(t)$ suitable for this depend on the specific application. It is clear that this model generalizes the linear constraint model.

In this model, a query language is provided based on the following operations. First, typical operations in a metric vectors space are allowed (sum of two vectors and scalar multiplication with a real number; and the standard dot product and thereof derived norm/length of a vector). Second, differential geometry operations (taking the derivative) are allowed:

- The velocity of a moving object $\text{vel}(p) = p'(t) = (p_1'(t), p_2'(t), \ldots, p_n'(t))$
- The acceleration of a moving object $\text{vel}'(p) = p''(t) = (p_1''(t), p_2''(t), \ldots, p_n''(t))$
- The speed of a moving object $\|\text{vel}(p)\|$
- The moving direction of a moving object $\frac{\text{vel}(p)}{\|\text{vel}(p)\|}$
- The distance between two moving objects p and q : $\|p - q\|$

A database system is constructed now as follows. First, a logical vector type is introduced. If τ is a function of type *time* $\to \mathbb{R}$, then the LV (logical vector) type is of the type $\tau^n = \tau \times \tau \times \cdots \times \tau$:

- A relation schema R is a finite set of pairs (A, T), where A is an attribute name and T of the LV-type.
- A database schema D is a finite set of relation schemas.
- A tuple is a total mapping from attribute names to the domains of the LV-types and a relation instance is a finite set of tuples over R.
- A database instance of D is a total mapping I from D such that for each $R \in D$, $I(R)$ is a relation instance of R.

A first-order formula $\varphi(x)$ with free variables defines a query Q as follows: if I is a constraint database, then the answer to Q on I is

$$Q(I) = \{a \mid I \models \varphi(a)\}.$$

- Variables and values of associated types or LV-types are terms
- Field operations on variables of type real or time are again terms

- Vector space operations on LV-types over the real type field are again terms
- If x is a vector and p is a term of the LV-type, then unit(x), vel(p), dir(p) are terms of the LV-type and len(p) is a term of the real type.

Let a be a time instant, p a value in the LV-type domain, and $x_1,\ldots,x_n$ real values, then $p(a;x_1,\ldots,x_n)$ is *true* if p is at $(x_1,\ldots,x_n)$ at the time instant a.

5.3.2.3 The Equations of Motion Model

Another way of approaching moving objects is storing objects through their *equation of motion* [27]. The idea behind this is to use Newton's Second Law, which states that once you know the total force acting on an object, you know its motion. Newton's Second Law connects force with acceleration through the object's mass. Let $x : \mathbb{R} \to \mathbb{R}^n : t \mapsto x(t) = (x_1(t), x_2(t), \ldots, x_n(t))$ describe the coordinate to a moving object at *time* t, let m be its mass, and let $F : \mathbb{R} \to \mathbb{R}^n : t \mapsto F(t) = (F_1(t), F_2(t), \ldots, F_n(t))$ describe the total force $F(t)$ acting on the object. The Second Law then states that

$$F(t) = m \frac{d^2 x(t)}{dt^2}.$$

Given two initial conditions $x_0 = x(t_0)$ and $v_0 = dx(t_0)/dt$, initial position and initial speed of the object, a single solution to this second-order differential equation can be found, not always analytically.

One can reduce the order of this equation by adding variables

$$\frac{d}{dt} X(t) = \frac{d}{dt} \begin{pmatrix} x(t) \\ v(t) \end{pmatrix} = \begin{pmatrix} v(t) \\ \frac{F(t)}{m} \end{pmatrix}.$$

The space in which the image of $X(t)$ lies is also referred to as the phase space.

A differential constraint over the variables is of the form

$$\dot{x}_i = f_i(x_1, \ldots, x_n, t),$$

where f_i is a multivariate polynomial in its variables. The author then describes an equation of motion as a finite set of triples where every triple contains a set of initial constraints, a set of differential constraints and an end time for which the triple is valid.

A trajectory is constructed from such a triple but first-order approximation, i.e. approximation with a linear piecewise curve. In the article, the author demonstrates this using Euler's method. The moving object database is then a triple consisting of a finite set of object identifiers, a mapping from this set to the set of all equations of motion and finally a time instant, which is an upper bound for existence of all the moving objects.

5.3.3 Modelling and Querying Moving Objects Databases

About a decade efforts attempt to achieve an appropriate kind of interaction between temporal and spatial database research. Spatiotemporal databases are the outcome of the aggregation of time and space into a single framework [1,53,56,68]. Substantial research work has been carried out focussing in modelling spatiotemporal databases, while recently new needs have been imposed by a series of ubiquitous applications as location-based services. This section presents a parallel line of research focussing on modelling trajectories as spatiotemporal objects (the so-called *moving objects*). Researchers in the field of Moving Objects Databases (MOD) have been studying the representation issues of trajectories into computer systems aiming at keeping track of object locations, as well as supporting location-aware queries. If, only time-dependent locations need to be managed (e.g. mobile phone users, cars, ships, etc.), then *moving point* is the basic abstraction; while, if the time-dependent shape or extent is also of interest (e.g. group of people, armies, spread of vegetation), then we are talking about *moving regions*.

A straightforward approach widely used in industry is to model a moving point by generating periodically a location-time point of the form (l,t), indicating that the object is at location l at time t, where l may be a coordinate pair (x,y). Points are stored in a database, and a database query language is used to retrieve the location information. This method is called *point-location management*, and it has several critical drawbacks, such as (1) does not enable interpolation or extrapolation, (2) leads to a critical precision/resource trade-off and (3) leads to cumbersome and inefficient software development.

In the literature of the MOD field, there are two main approaches to model trajectory data: one for querying current and future positions of the moving objects in [64,76,77] and the second for querying past trajectories of the moving objects in [22,25,30,44].

Querying current and future positions must consider the problem of managing the positions of a set of entities in a database. However, to keep the location information up-to-date, we encounter an unpleasant trade-off. If updates are sent and applied to the database very often, the error in location information in the database is kept small, yet the update load becomes very high. Indeed, keeping track of a large set of entities is not feasible. Conversely, if updates are sent less frequently, the errors in the recorded positions relative to the actual positions become large. This problem was explored by Wolfson et al. [64,76,77], who have developed a model, called MOST, that allows one to store the motion vector rather than the objects' positions in the database, avoiding a very high volume of updates. In Wolfson and colleagues' work, the location of a moving object is simply given as a linear function of time, which is specified by two parameters: the position and velocity vector of the object. Thus, without frequent update messages, the location server can compute the location of a moving object at a given time t through linear interpolation: $y(t) = y_0 + v(t - t_0)$ with time $t > t_0$. An update message is only issued when the parameter of the linear function, i.e. v, is changed. This update approach is called *dead-reckoning*. It offers a great performance benefit in linear mobility patterns, but performance suffers

5 Trajectory Data Models

when the randomness of the mobility pattern increases. In addition, Wolfson et al. group incorporates a new concept of dynamic attributes, which change over time; hence the results of queries also change over time, leading to a notion of continuous queries. The related query language, called future temporal logic (FTL), allows one to specify temporal relationships between objects in queries. This approach is restricted to moving point objects, and is dealing with current and expected near future movement. FTL has the following SQL/OQL type syntax:

RETRIEVE target-list
WHERE condition-list

The condition part is specified as a FTL formula. FTL formulas are interpreted over future histories specifying the object locations. Static attributes remain unchanged, while dynamic attributes change according to their functions. FTL employs a variety of spatial, temporal predicates and operators. Later we present three representative examples illustrating the FTL query language:

Q1: Retrieve names of red colour objects that will be inside the region P within 10 units of time.

RETRIEVE O.name
WHERE O.colour = red AND Eventually-within-10 (INSIDE(O,P))

Q2: Retrieve names of objects that will be within a distance of 10 from a truck for the next five units of time.

RETRIEVE O.name
WHERE Always-for-5 (DISTANCE(O,O')= 10) AND O'.type = truck

Q3: Retrieve all objects that enter a tunnel in the next 5 units of time and stay inside it for the subsequent 10 time units.

RETRIEVE O.type
WHERE Not Inside(O,P) AND Eventually-within-5(Always-for-10 (Inside(O,P)) AND P.type = tunnel

The need for capturing complete histories of objects' movement has promoted the investigation of continuously moving objects. Clearly as location data may change over time, the database must describe not only the current state of the spatial data but also the whole history of this development. Thus it should allow to go back in time at any particular instant, to retrieve the state of the database at that time, to understand the evolution, to analyze when certain relationships were fulfilled and so forth. This approach was developed by Guting and colleagues [22, 25, 30, 44]. They described a new approach where moving points and moving regions are viewed as three-dimensional (2D space + time) or higher-dimensional entities whose structure and behaviour is captured by modelling them as abstract data types. Such types and their operations for spatial values changing over time can be integrated as base (attribute) data types into object-relational, object-oriented or other extensible database management system. More specifically, they introduced a type constructor τ, which transforms any given atomic data type a into a type $\tau(a)$ with semantics $\tau(a) = time \rightarrow a$. In this way, the two basic types defined, namely *mpoint* and *mregion*, may also be represented as $\tau(point)$ and $\tau(region)$, respectively. They provided an algebra with data types such as moving point, moving line and moving region together with a comprehensive set of operations. All the types that are produced by application of the type constructor τ on other data types are functions over an infinite domain; hence each value is an infinite set.

It is important to note that in MOD modelling, the trajectory of a moving point can be described either as a curve or as a polygonal line in three-dimensional space. In the first case, a curve is defined as a certain kind of infinite set of points without fixing any finite representation. In the second case, the definition uses a finite representation of a polyline, which in turn defines the infinite points set making up the trajectory of the moving point. In Erwig et al. [22], the difference between these two levels of modelling is discussed at some depth, and the terms abstract and discrete modelling are introduced for them. As an extension to the abstract model in [24,30] introduced the concept of *spatiotemporal predicates*. The goal was to investigate temporal changes of topological relationships induced by temporal changes of spatial objects. A corresponding query language incorporating these concepts was presented in [23]. In [25], the authors presented the definition of the discrete representation of the above-discussed abstract data types. They introduced the concept of sliced representation, the basic idea of which is to decompose the temporal development of a moving value into fragments called *slices* such that within a slice this development can be described by some kind of simple function. Algorithms implementing the rather large set of operations defined in [30] are studied in [25,44].

The final outcome of this work was a system that implements the above-described moving objects data model and query language completely integrated into a DBMS environment [4]. More specifically, the prototype has been developed as an algebra module in SECONDO's extensible environment [18]. Further we provide three representative queries exemplifying the resulted SQL-like query language [4]:

Q1: Where exactly were the trains during period P?

SELECT Id, Line, trajectory(Trip atperiods P) AS Stretch
FROM Trains
WHERE Trip present P;

Q2: At what times have trains passed through (underground) the region R?

SELECT Id, Line, deftime(Trip at R) AS Times
FROM Trains
WHERE Trip passes R;

Q3: Where have the trains passing through station S been at time T (as far as they are moving at this time).

SELECT Id, Line, val(Trip atinstant T) AS Pos
FROM Trains, Stations
WHERE Trip passes Loc AND SName contains S AND Trip present T

Another approach following the paradigm of moving objects was presented in [52]. This research focussed on the representation and querying of continuously as well as discretely moving objects similar to those presented in [30]. From a theoretical point of view, a data type-oriented model (STAU) that supports the representation of objects both under object-oriented and object-relational platforms was introduced. From a technical point of view, two data cartridges under ORACLE object-relational DBMS were developed. The first cartridge provides pure temporal functionality implementing TAU temporal types [38]. The second cartridge supports a palette of moving object data types, which has been implemented by merging the temporal cartridge with Oracle's spatial cartridge. The resulted system supports a wide set of object methods that extends the Oracle PL/SQL query language with spatiotemporal semantics. Indicative examples of the aforementioned query language include:

Q1: When did John leave the rectangular area defined by (x_1,y_1) lower left and (x_2,y_2) upper right co-ordinates?

SELECT h.route.f_leave(SDO_GEOMETRY(2003, NULL, NULL,
SDO_ELEM_INFO_ARRAY(1,1003,3), SDO_ORDINATE_ARRAY(x_1,y_1,x_2,y_2)))
FROM humans h
WHERE h.id = John

Q2: What is John's speed at 24/11/2007-10:45:30?

SELECT h.route.f_speed(tau_tll.d_timepoint_sec(2007,11,24,10,45,30))
FROM humans h
WHERE h.id = John

Q3: Find John's friends that are located within 1,000 m distance from his current location.

SELECT f.id
FROM humans h, friends f
WHERE h.id = John AND h.route.f_within_distance(1000, f.route, tolerance, NOW))

By assuming that a trajectory is modelled in its finest spatial granularity (exact location), all the previously mentioned data models provide support at two levels.

First, they provide a mechanism to split a trajectory into sub-paths, according to some variables such as the sampling rate and the most appropriate update time. Second, MOD models usually imply a linear interpolation between the exact locations. Considering the first issue, sliced representation [25] is adopted as the solution for the model proposed in [30]. In [54], the authors utilize sliced representation and develop a moving type that associates a period of time with the description of a simple function that models the behaviour of the moving object in that specific time period. Considering the second issue, linear interpolation is considered sufficient for the querying purposes of a MOD. However, other types of interpolations could be as well important either for making motions more realistic or for sub-serving the tasks of privacy and/or modelling in various granularities. The model in [54] provides an extensible mechanism to support different kinds of interpolation, currently implementing linear and arc sub-motions. This model was recently extended [10, 55] to support not only historical queries but also dynamic ones.

Following the modelling primitives described earlier, several solutions have been proposed to address specific query processing issues in MODs. Research in the field is driven by related work performed in the domain of (stationery) spatial databases. For instance, queries of the form *'find all objects located within a given area during a certain time interval'* generalize the spatial range query of the form *'find all objects within a given area'*. Many different types of the so-called *coordinate-based queries* [59] have been proved to be useful for MODs: Queries of the form *'find all objects' locations within a given area at a certain time instance'*, called *timeslice* queries, constitute a special type of range queries where the temporal extent is set to zero. Another straightforward extension of pure spatial queries includes the *nearest neighbour* queries of the form *'find the nearest moving object to a query object at a certain time instance (or during a certain time interval)'*. As discussed in [26], in the case of time-interval nearest neighbour queries, the query object can be either a two-dimensional point or another trajectory, while the query may return either the nearest to the query object in any time during a time interval or in every time instance of the query time interval (historical continuous queries). The last extension of spatial queries already discussed in the spatiotemporal trajectory literature deals with trajectory join queries [5, 6], which are categorized in the so-called *distance join* and *k-closest pairs* queries. The former is defined as follows: Given two sets of trajectory data sets $Q = q_1,..,q_n$ and $T = t_1,..,t_m$ compute all pairs (q_i,t_j), where q_i and t_j have distance no more than a particular threshold at a given time stamp. In correspondence to the classic closest-pair problem of computational geometry, the latter finds the k closest pairs of trajectories between the two data sets at the given time stamp, i.e., the pairs of (q_i,t_j) that have the k smallest in-between distances at the given time stamp. Both queries can be generalized in the *time-relaxed* context where the temporal dimension is of no interest; as such the latter query type is transformed to *'find the k closest pairs of trajectories between the two data sets at any time stamp'*. Another useful co-ordinate-based query (Fig. 5.6) in MOD derives from the so-called *trajectory similarity problem* and aims to find 'similar' trajectories of moving objects. To handle such queries efficiently, MOD systems

5 Trajectory Data Models   141

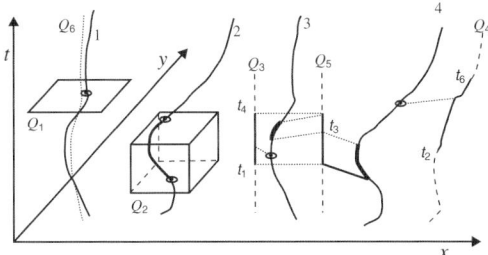

Fig. 5.6 Coordinate-based queries: timeslice (Q1), range (Q2), point nearest neighbor (Q3) trajectory nearest neighbor (Q4), historical continuous point nearest neighbor (Q5), most similar trajectory (Q6)

should incorporate query processing methods for supporting *most similar trajectory (MST)* search also discussed in [15, 70, 74].

According to the classification in [57, 59], apart from co-ordinate-based queries, the so-called *trajectory-based queries* are also of great interest. In contrast to co-ordinate-based, trajectory-based queries require the knowledge of the complete – or at least a subset of the – object's trajectory to be processed. Such queries consider topological relations (e.g. enter, leave, etc.) and may provide derived information about an object's navigation (e.g. average speed, travelled distance, etc.). Furthermore, the combination of a range with topological queries produces another class of queries called *combined queries*. As an example [59], consider the following query *'What were the trajectories of objects after they left Tucson street between 7 and 8 a.m. today, in the next hour'*, which first locates the trajectories contained in an inner range query window (*Tucson street, between 7 and 8 a.m. today*) and then retrieve those parts of objects' trajectories contained in an outer query window (*in the next hour*).

5.4 Open Issues

As presented in Sect. 5.3.3, consequent quality work has been done regarding trajectory modelling and querying. However, open issues remain particularly regarding conceptual modelling of trajectories, uncertainty, multiple representation of trajectories, continuously acquired trajectories and query capabilities. They are described further.

5.4.1 Conceptual Modelling of Trajectories

Most of the works on moving objects have paid very little attention on the conceptual description of the trajectory. In those models, the trajectory of a given object is a by-product, so to speak, of capturing the object's mobility. Indeed describing an object type as a moving point allows representing the position of the object all along its lifespan. It does not allow the system to be aware of the semantic segmentation of the object's paths into semantically meaningful trajectories. To be able to associate a trajectory or a list of trajectories to the object we need more than moving points. Trajectories should be promoted as a modelling construct i.e. be first-class data, rather than computable derived data. Moreover, the specification of the trajectory construct should be done at the conceptual level to fix a purely conceptual view of the concept and to ensure its maximal flexibility. Indeed, when looking for a conceptual model for trajectories, we have to focus on trajectory characteristics that are generic, i.e. independent of any specific application domain, while being relevant to the application realm and not driven by considerations pertaining to their implementation in a computer-based system. To propose a conceptual solution for the trajectory concept, the following issues need to be tackled.

5.4.1.1 Conceptual Description of Trajectories

From a conceptual point of view, the concept of trajectory denotes the evolving position of some object in some space, from an initial position to a final position. A trajectory has two facets:

- A spatiotemporal facet: it allows to record the positions of the moving object.
- A semantic facet: it allows to associate application-dependant information or characteristics to the whole trajectory as well as to any of its subparts. For example, a point-based trajectory for a person could store the activity the person is doing between two defining points not necessarily consecutive (visiting, walking to work, etc.) and the transport means. Obviously, the trajectory of a car will bear completely different data: a car trajectory could store for each defining point its road distance from the previous point, the duration of *en route* stops, the amount of highway fees paid and the gasoline consumption over the last segment.

5.4.1.2 Constraints of Trajectories with Their Environment

In the same way as we need to associate semantic data to trajectories and/or to their subparts, we want to be able to describe the constraints holding between the trajectories or their subparts and their environment [75]. As trajectories are spatiotemporal object types, semantic as well as both topological and temporal constraints should be considered. For example, a topological constraint might describe the fact that the whole trajectory of this person is included in the area delimited by the old city part of Lausanne but also that this particular part of the trajectory is equal to the geometry of the stairs going to the cathedral.

5.4.1.3 Spatial, Temporal or Spatiotemporal Operators

To manipulate trajectories, a set of operators should be identified. As trajectories are spatiotemporal objects, the data model must offer traditional temporal and spatial operators like the ones proposed by Allen [2] in the temporal dimension and

Egenhofer [20] in the spatial dimension, respectively, as well as trajectory-specific operators (such as, for instance, how to 'sum-up' two trajectories). Indeed, promoting trajectories as first class modelling constructs has direct consequences on the operators applying to trajectories. It implies evolving from operations on moving objects as defined by e.g. [30] to manipulation of trajectories as a whole. Thus investigations should be done to define an underlying algebra for trajectories.

5.4.2 Uncertainty

The representation of moving objects within a MOD is inherently imprecise due to errors introduced from different sources like measurement, sampling and pre-processing.

So far, the related research [3, 58, 71] emphasizes on sampling errors introduced by the interpolation process, which is utilized to 'predict' the moving objects positions within the sampled points. The usually adopted interpolation method is linear interpolation, which is the simplest one. More advanced techniques, like polynomial splines, should be also considered so as to better approximate the actual movements.

Interpolation uncertainty, although important, is only one source of uncertainty. Further uncertainty sources should be considered and their impact on the represented trajectory with respect to the actual trajectory should be investigated. Examples of such sources are compression, simplification and matching of trajectories to the network (in case of a network-constrained movement).

The uncertainty so far refers to the spatial dimension, i.e. how well the stored positions of the moving object represent its actual positions. Spatial information, however, is not the only aspect of a movement. Time and speed, for example, constitute other aspects that might contain uncertainty in their values.

The idea of restricting uncertainty by exploiting the underlying network has started to be addressed in the literature [3]. However, a further uncertainty factor is imposed here that of finding the position of the moving object within the network, i.e. network matching. Investigating how matching affects the quality of trajectories, and how the uncertainty it imposes interacts with the inherent uncertainty of trajectories seems a promising research line. Furthermore, the uncertainty area of an object within a network depends on the geometry of the network, which, in the general case, does not solely consists of straight lines. More complex geometries, e.g. circles or spirals, should be investigated and the shape of uncertainty areas should be allocated.

To conclude, uncertainty in the representation of a moving object is an important issue within a moving object database, since the adopted representation is the basis for other DB operations like querying and indexing. Two critical questions arise: what is an effective representation schema for trajectories under uncertainty and how the uncertainty is propagated into other MOD operations. Since uncertainty is a reality in MOD, the efforts should be towards its limitation so as to provide the end user with reliable results.

5.4.3 Streaming Models

The streaming model is completely suitable for moving object data since they encounter frequent updates, their volume is unexpectedly varied, and they are being processed under real-time conditions with several continuous spatiotemporal queries. Thus, MODs perfectly fit with the stream concept, and it sounds reasonable to go towards a streaming procedure that feeds a trajectory database or a trajectory data warehouse. Streaming spatiotemporal data has addressed a considerable research attention in the past few years [32, 45, 46, 51]. Existing work in streaming models include [51], which attempts to model the management of moving objects under the assumption that the trajectories are continuous, time-varying and possibly unbounded data streams, proposing a basic framework for managing trajectory streams with the introduction of constructs for advanced query capabilities in an SQL-like language. However, modelling of moving objects must be further studied, in addition to the introduction of algebraic constructs for windows and to the proposal of syntax rules for query language.

5.4.4 Multiple Representation

Multiple representation is an important issue regarding the modelling of trajectories. Multiple representation means that we want to store or to be able to retrieve several representations for the same phenomenon in the database. This is an important requirement as each application has its own perception of the real world and its own data processing tasks leading to specific requirements both in terms of what information is to be kept and in terms of how information is to be represented. Spatial and temporal applications show additional requirements in terms of multiple representation as they need flexibility also in the perception and representation of spatial and temporal features.

In the context of trajectories, multiple representation may result from the description of the same trajectory according to several viewpoints but more importantly according to different spatial and temporal granularity. Viewpoint, here, should be understood as the expression, by a group of users, of their specific interests in data management. Granularity refers to the notion that the world is perceived at different level of details i.e. in the temporal dimension using more or less time steps like in [38] and in the spatial dimension considering a smaller or bigger spatial resolution. Complex applications naturally include the need for multiple representations of trajectories. Indeed different tasks often require different granularities: for instance, considering the trajectory of a person travelling from home in Lausanne to work in Geneva, some tasks might be only interested to analyse the trajectory from the starting point in Lausanne to the arrival point in Geneva and then use a coarse spatial and temporal granularity. On the contrary, another task might need a more detailed description: at 7.40 a.m. the person has left home to walk to the bus station for 10 min, then has taken the bus to the train station where the person has been waiting for 5 min, then travelled for 30 min, etc. In this example, the same trajectory is modelled at different levels of spatial and temporal granularity. Moreover, the same trajectory may show parts in different granularity, e.g. more detailed data for critical sections of the trajectory: for instance, a detailed description of the trajectory between the person's home and the train station will be kept while from Lausanne train station to Geneva train station no specific detail is necessary.

Although multiple representation has received a lot of attention in the spatiotemporal database community, multiple representation applied to the description of moving objects and their trajectories is still an open issue on which few propositions exists [14, 33]. In this area, research work has to be done to provide for a model describing multiple representations of the same trajectory including conversion operators to shift among granularities and an adapted query language. Proposed operators for granularity change exist, but how to maintain multiple representations with maximal flexibility has rarely been addressed.

5.4.5 Query Capabilities

Open issues regarding the query types supported by MODs include novel query types, meaningful only in the spatiotemporal domain, as well as the expansion of query types from other domains. In particular, there exist a significant number of spatial queries not yet adopted in the spatiotemporal framework, and particularly in trajectory databases. For example, the recently proposed group nearest neighbour query [48] can be also applied in the spatiotemporal domain; nevertheless, its employment is not straightforward at all, since a trajectory GNN query would had to clarify several issues concerning its definition (a) the type of the query objects (static or continuously moving), (b) the time interval during which the GNN is requested, (c) the way of determining the distance of GNN from the query objects, since distances from all query objects could be calculated on exactly the same time stamp or considering the entire query period. In addition, such type of queries involving the calculation of the nearest distance of trajectories could be extended by employing the notion of network distance discussed in the domain of spatial network databases (SNDB) [47].

Regarding the trajectory similarity topic, the majority of the existing approaches, being mainly inspired by the time series analysis domain, propose generic similarity metrics for two-dimensional data [15, 74] in order to answer queries requesting about the most similar to a given trajectory. However, the notion of 'textitmost similar' in the trajectory domain can be considered through several aspects, since trajectories have spatial, temporal, spatiotemporal and other derived features. Consider for example the following queries:

- Query 1 (*spatial similarity*): Find objects whose route (i.e. the projection of trajectory on two-dimensional plane) is quite similar to that of object id= 132 (irrespective of time).

- Query 2 (*spatiotemporal similarity*): Find objects that follow a route similar to that of object id= 132 during the same time interval, e.g. 3–6 PM.
- Query 2a (*speed pattern-based spatiotemporal similarity*): Find objects that follow a route similar to that of object id= 132 and, additionally, move with a similar speed pattern.
- Query 3 (directional similarity): Find objects that follow a direction similar to a given direction pattern, e.g. NE during the first half of the route and subsequently W.

To the best of our knowledge, there is no work dealing with the different similarity types that can be defined based on these underlying parameters of the trajectories. However, such queries are expected to be at least as popular as those already examined, which mainly deal with the spatial similarity between trajectories.

5.4.6 Conclusion

The objective of this chapter was to provide an extensive discussion of the state-of-the-art on data modelling of trajectories. We have initiated our discussion by describing trajectory as the record of time-varying phenomenon. In terms of data representation, trajectories are described as a collection of sample points that need to be linked. To find the suitable curve connecting the sample points, interpolation methods are used. As shown, there are two kinds of methods that may apply to this problem, and the trajectory accuracy depends basically on the number of sample points and the method used. Interpolation only gives a guess of a probable trajectory, leaving a certain amount of uncertainty that needs to be taken care of. Two approaches have been presented to treat uncertainty of trajectories.

In terms of data modelling, many research efforts have been done in modelling spatiotemporal applications, among them some are specifically defined for moving objects and their trajectories, some are not. Indeed, since trajectory applications are a sub-domain of spatiotemporal applications, we have analysed these off-the-shelf spatiotemporal models to deal with trajectory representation. Similarly, we have analysed constraint database modelling that provides a method to represent infinite geometrical entities in databases and thus could be specialized to model moving points and trajectory data. Finally, we have presented important researches in the field of MOD that are studying representation issues of trajectories as well as their querying. From a theoretical point of view, constraint and moving objects models approaches are less conceptual and more implementation-oriented than spatiotemporal data models are. Former approaches focus more on definition of mathematical models, abstract data types, algorithms for set of operations and query answering approaches.

For completeness of the discussion, we have shown in Sect. 5.4 open issues in trajectory modelling that we found relevant. The first open issue concerns the definition of an adequate conceptual representation of trajectories not as a by-product of capturing objects' mobility but as a first class-construct. Effective representation of trajectories taking into account uncertainty and its propagation to operations are also important issues that need to be addressed. Besides, the high-data volume and frequent updates in the context of streaming spatiotemporal data model has not received so far the attention it requires. Another important open issue regarding the modelling of trajectories is to deal with several representations for the same trajectory in the database (i.e. multiple representation). Last but not least, queries capabilities must be improved in the context of trajectory databases, for instance queries based on: derived trajectory information (e.g. speed), nearest neighbour or trajectory similarity.

References

1. T. Abraham and J.F. Roddick. Survey of spatio-temporal databases. *Geoinformatica*, 3(1):61–99, 1999.
2. J.F. Allen. Maintaining knowledge about temporal intervals. *Communications ACM*, 26(11):832–843, 1983.
3. V.T. de Almeida and R.H. Güting. Supporting uncertainty in moving objects in network databases. In *Proceedings of the 13th International Symposium on Geographic Information Systems (GIS'05)*, pp. 31–40, ACM, 2005.
4. V.T. Almeida, R.H. Güting, and T. Behr. Querying moving objects in secondo. In *Proceedings of Mobile Data Management (MDM'06)*, p. 47, 2006.
5. P. Bakalov, M. Hadjieleftheriou, E.J. Keogh, and V.J. Tsotras. Efficient trajectory joins using symbolic representations. In *Proceedings of Mobile Data Management (MDM'05)*, pp. 86–93, 2005.
6. P. Bakalov, M. Hadjieleftheriou, and V.J. Tsotras. Time Relaxed Spatiotemporal Trajectory Joins. In *Proceedings of the 13th Annual International Workshop on Geographic Information Systems (GIS'05)*, pp. 182–191, 2005.
7. Y. Bédard. Visual modeling of spatial databases: Towards spatial pvl and uml. *Geomatica*, 53:169–185, 1999.
8. Y. Bédard, S. Larrivée, M.-J. Proulx, and M. Nadeau. Modeling Geospatial Databases with Plug-Ins for Visual Languages: A Pragmatic Approach and the Impacts of 16 years of Research and Experimentations on Perceptory. In *Conceptual Modeling for Advanced Application Domains*, Vol. 3289, pp. 17–30. Springer, Berlin Heidelberg New York, 2004.
9. J. Bochnak, M. Coste, and M. Roy. *Géométrie Algébrique Réelle*. Springer, Berlin Heidelberg New York, 1987.
10. Boosting location-based services with a moving object database engine. In *Proceedings of the 5th Workshop on Data Engineering for Wireless and Mobile Access (MobiDE'06)*.
11. K. Borges, C. Davis, and A. Laender. Omt-g: An object-oriented data model for geographic applications. *GeoInformatica*, 5:221–260, 2001.
12. J. Brodeur, Y. Bédard, and M.-J. Proulx. Modelling Geospatial Application Database Using Uml-Based Repositories Aligned with International Standards in Geomatics. In ACM, editor, *Proceedings of the 8th International Symposium on Geographic Information Systems (GIS'00)*, pp. 39–46, 2000.
13. E. Camossi, M. Bertolotto, E. Bertino, and G. Guerrini. A Multigranular Spatiotemporal Data Model. In *Proceedings of the 11th International Symposium on Geographic Information Systems (GIS'03)*, pp. 94–101, ACM, 2003.
14. E. Camossi, M. Bertolotto, and E. Bertino. A flexible approach to spatio-temporal multigranularity in an object data model. *International Journal of Geographical Information Science*, 20(5), 2006.

15. L. Chen, T. Özsu, and V. Oria. Robust and Fast Similarity Search for Moving Object Trajectories. In F. Ozcan (ed.), *Proceedings of the International Conference on Management of Data (SIGMOD'05)*, pp. 491–502. ACM, 2005.
16. H. Darwen. Valid Time and Transaction Time Proposals: Language Design Aspects. In *Temporal Databases: Research and Practice, LNCS 1399*, pp. 195–210, 1998.
17. C. Date, H. Darwen, and N. Lorentzos. *Temporal Data and the Relational Model*. Model, Morgan Kaufmann, 2003.
18. S. Dieker and R.H. Güting. Plug and play with query algebras: secondo - a generic dbms development environment. In *Proceedings of the International Symposium on Database Engineering & Applications (IDEAS '00)*, pp. 380–392. IEEE Computer Society, 2000.
19. M.J. Egenhofer. Approximations of geospatial lifelines. 2003.
20. M.J. Egenhofer and R.D. Franzosa. Point set topological relations. *International Journal of Geographical Information Systems*, 5:161–174, 1991.
21. B. El-Geresy and C. Jones. *Five Questions to Answer in Time: A Critical Survey of Approaches to Modelling in Spatio-Temporal GIS*, Chap. 3. GIS and Geocomputation-Innovations in GIS 7. Taylor & Francis, London, 2000
22. M. Erwig, R.H. Güting, M. Schneider, and M. Vazirgiannis. Spatio-temporal data types: An approach to modeling and querying moving objects in databases. *GeoInformatica*, 3(3):269–296, 1999.
23. M. Erwig and M. Schneider. Developments in Spatio-Temporal Query Languages. In *Proceedings of 10th International Conference and Workshop on Database and Expert Systems Applications (DEXA'99)*, pp. 441–449, 1999.
24. M. Erwig and M. Schneider. Spatio-temporal predicates. *IEEE Transaction Knowledge Data Engeneering*, 14(4):881–901, 2002.
25. L. Forlizzi, R.H. Güting, E. Nardelli, and M. Schneider. A Data Model and Data Structures for Moving Objects Databases. In *Proceedings of the International Conference on Management of Data (SIGMOD'00)*, pp. 319–330, 2000.
26. E. Frentzos, K. Gratsias, N. Pelekis, and Y. Theodoridis. Nearest Neighbor Search on Moving Object Trajectories. In *Proceedings of 9th International Symposium on Advances in Spatial and Temporal Databases (SSTD'01)*, Vol. 3633. *Lecture Notes in Computer Science*, pp. 328–345. Springer, Berlin Heidelberg New York, 2005.
27. F. Geerts. Moving Objects and their Equations of Motion. In *Proceedings of the 1st International Symposium on Applications of Constraint Databases*, volume 3074 of *Lecture Notes in Computer Science*, pp. 41–52. Springer, Berlin Heidelberg New York, 2004.
28. T. Griffiths, A. Fernandes, N. Paton, and R. Barr. The tripod spatio-historical data model. *Data and Knowledge Engineering*, 49:23–65, 2004.
29. S. Grumbach, M. Koubarakis, P. Rigaux, M. Scholl, and S. Skiadopoulos. *Spatio-temporal Models and Languages: An Approach Based on Constraints*, Chap. 5, pp. 177–201, 2003.
30. R.H. Güting, M.H. Böhlen, M. Erwig, C.S. Jensen, N.A. Lorentzos, M. Schneider, and M. Vazirgiannis. A foundation for representing and quering moving objects. *ACM Transactions on Database System*, 25(1):1–42, 2000.
31. R. Guting, M. Bohlen, M. Erwig, C. Jensen, M. Schneider, N. Lorentzos, E. Nardelli, M. Schneider, and J. Viqueira. Spatio-Temporal Models and Languages: An Approach Based on Data Types. In *Spatio-Temporal Databases: The Chorochronos Approach, LNCS 2520*, Chap. 4, pp. 117–176, 2003.
32. M.A. Hammad, W.G. Aref, and A.K. Elmagarmid. Stream Window Join: Tracking Moving Objects in Sensor-Network Databases. In *Proceedings of 15th International Conference on Scientific and Statistical Database Management (SSDBM'03)*, pp. 75–84, 2003.
33. K. Hornsby and M.J. Egenhofer. Modeling moving objects over multiple granularities. *Annual Mathematics Artificial Intelligence*, 36(1–2):177–194, 2002.
34. B. Huang and C. Claramunt. Stoql: An ODMG-Based Spatio-Temporal Object Model and Query Language. In *Proceedings of the 10th International Symposium on Spatial Data Handling (SDH'02)*, pp. 225–237, 2002.
35. ISO/IEC. *Information Technology – Database languages – SQL – Part 7: Temporal (SQL/Foundation). ISO/IEC 9075-2 Working Draft*. ISO, 2001.

36. ISO/TC211. *Geographic Information and Temporal Schema. ISO 19108:2002*. ISO, 2002.
37. ISO/TC211. *Geographic Information and Spatial Schema. ISO 19107:2003*. ISO, 2003.
38. I. Kakoudakis. *The Tau Temporal Object Model*, M.Sc. Thesis, Umist, 1996.
39. P.C. Kanellakis, G.M. Kuper, and P. Revesz. Constraint query languages. *Journal of Computer and System Sciences*, 51:26–52, 1995.
40. V. Khatri, S. Ram, and R. Snodgrass. Augmenting a Conceptual Model with Geospatiotemporal Annotations. *IEEE Transactions on Knowledge and Data Engineering*, 16:1324–1338, 2004.
41. B. Kuijpers, J. Paredaens, and D.V. Gucht. Towards a theory of movie database queries. In *Proceedings of the 7th International Workshop on Temporal Representation and Reasoning (TIME'00)*, pp. 95–102. IEEE Computer Society, 2000.
42. G. Kuper, L. Libkin, and J. Paredaens. *Constraint Databases*. Springer, Berlin Heidelberg New York, 2000.
43. S. Larrivée, Y. Bédard, and J. Pouliot. How to Enrich the Semantics of Geospatial Databases by Properly Expressing 3d Objects in a Conceptual Model. In *Proceedings of the Workshops On The Move to Meaningful Internet Systems*, number 3762 in LNCS. Springer, Berlin Heidelberg New York, 2005.
44. J.A.C. Lema, L. Forlizzi, R.H. Güting, E. Nardelli, and M. Schneider. Algorithms for moving objects databases. *The Computer Journal*, 46(6):680–712, 2003.
45. M.F. Mokbel, X. Xiong, W.G. Aref, S.E. Hambrusch, S. Prabhakar, and M.A. Hammad. Place: A Query Processor for Handling Real-Time Spatio-Temporal Data Streams. In *Proceedings of 30th International Conference on Very Large Data Bases (VLDB'04)*, pp. 1377–1380, 2004.
46. M.F. Mokbel, X. Xiong, M.A. Hammad, and W.G. Aref. Continuous query processing of spatio-temporal data streams in place. *GeoInformatica*, 9(4):343–365, 2005.
47. D. Papadias, J. Zhang, N. Mamoulis, and Y. Tao. Query Processing in Spatial Network Databases. In *Proceedings of 29th International Conference on Very Large Data Bases (VLDB'03)*, pp. 802–813, 2003.
48. D. Papadias, Q. Shen, Y. Tao, and K. Mouratidis. Group nearest neighbor queries. In *Proceedings of the 20th International Conference on Data Engineering (ICDE'04)*, pp. 301–312. IEEE Computer Society, 2004.
49. J. Paredaens, G. Kuper, and L. Libkin, editors. *Constraint databases*. Springer, Berlin Heidelberg New York, 2000.
50. C. Parent. A Framework for Characterizing Spatio-Temporal Data Models. In S.S. Y. Masunaga (ed.), *Advances in Multimedia and Databases for the New Century*, pp. 89–97. World Scientific, Singapore, 2000.
51. K. Patroumpas and T.K. Sellis. Managing Trajectories of Moving Objects as Data Streams. In J. Sander and M.A. Nascimento, editors, *Proceedings of 2nd International Workshop on Spatio-Temporal Database Management (STDBM'04)*, pp. 41–48, 2004.
52. N. Pelekis. *STAU: A Spatio-Temporal Extension to ORACLE DBMS*. Ph.D. Thesis, 2002.
53. N. Pelekis, B. Theodoulidis, I. Kopanakis, and Y. Theodoridis. Literature review of spatio-temporal database models. *Knowledge Engineering Review*, 19(3):235–274, 2004.
54. N. Pelekis, B. Theodoulidis, Y. Theodoridis, and I. Kopanakis. An Oracle data cartridge for moving objects, laboratory of information systems, department of informatics, university of piraeus, unipi-isl-tr-2005-01, 2005. http://isl.cs.unipi.gr/db/ publications.html.
55. N. Pelekis, Y. Theodoridis, S. Vosinakis, and T. Panayiotopoulos. Hermes – A Framework for Location-Based Data Management. In *Proceedings of 10th International Conference on Extending Database Technology (EDBT'06)*, pp. 1130–1134, 2006.
56. D.J. Peuquet. Making space for time: Issues in space-time data representation. *Geoinformatica*, 5(1):11–32, 2001.
57. D. Pfoser. Indexing the trajectories of moving objects. *IEEE Data Engeneering Bullettin*, 25(2):3–9, 2002.
58. D. Pfoser and C.S. Jensen. Capturing the uncertainty of moving-object representations. In R.H. Güting, D. Papadias, and F.H. Lochovsky, (eds.), *Proceedings of the 6th International Symposium on Advances in Spatial Databases (SSD'99)*, Vol. 1651. *Lecture Notes in Computer Science*, pp. 111–132. Springer, Berlin Heidelberg New York, 1999.

59. D. Pfoser, C.S. Jensen, and Y. Theodoridis. Novel Approaches in Query Processing for Moving Object Trajectories. In *Proceedings of 26th International Conference on Very Large Data Bases (VLDB'00)*, pp. 395–406, 2000.
60. R. Price, N. Tryfona, and C. Jensen. Extended spatiotemporal uml: Motivations, requirements, and constructs. In *Journal of Database Management*, 11:14–27, 2000.
61. R. Price, N. Tryfona, and C. Jensen. *Extending UML for Space- and Time-Dependent Applications*. Idea Group Publishing, 2002.
62. S. Ram, R. Snodgrass, V. Khatri, and Y. Hwang. *DISTIL: A Design Support Environment for Conceptual Modeling of Spatio-temporal Requirements*, pp. 70–83. 2001.
63. P. Rigaux, M. Scholla, L. Segoufin, and S. Grumbach. Building a constraintbased spatial database system: Model, languages, and implementation. *Information Systems*, 28:563–595, 2003.
64. A.P. Sistla, O. Wolfson, S. Chamberlain, and S. Dao. Modeling and Querying Moving Objects. In *Proceedings of the 13th International Conference on Data Engineering (ICDE'97)*, pp. 422–432. IEEE Computer Society, 1997.
65. R. Snodgrass, M. Böhlen, C. Jensen, and N. Kline. *Adding valid time to SQL/Temporal. ANSI X3H2-96-501r2, ISO/IEC JTC1/SC21/WG3 DBL MAD-146r2*, 1996.
66. R. Snodgrass, M. Böhlen, C. Jensen, and A. Steiner. *Adding transaction time to SQL/Temporal: Temporal change proposal. ANSI X3H2-96-152r, ISO-ANSI SQL/ISO/IECJTC1/SC21/WG3 DBL MCI-143*. ISO, 1996.
67. R. Snodgrass, M. Böhlen, C. Jensen, and A. Steiner. Transitioning Temporal Support in tsql2 to sql3. In *Temporal Databases: Research and Practice, LNCS 1399*, pp. 150–194, 1998.
68. *Spatio-Temporal Databases: The CHOROCHRONOS Approach*, Vol. 2520 of *Lecture Notes in Computer Science*. Springer, Berlin Heidelberg New York, 2003.
69. J. Su, H. Xu, and O.H. Ibarra. Moving Objects: Logical Relationships and Queries. In C.S. Jensen, M. Schneider, B. Seeger, and V.J. Tsotras, editors, *Proceedings of 7th International Symposium on Advances in Spatial and Temporal Databases (SSTD'01)*, volume 2121 of *Lecture Notes in Computer Science*, pp. 3–19. Springer, Berlin Heidelberg New York, 2001.
70. Y. Theodoridis. Ten benchmark database queries for location-based services. *The Computer Journal*, 46(6):713–725, 2003.
71. G. Trajcevski, O. Wolfson, K. Hinrichs, and S. Chamberlain. Managing uncertainty in moving objects databases. *ACM Transactions Database System*, 29(3):463–507, 2004.
72. N. Tryfona and C. Jensen. Conceptual data modeling for spatiotemporal applications. *GeoInformatica*, 3:245–268, 1999.
73. N. Tryfona, R. Price, and C. Price. *Spatiotemporal Conceptual Modeling.*, chapter 3, pp. 79–116, Berlin, 2003.
74. M. Vlachos, D. Gunopulos, and G. Kollios. Discovering Similar Multidimensional Trajectories. In *Proceedings of the 18th International Conference on Data Engineering (ICDE'02)*, pp. 673–684. IEEE Computer Society, 2002.
75. N.V. de Weghe, F. Witlox, A.G. Cohn, T. Neutens, and P.D. Maeyer. Efficient storage of interactions between multiple moving point objects. In *OTM Workshops (2)*, pp. 1636–1647, 2006.
76. O. Wolfson, B. Xu, S. Chamberlain, and L. Jiang. Moving Objects Databases: Issues and Solutions. In *Proceedings of the 10th International Conference on Scientific and Statistical Database Management (SSDBM'98)*, pp. 111–122, IEEE Computer Society, 1998.
77. O. Wolfson, A.P. Sistla, S. Chamberlain, and Y. Yesha. Updating and querying databases that track mobile units. *Distributed and Parallel Databases*, 7(3):257–387, 1999.
78. J. Zhang and M. Goodchild. *Uncertainty in Geographical Information*. Taylor & Francis, New York, 2002.
79. E. Zimanyi, C. Parent, and S. Spaccapietra. *Conceptual Modeling for Traditional and Spatio-Temporal Applications – The MADS Approach*. Springer, Berlin Heidelberg New York, 2006.

Chapter 6
Trajectory Database Systems

E. Frentzos, N. Pelekis, I. Ntoutsi, and Y. Theodoridis

6.1 Introduction

In this chapter, we deal with trajectory database management issues and physical aspects of trajectory database systems, such as indexing and query processing. Our emphasis is on historical databases handling past positions of moving objects represented as trajectories. This is because only such databases can be used in the context of trajectory data warehouses, which is the core subject of this book.

Outlining the main topics that we will discuss in this chapter, we include operational trajectory database engines, indexing techniques for moving object trajectories, query processing, querying under the presence of uncertainty, the mapmatching problem, and, finally, issues on trajectory compression. All topics are presented under a two-stage approach: we first discuss the state of the art, illustrating the most popular proposals on each topic, and then we provide directions and hints for future work in each particular topic.

6.2 Trajectory Database Engines

As already stated in the previous chapter, the research area of Moving Objects Databases (MODs) has addressed the need for representing movements of objects (i.e., trajectories) in databases to perform ad hoc querying, analysis, as well as data mining on them. During the last decade, there has been a lot of research ranging from data models and query languages to implementation aspects, such as efficient indexing, query processing, and optimization techniques. The realization of data models proposed in the literature as well as packaging corresponding functionality to specific technical solutions results in MOD engines. In the literature, one can

---

E. Frentzos
Computer Technology Institute (CTI) and Department of Informatics, University of Piraeus, Greece, e-mail: efrentzo@unipi.gr

find at least two MOD engines developed to realize the model proposed by Güting et al. [25], namely the SECONDO prototype [3] and the HERMES engine [43, 44]. These will be reviewed in the following paragraphs.

6.2.1 SECONDO

The first development concerns a follow-up on the study of abstract moving object data types and algorithms for the operations defined in [25]. Whereas [19] just provides a succinct look into this issue, Lema et al. [34] present a systematic study of algorithms for a subset of the methods introduced in [25]. The final outcome of this work was a research prototype, which has been recently demonstrated in [3]. The prototype has been developed as an algebra module in the extensible DBMS environment SECONDO [17, 26]. The module uses the *sliced representation*, described in the previous chapter, representing a time-dependent value as a sequence of simple temporal functions. Having defined the physical storage of each of the objects in the type system, the next step is the development of the temporal counterparts of operations defined in the ROSE algebra. For example, an operation answering whether a point resides or not inside a region (i.e., *inside[point region]: bool*) is transformed, by an approach called *lifting*, to an operation returning a time-varying Boolean representing the periods where a moving point is inside the region (i.e., *inside[mpoint region]: mbool*). Finally, special operators for moving types are offered such as projections into time and range of values, intersections with values or sets of values from time and range of values, and methods that determine rate of change. The above-described functionality has been embedded into SQL and slightly adapted so that queries can be written directly as PROLOG terms, as this is the development language of SECONDO.

SECONDO is a DBMS prototype platform especially adjusted for extension by algebra modules for nonstandard applications. It does not support a predetermined data model, but rather is open for implementation of new models. It consists of (a) a *kernel*, which offers query processing over a set of implemented type system algebras, (b) an *optimizer*, which implements the essential part of an SQL-like language, and (c) an extensible GUI where new data types and models can provide specialized viewers for moving objects (see Fig. 6.1a). As such, to realize the spatiotemporal algebra introduced in [25], two modules have been built in SECONDO: the first provides all the spatial data types and operations (i.e., the ROSE algebra module [24]) and the second provides the above-mentioned moving object algebra module. More specifically, the *optimizer* provides cost-based optimization of conjunctive queries producing an execution plan and the *kernel* evaluates the query plan, also called an executable query, or a query at the executable level, which is just a term of the implemented algebras. Query processing is performed as follows (Fig. 6.1): the *command manager* receives an executable query, parses it, and passes the result to the *query processor*. The query processor then evaluates the query by building an operator tree and then traverses it, calling operator implementations from the algebras. SECONDO stores (and retrieves) moving objects into a database with the help of the

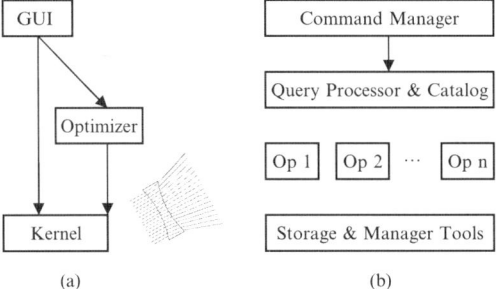

Fig. 6.1 SECONDO system [3]: the three major components and a rough architecture of the kernel

storage manager, while the objects are managed by the *catalog*. The reader who is interested in details about this process is referred to [17].

6.2.2 Hermes

Hermes, a database engine for handling objects that change location, shape, and size either discretely or continuously in time has been recently proposed by Pelekis et al. in [43, 44]. Hermes provides spatiotemporal functionality to state-of-the-art object-relational database management systems (ORDBMS). The prototype has been designed as an extension of STAU [42, 45], which is providing a system extension to Oracle ORDBMS [39] data management infrastructure for historical MODs. The system can be used either as a pure temporal or a pure spatial system, but its main functionality is to support the modeling and querying of continuously moving objects. Such a collection of data types and their corresponding operations are defined, developed, and provided as an Oracle data cartridge, called Hermes Moving Data Cartridge (Hermes-MDC), which is the core component of the Hermes system architecture. Embedding the functionality offered by Hermes-MDC in Oracle DML [39], one obtains an expressive and easy to use query language for moving objects. In particular, Hermes-MDC defines a palette of moving object data types, illustrated in the UML class diagram of Fig. 6.2 [44].

The usefulness and applicability of the server-side extensions provided by Hermes has been demonstrated by developing an application on top of this framework [43], which builds and visualizes the results of a palette of MOD queries that have been proposed in the literature [64] as an advanced location-based services (LBS) benchmarking framework for the evaluation of MOD engines. Among others, Hermes functionality includes the following:

– *Queries on stationary reference objects*; examples include distance-based or nearest-neighbor queries (e.g., find nearby or closest landmarks, respectively, with respect to one's current location) and topological queries (e.g., find those who have crossed this area during the past hour)

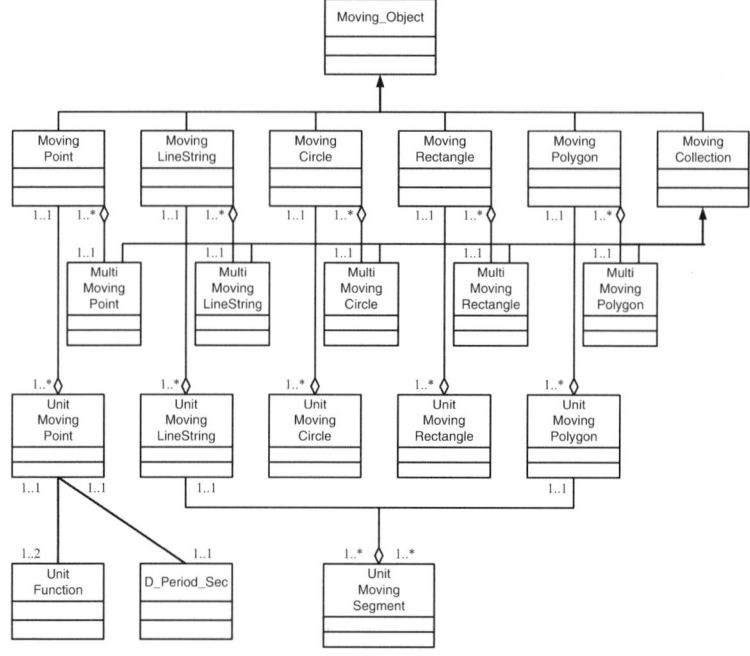

Fig. 6.2 Hermes-MDC class diagram [44]

- *Queries on moving reference objects*; examples include distance-based (e.g., find those who have passed close to me this evening) and similarity-based queries (e.g., find the three most similar trajectories to the one I followed yesterday)
- *Queries involving unary operators*, such as traveled distance or speed (e.g., find the average speed of the trajectory I have followed during weekend)

6.3 Trajectory Indexing

Like in traditional databases, querying in MODs could be very expensive because of the nature of data and the complexity of query processing algorithms. Given also that location aware devices are almost ubiquitous in these days, trajectory databases will, sooner or later, face enormous volumes of data. It consequently arises that performance, in the presence of vast data sizes, will be a significant problem for trajectory databases. Since ordering is far from nature of the geographic (multidimensional) data, traditional indexes like B-trees are not useful in spatial (and consequently in spatiotemporal) databases. In the domain of spatial databases, the R-tree proposed by Guttman [27] is "almost ubiquitous", with applications ranging from geographical information systems (GIS) and computer-aided design (CAD) to image and multimedia management systems [35]. The R-tree can be considered as

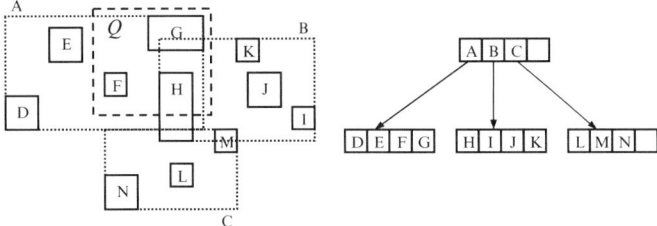

Fig. 6.3 An example of spatial data, their MBBs, a range query, and the corresponding R-tree [35]

an extension of the B-tree in n-dimensional spaces. Similar to the B-tree, R-tree is a height balanced tree with the index records in its leaf nodes containing pointers to the actual data objects. Leaf node entries are of the form (*id, MBB*), where *id* is an identifier that points to the actual object and minimum bounding box (*MBB*) is a n-dimensional interval. Nonleaf node entries are of the form (*ptr, MBB*), where *ptr* is a pointer to a child node and *MBB* is the bounding box that covers all child nodes. A node in the tree corresponds to a physical disk page (or disk block, which is the fundamental element on which the actual disk storage is organized) and contains between m and M entries (M is the node capacity and m is a tuning parameter – usually m is set to $M/2$, which guarantees that the space utilization is at least 50%). Contrary to the B-tree, node MBBs belonging to the same tree level are allowed to overlap. Figure 6.3 illustrates a set of spatial objects and the corresponding R-tree.

In the domain of *spatiotemporal indexing*, R-tree variations and extensions include, among others, three-dimensional R-trees [66], TB-trees and STR-trees [49], FNR-trees [20] and MON-trees [2], while SETI [11] is a hybrid R-tree-based and partition-based technique. Since our interest in this chapter focuses on historical MODs, we restrict our discussion to indexing techniques recording past locations. The reader interested in indexing current locations and motion vectors can find very interesting works in [52, 55, 56, 62, 74].

Taking into consideration the fact that motivation behind MODs usually comes from emerging applications such as fleet management and LBS solutions, trajectory indexing techniques are classified into those organizing motion in either unrestricted space or fixed networks. In the latter case, the underlying infrastructure is not only an additional information that somehow has to be integrated in the index, but also affects fundamental concepts, such as the notion of distance (i.e., network vs. Euclidean distance).

6.3.1 Indexing Trajectories in Unrestricted Space

On the subject of indexing moving object trajectories in unrestricted space, the three-dimensional R-tree [66] was proposed as a straightforward extension of the R-tree in the three-dimensional space formed by the $2 + 1$ (spatial and temporal, respectively) dimensions. It treats time as an extra spatial dimension and is capable of answering coordinate-based queries, as they are defined in the previous chapter. Although originally designed to index multimedia data, the proposal by Pfoser et al. [49] enables it to support trajectories as well. Obviously, the three-dimensional R-tree indexes collections of line segments in the three-dimensional (spatiotemporal) space, only concerning about the processing of the traditional coordinate-based queries, being at the same time inefficient to handle trajectory-based queries (also discussed in the previous chapter) whose processing requires the extraction of a part of – or even, the complete – moving object's trajectory.

The trajectory bundle tree (TB-tree), proposed by Pfoser et al. [49], tries to overcome this inefficiency. The TB-tree is a height-balanced tree with the index records in its leaf nodes based on the three-dimensional R-tree. However, it turns out to be fundamentally different from other spatiotemporal access methods mainly due to its insertion and split strategy. Its insertion algorithm is not based upon the spatial and temporal relations of moving objects but it relies only on the moving object identifier (id). When a new line segment is inserted, the algorithm searches for the leaf node containing the last entry of the same trajectory, and simply inserts the new entry in it, thus forming leaf nodes that contain line segments from a single trajectory. If the leaf node is full, then a new one is created and inserted at the right end of the tree. For each trajectory, a double-linked list connects the leaf nodes that contain its portions together (Fig. 6.4), resulting in a structure that can efficiently answer trajectory-based queries. Pfoser et al. [49] propose also the STR-tree, which tries to combine the desired properties of both TB and three-dimensional R-tree; however as presented in the respective experimental study, same as the three-dimensional R-tree, it is also inefficient on trajectory-based queries. Zhu et al. [78] extend the work in [49] proposing the octagon-prism tree (OP-tree), which indexes trajectories by using octagon approximations instead of MBBs. On the basis of the conducted experiments, OP-trees are shown to outperform the original TB-tree both on range and trajectory-based queries.

Unfortunately, in spite of its clear advantages on trajectory-based query processing, the TB-tree (and its variation, the OP-tree) has a crucial drawback: because of its insertion strategy, new trajectory data are always inserted at the right "end" of the tree, leading its performance to heavily depend on the order of data insertion. This insertion strategy may not lead to problematic behavior only under the assumption that trajectory data are inserted in the index in pure chronological order: the insertion

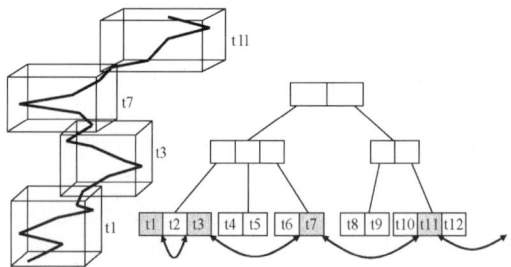

Fig. 6.4 The TB-tree structure

strategy will organize temporally close line segments to be also close in the index. However, in real-world applications, this assumption is not guaranteed to be true. For example, in an application where the insertions occur in real time, if the moving object enters an area where the position transmission system does not function, its trajectory could be stored locally in the object memory and be transmitted to the central server – where the index operates – at a later time; meanwhile, other moving objects could have transmitted their positions, increasing the temporal overlapping between the tree nodes, which subsequently leads to the deterioration of the index performance.

SETI [11] is a hybrid structure, indexing trajectories at two levels to disjoint the spatial from temporal indexing. Acknowledging that trajectory data sets continually expand in the temporal dimension while the spatial boundaries remain static or at least rarely change, SETI partitions the two-dimensional space into disjoint hexagon cells that remain static during the structure's lifetime, while other adaptive spatial partitioning strategies can also be used. Each cell logically contains only those trajectory segments that are completely within the cell, while in the case of a trajectory segment that crosses the cell boundary is split and subsequently inserted into both cells. Actually, trajectory segments are inserted into a data file; each page of the data file contains segments from only one cell. Then, a temporal index (e.g., a one-dimensional R-tree) indexing the time intervals of each particular cell in the data file is assigned to the corresponding cell. Figure 6.5 summarizes the SETI structure.

The insertion and searching algorithms follow a multistep approach composed of spatial filtering, temporal filtering, and refinement. In particular, during each insertion, the algorithm locates the cell into which the segment has to be inserted (considering also possible splits between cells), and then inserts it in the corresponding page of the data file, updating at the same time the corresponding entry of the one-dimensional R-tree if it is necessary. Although as presented in the experimental study of [11] SETI clearly outperforms the three-dimensional R-tree and the TB-tree in time-interval and time-slice queries, it cannot be used to process trajectory-based queries. This is due to the fact that trajectory line segments are organized inside the index, based only on their spatial and temporal relations; as such successive line segments of the same trajectory may be placed in different disk pages. Therefore, in the worst case scenario the retrieval of a single trajectory would require to read one disk page for each trajectory line segment. Moreover, authors do not provide any

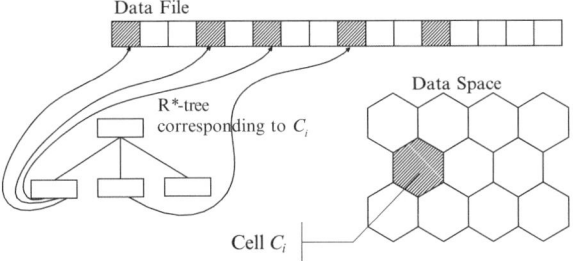

Fig. 6.5 The SETI structure

nearest-neighbor query processing algorithm, while the development of an efficient one is not a straightforward task.

6.3.2 Indexing Trajectories in Fixed Networks

The first proposal concerning the indexing of trajectories in fixed networks was presented in [20], introducing the FNR-tree based on the original R-tree. Instead of using a single R-tree to index object trajectories, the FNR-tree utilizes a forest of R-trees. More specifically, the FNR-tree is a two-stage access method, consisting of a two-dimensional R-tree, which organizes a set of one-dimensional R-trees (Fig. 6.6). The two-dimensional R-tree is used to index the spatial data of the network, whereas each one of the one-dimensional R-trees corresponds to a leaf of the two-dimensional R-tree, and indexes respective time intervals. As long as there are no structural changes in the spatial network, the two-dimensional R-tree remains fixed, whereas one-dimensional R-trees change as objects move. The insertion and range query processing algorithms presented in [20] are much alike those of SETI, consisting also of the same three steps of spatial filtering, temporal filtering, and refinement. The experimental study presented in [20] shows that the FNR-tree outperforms the three-dimensional R-tree by several orders of magnitude considering simple range queries, while it demonstrated a weakness in the case of time-slice queries with the entire spatial extent. Same as SETI, there is neither obvious way for the FNR-tree to support efficient trajectory-based query processing nor nearest-neighbor query processing.

Exploiting the same property of a spatial network, a variation of the FNR-tree, called MON-tree, has been proposed in [2]. In this index, instead of using one-dimensional R-tree for every leaf node of the two-dimensional R-tree, the MON-tree utilizes a two-dimensional R-tree for every polyline of the spatial network. The MON-tree is shown to significantly outperform the three-dimensional R-tree and

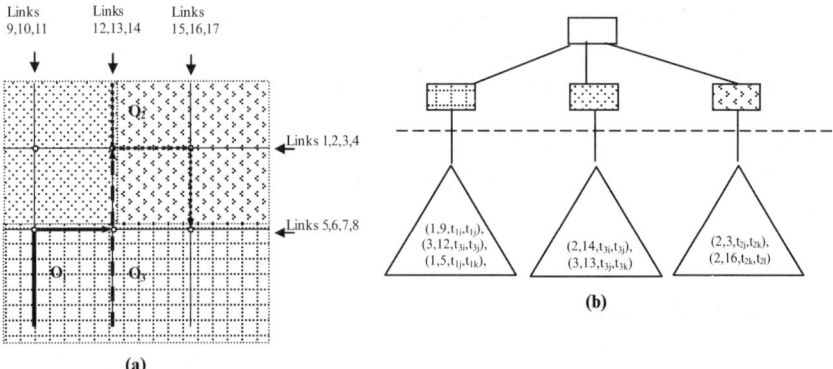

Fig. 6.6 An FNR-tree example: (a) trajectories of three objects on a road network and (b) the corresponding FNR-tree components

the FNR-tree, in time-interval and time-slice queries, while it shows the same disadvantage with the previously described schemes, being unable to efficiently process trajectory-based queries.

Another interesting methodology on the same subject of indexing of objects moving on networks is presented in [48]. This approach suggests the mapping of the underlying network from two- to one-dimension by sorting the network edges according to their Hilbert values. Hilbert values is an approach for ordering the two-dimensional space; they are determined by applying a Hilbert curve covering the two-dimensional space, mapping every two-dimensional to a one-dimensional point [73]. Then, the problem of indexing three (i.e., 2 spatial +1 temporal) dimensions is reduced to the problem of indexing two (i.e., 1 spatial +1 temporal) dimensions, which can be efficiently handled by employing any existing simple spatial index as the well known R-tree, also supported by existing DBMS. After that, each range query has to be mapped accordingly to the reduced one-dimensional space, producing thus a number of two-dimensional (spatial and temporal) rectangles, which are subsequently posed against the R-tree. The technique also uses an R-tree to index the underlying network so as to speed up the query mapping process. The experimental study presented in [48] shows that the proposed method clearly outperforms the three-dimensional approach (e.g., three-dimensional R-tree, treating time as an extra spatial dimension) as the query size increases. However, same as previous, there is no obvious way on how this approach [48] can process nearest-neighbor and trajectory-based queries.

6.4 Trajectory Query Processing and Optimization

Since spatiotemporal query types are guided by existing work on the domain of spatial querying, it is expected that the majority of the proposed algorithms for trajectory query processing will also be an extension of algorithms already employed in the context of spatial databases. For example, the spatiotemporal range query algorithm involving both spatial and temporal components, in the R-tree-like structures storing historical trajectory information, is a straightforward generalization of the original *FindLeaf* algorithm presented in [27] in the three-dimensional space. For a more detailed discussion on the definitions of spatiotemporal query types, the interested reader may refer to the previous chapter, which also contains comprehensive examples. Following also from the previous chapter, the query types we will deal with in the context of query processing are *range, nearest neighbor, join, similarity, and trajectory-based* queries.

6.4.1 Range Search

The majority of the aforementioned spatiotemporal indexes provide range search algorithm exploiting both the spatial and temporal dimensions. As already mentioned, since most of them are based on the well-known R-tree, the respective range search algorithm follows the one presented in [27]. Following the example illustrated in Fig. 6.3 for spatial objects, consider a range query Q executed against the two-dimensional R-tree. The algorithm starts by visiting the tree root, checking whether the MBBs of the root entries are overlapping Q. If a node entry MBB overlaps Q, the algorithm follows the pointer to the corresponding child node (in our case entries A and B), where it repeats recursively the same task. If the algorithm reaches a leaf node, leaf entries are examined against Q and if their MBB overlap, the algorithm reports their ids (objects F and G when the algorithm visits leaf node A, and object H when in node B). The extension of the above algorithm in the spatiotemporal domain is a straightforward task, where each two-dimensional MBB is simply replaced by the respective three-dimensional MBB of actual objects, nodes or queries.

Regarding two-stage structures, such as SETI, FNR, and MON-tree, range search is generally a three-step task. It consists of a spatial filtering process, which is followed by a temporal filtering and a subsequent refinement step joining the results of the spatial and temporal filtering. Spatial and temporal filtering are performed through the respective spatial and temporal components; if this is a one- or two-dimensional R-tree, the algorithm is essentially the same with the one previously presented for simple R-trees. The refinement step is necessary, since objects retrieved from the spatial filtering are approximated by MBBs, therefore, we cannot determine whether the spatial object is actually inside the query before it has been retrieved (something that happens only after the temporal filtering step). To make this more comprehensive, consider a range query over SETI partially overlapping an index cell; then, line segments inside it may or may not be actually inside the query, something that can be determined only after the first two steps that retrieve the actual trajectory components (i.e., line segments). Generally speaking, such an approach is much more efficient since these indexes exploit the fact that the spatial domain remains unchanged, while the time domain evolves monotonically; as a result, all these approaches outperform the R-tree by several orders of magnitude [2, 11, 20].

Moreover, the spatiotemporal domain includes several approaches trying to optimize the range search procedure based on several properties of the real spatiotemporal applications. For example, the work presented in [47] uses the restrictions placed on the movement of objects by the existing infrastructure to improve the performance of spatiotemporal queries executed against a spatiotemporal index. The strategy followed does not affect the structure of the index itself. Instead, they adopt an additional preprocessing step before the execution of each query. In particular, provided that the infrastructure is rarely updated, it can be indexed by a conventional spatial index such as the R-tree. On the other hand, a general-purpose spatiotemporal index, such as the TB-tree [49] or the three-dimensional R-tree [66], can be used to index trajectories of moving objects. Then, a preprocessing step of the query divides the initial query window in a number of smaller windows, from which the regions covered by the infrastructure have been excluded (see Fig. 6.7). Each one of the smaller queries is executed against the (general-purpose spatiotemporal) index

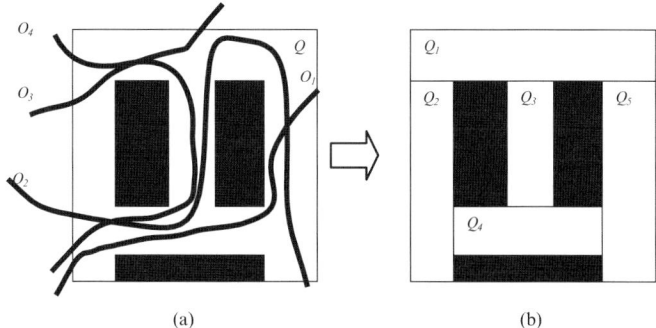

Fig. 6.7 The initial query window Q (**a**) is decomposed into a number of smaller query windows $Q_1, Q_2, \ldots$ (**b**) with respect to infrastructure elements (drawn in *black*)

returning a set of candidate objects, which are finally refined with respect to the initial query window.

In the evaluation presented in [47], the performance of two spatiotemporal indexes (TB- and three-dimensional R-tree) was compared, using either the described query preprocessing step (i.e., dividing the initial window in smaller windows) or not, and it was shown that the query performance was improved for both indexes when this step was used.

Recently, work has also been done on how to optimally split trajectories for the purpose of improving range query performance [28, 30, 53]. Hadjieleftheriou et al. [28] use a partially persistent structure, the PPR-tree, trying to confront the problem of the dead space generated by MBB approximations of moving object trajectories. Dead space is termed as the amount of space in an MBB approximation, which does not actually covers any object contained inside it. They introduce "artificial object updates" partitioning the trajectories into smaller elements, thus reducing the dead space; they use nonlinear functions to describe the moving objects' trajectories, which are initially indexed by the PPR-tree. This work is extended in [30] where a multiversion R-tree, such as the one proposed in [62] is used instead of the PPR-tree, leading to an indexing scheme with improved performance. Moreover, the proposed algorithms for handling the problem of the dead space introduced in MBBs can be used in combination with any spatiotemporal data archive as the R-tree and its variants.

6.4.2 Nearest-Neighbor Search

Nearest-neighbor (NN) search has been in the core of spatial and spatiotemporal database research during the last decade. The literature on NN query processing algorithms mainly deals with either stationary [14, 31, 54] or moving query points over static data sets [57, 60] or data sets constituting by current or future (predicted) locations [7, 33, 37, 58, 75, 77]. Apparently, these types of queries do not cover NN search on historical trajectories, which is the subject of this work; the only relative proposal is presented in [21], which investigates mechanisms to perform NN search on R-tree-like structures storing historical information about moving object trajectories. The depth-first and best-first algorithms proposed in [21] vary with respect to the type of the query object (stationary or moving point) as well as the type of the query result (historical continuous or not), thus resulting in four types of NN queries, which are thoroughly discussed in the previous chapter. The proposed algorithms where implemented on two members of the R-tree family for trajectory data (the TB-tree and the three-dimensional R-tree) demonstrating their scalability and efficiency through an extensive experimental study using synthetic and real data sets.

6.4.3 Trajectory Joins

Distance join has not been considered extensively in the domain of spatiotemporal databases. The limited existing work on this subject considers joining of moving objects trajectories utilizing dedicated index structures [5, 6] or general-purpose indexes [4].

Bakalov et al. [5] consider the problem of evaluating all pairs of similar trajectories between two data sets. According to [5], two trajectories are considered similar during a given time interval, when, given a distance function, all distances between timely corresponding trajectory positions are within the given threshold. Then an approximation technique is used to reduce trajectories to symbolic representations (strings) so as to lower the dimensionality of the original (three-dimensional) problem to one. Using the constructed strings, a special lower-bounding metric supports a pruning heuristic used to reduce the number of candidate pairs to be examined. The overall scheme is subsequently indexed by a structure based on the B-tree, requiring also minimal storage space. The same work is extended in [6] to support time-relaxed spatiotemporal trajectory joins.

Another variation on the subject of joining trajectories is the closest-point-of-approach recently introduced in [4]. Closest-point-of-approach requires finding all pairs of line segments between two trajectories such that their distance is less than a predefined threshold. The work presented in [4] proposes three approaches : the first utilizes packed R-trees treating trajectory segments as simple line segments in the $d+1$-dimensional space, and then employs the well-known R-tree join algorithm [32], which requires carefully controlled synchronized traversal of the two R-trees; The second is based on a plane-sweep along the temporal dimension algorithm; and the third is an adaptive algorithm, which naturally alters the way in which it computes the join in response to the characteristics of the underlying data.

6.4.4 Similarity Search

Similarity search has been well studied in the time series analysis domain; consequently, techniques addressed there are usually extended in the spatiotemporal

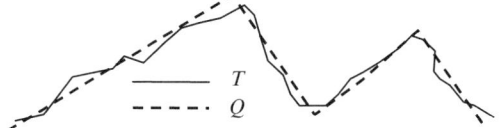

Fig. 6.8 Two similar trajectories T and Q

domain, in which trajectories as T and Q presented in Fig. 6.8 are considered. Historically, similarity search has been based on the *Euclidean distance* between time series, nevertheless, having several disadvantages which the following proposals are trying to confront. In particular, in order to compare sequences with different lengths, Berndt and Clifford [8] used the *dynamic time warping* (DTW) technique that allowed sequences to be stretched along the time axis so as to minimize the distance between sequences. Although DTW incurred a heavy computation cost, it was more robust against noise. *Longest common subsequence* (LCSS) measure [70] matches two sequences by allowing them to stretch, without rearranging the sequence of the elements, but allowing some elements to be unmatched (which is the main advantage of the LCSS measure compared with Euclidean distance and DTW). Therefore, LCSS can efficiently handle outliers and different scaling factors. Authors introduce two similarity measures, namely S_1 and S_2, allowing time stretching and translations, respectively, which were proved to be very robust to the presence of noise and provided an intuitive notion of similarity between trajectories by giving more weight to the similar portions of the trajectories. In [12], a distance function, called *edit distance on real sequences* (EDR), was introduced. EDR distance function is based on the edit distance, which is the number of insert, delete, or replace operations that are needed to convert trajectory T into Q. In the respective experimental study presented in [12], EDR was shown to be more robust than DTW and LCSS over trajectories with noise.

To speed up the similarity search between trajectories, both [70] and [12] rely on dedicated index structures, thus achieving pruning of over 90% of the total number of indexed trajectories.

6.4.5 Trajectory-Based Querying

Trajectory-based querying is mainly discussed in [49] and [78], where dedicated index structures (TB-tree and OP-tree, respectively) are proposed to efficiently support this type of queries. Regarding the aforementioned structures, trajectory-based querying is a rather straightforward task to perform: having located one leaf node containing entries of a specific trajectory, one may recursively follow the pointers to the previous and the successive node containing entries of the same trajectory (recall Fig. 6.4 for the TB-tree case), until the spatial or temporal query criterion has been verified or the entire moving object trajectory has been retrieved.

Regarding the rest of the index structures, which do not consider trajectory preservation, the processing of trajectory-based queries can be performed by employing the algorithm proposed in [49] regarding the three-dimensional R-tree and the STR-tree. As such, having retrieved an initial segment belonging to the trajectory under consideration, the algorithm tries to find its connecting segment, first, in the same leaf node, and, second, in other leaf nodes. Searching in other leaf nodes is conducted as a range search, with the endpoint of the segment in question as a predicate. Arriving at the leaf level, the algorithm checks whether a segment is connected to the segment in question in the specified way (backward of forward connected). Using this recursive approach, successive segments of the trajectory are retrieved, until the spatial or temporal query criterion has been verified, or the entire moving object trajectory has been retrieved. However, this simple algorithm incurs heavy computation cost even in the presence of a buffer, since the worst case scenario corresponds to a case where every trajectory segment is stored in different disk page.

6.4.6 Spatiotemporal Query Optimization

The determination of the best execution plan for a query requires estimating the number of data items that it retrieves, as well as its cost, in terms of I/O and CPU effort. Like traditional databases, spatial query optimization tools include cost-based models, exploiting analytical formulas for selectivity and cost of a query, and histogram-based techniques. On the other hand, although the domain of spatiotemporal databases has been in the center of the research interest for several years developing many novel indexing techniques most of them based on the R-tree, the work conducted for estimating the selectivity of trajectories as well as developing cost models for such indexing schemes is very limited. Specifically, on the subject of selectivity estimation in spatiotemporal databases, research includes [15, 29, 61], all of them estimating the selectivity of several spatiotemporal predictive queries. Apparently, none of them covers the domain of historical trajectory databases; therefore, the interested reader is referenced to the cited papers.

Although models for the prediction of the R-tree performance have been extensively examined during the last decade, they cannot be straightforwardly applied in the spatiotemporal domain. For example, the traditional analysis on R-trees cost models, such as [65, 67], relies on the assumption that the extent of the data inserted in the tree is equally distributed along each dimension, i.e., resulting in square node rectangles. Though this is a reasonable assumption concerning spatial objects, in the spatiotemporal domain, the temporal dimension behaves differently from the two spatial ones. For example, in the widely used three-dimensional R-tree, when an object updates its position rarely, its trajectory's line segments will tend to be elongated in the time dimension, resulting in elongated (in the temporal dimension) leaf nodes.

To resolve this problem, Tao and Papadias [59] examine the R*-tree split algorithm and propose an *extent regression function* (ERF), which computes the node extents as a function of the number of node splits. In particular, using each level's and axis' *length distribution function* (which at the leaf level derives from the actual data), they calculate the introduced *extent regression function* $\mathrm{ERF}_i(t)$ for each tree level at the ith dimension having as parameter t the total number of splits performed along the ith dimension in this tree level. The average extent $s_{i,j}$ of a level-i node along the jth dimension is calculated using the computed ERFs adopting also a technique that estimates the number of splits performed along the jth dimension at the ith tree level by minimizing an objective function under constraints. Finally, having estimated off-line, and without accessing the tree, the average values of $s_{j,i}$ at each tree level, they provide the following generalized formula regarding the expected number of node accesses $C_W(R,q)$ for a query window q:

$$C_W(R,q) = \sum_{j=1}^{1+\log_f(N/f)} \left\{ \frac{N}{f^j} \prod_{i=1}^{d}(s_{i,j}+q_i) \right\}, \quad (6.1)$$

where N is the data set cardinality, f is the fanout of tree nodes, j is the respective level of the R-tree, and q_i is the extent of query q along the ith dimension (a formula that origins itself in the spatial database domain [65]). The experimental evaluation presented in [59] shows that the proposed model provides accurate estimates for the expected number of node accesses in all settings, while other tested cost models (such as [65]) completely fail. Although the model is not developed only for spatiotemporal data, it is capable to predict the performance of a three-dimensional R*-tree since it supports tree nodes being elongated in the temporal dimension. However, it cannot be used to other R-tree variants since the calculation of the ERF is based on the R*-tree splitting algorithm.

6.5 Dealing with Location Uncertainty

The recorded location of a moving object does not always represent its precise location mainly due to GPS erroneous measurements and sampling errors. This problem, known as *location uncertainty*, affects several aspects of a MOD like representation, querying, and indexing. So far, related research emphasizes on representation issues, i.e., how the notion of uncertainty is incorporated into the representation of moving objects within a MOD (see Chap. 5 for more details). Lately, however, several approaches have raised that deal with querying and indexing under uncertainty. In this section we present these approaches.

Pfoser and Jensen [46] constrain the uncertainty area of the moving object between two consecutive sampled positions to be the intersection of the uncertainty areas of the samples. In addition, they illustrate how their uncertainty model can be used for query processing purposes in conjunction with a moving point index that supports range queries. Supported queries are the so-called *probabilistic range queries* (PRQ) (i.e., "Retrieve the moving-object positions that were inside query rectangle A at some time between time points B and C with a probability of at

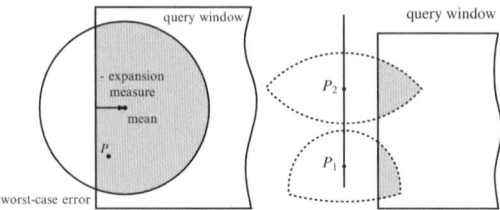

Fig. 6.9 Filter step (*left*) – refinement step (*right*) [46]

least $X\%$"). The standard filter-and-refinement method, borrowed from the spatial query processing domain, is adopted for query processing purposes. More specifically, the authors expand the query window A so as to retrieve all line segments containing positions that lie in A with a probability higher or equal to $X\%$; the expansion is determined using the probability $X\%$ and the worst-case sampling error (represented as a circle). This comprises the filtering step (left part of Fig. 6.9), which usually returns a superset of the qualifying positions. In the refinement step (right part of Fig. 6.9), the positions contained in the retrieved line segments that actually lie within the query rectangle A with probability at least $X\%$ are identified. Instead of the worst-case sampling rate that is used during the filtering step, the sampling error (represented as an ellipse), unique for each position, is used during the refinement step to evaluate positions in time.

Actually the emphasis in this work is on reducing uncertainty in-between sampled positions, rather than querying and indexing. The rest approaches [13, 68, 69], as will be shown below, adopt some simpler uncertainty model and give emphasis on efficient and effective query processing.

Trajcevski et al. [69] associate an uncertainty threshold r to the whole trajectory. Each point (x,y,t) of the trajectory is associated with an r-uncertainty area, which is actually a horizontal disk with radius r and center (x,y,t); (x,y) is the expected position at time t. Thus, the trajectory is modeled as a cylindrical volume in three-dimensional space around the given trajectory polyline. In this work, two categories of operators for querying moving objects under uncertainty have been introduced, namely point queries and spatiotemporal range queries, both referring to a single trajectory. *Point queries* either refer to the location of the moving object at a specific time point or to the time point(s) at which the moving object is expected to be at a specific location. *Spatiotemporal range queries* extend the traditional spatiotemporal range queries by also considering the uncertainty that is inherent in the database locations of the moving objects.

Location uncertainty affects the defined queries in both their temporal and spatial aspects: Regarding the temporal effect, one may query for the objects that are inside the query region *sometime* during the time interval or for those that are inside the query region *always* during the time interval. Intuitively, the "sometime" operator corresponds to cases where the moving object appears within the query region for some time during the query time, whereas the "always" operator corresponds to cases where the moving object lies within the query region during the whole

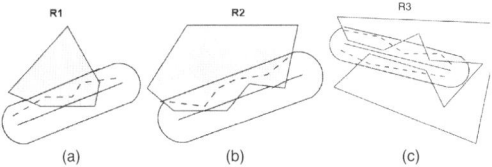

Fig. 6.10 Possible positions of a moving point with respect to region R_i: (**a**) Possibly–Sometime–Inside, (**b**) Possibly–Always–Inside, and (**c**) Always–Possibly–Inside [69]

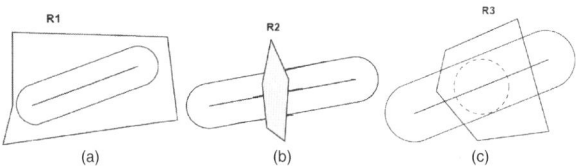

Fig. 6.11 Definite positions of a moving point with respect to region R_i: (**a**) Definitely–Always–Inside, (**b**) Definitely–Sometime–Inside, and (**c**) Sometime–Definitely–Inside [69]

query time. Regarding the spatial effect, one may query for the objects that are "possibly inside" or "definitely inside" the query region. Intuitively, the "possibly inside" operator corresponds to cases where part of the uncertainty area of the moving object appears within the query region, whereas the "definitely inside" operator corresponds to cases where the whole uncertainty area of the moving object lies within the query region. Combining the temporal and spatial effects, the following types of spatiotemporal queries under uncertainty arise: Possibly–Sometime–Inside, Possibly–Always–Inside, Always–Possibly–Inside (Fig. 6.10), Definitely–Always–Inside, Definitely–Sometime–Inside, Sometime–Definitely–Inside (Fig. 6.11).

For the purposes of query processing, Trajcevski et al. [69] assume that a three-dimensional indexing schema is provided by the underlying ORDBMS. The insertion of each trajectory is achieved by enclosing the respective trajectory volume between t_i and t_{i+1} in a MBB. The standard filter-and-refinement method is adapted for query processing: During the filtering step, the trajectories that have at least one of their MBBs intersecting with the query polygon are retrieved. For the refinement step, the method relies on the areas of geometry and motion planning [40]. Although in this work a simple uncertainty model (simpler than [46]) is considered, an interesting set of queries over uncertain trajectories is presented.

Trajcevski [68] provides a methodology for answering PRQ under uncertainty. The queries treated there are of the form: *What is the probability that a given moving object was/will be inside a given region sometimes/always during a given time interval?* This probability is given by the fraction of the intersection area between the trajectory volume and the query region, with the whole trajectory volume. In [72], Wolfson et al. introduce a probabilistic model for processing PRQ in motion databases. The output of this type of query consists of pairs of the form (o_i, p_i), where p_i is the probability that the object o_i intersects the query region R at time t. The uncertain position of the moving object is represented through a density function. Query predicates are distinguished into two parts: the static part, C_1, which refers to the static attributes of the objects, e.g., color, type, etc., and the dynamic part, C_2, which refers to the location attributes. The idea is to first retrieve the set of objects satisfying the predicates of the static part, i.e., C_1, and then proceed with the dynamic part, i.e., C_2. So, after the retrieval of the set of objects satisfying C_1, the routes of the resulted objects are retrieved. Then, for each route r and each atomic predicate p appearing in C_2, the list of the intervals of route r with the region defined by p is retrieved – any spatial indexing schema can be used toward this aim. Finally, the list of intervals of route r with all predicates of query q is computed. For each route r and for each object o traveling on r, the probability that it satisfies query q is given by:

$$\sum_{i=1}^{k} \int_{u_i}^{v_i} f_o(x) dx, \tag{6.2}$$

where k is the number of intervals of r with q, $[u_i, v_i]$ are the limits of each interval I_i, and $f_o(x)$ is the density function.

Cheng et al. [13] study the execution of probabilistic range queries (PRQ) and probabilistic nearest-neighbor queries (PNNQ) under uncertainty. They adopt a generic uncertainty model that for each time point associates an object with an uncertainty region. The position of the object is modeled through a probability density function, which is zero outside the uncertainty region. The algorithm for *probabilistic range queries* processing integrates the probability distribution function in the overlapping area defined by the query region and the object's uncertainty region. Processing a *nearest-neighbor query* involves evaluating the probability of each object being closest to a query point q. The adopted solution consists of four steps: projection, pruning, bounding, and evaluation phases. During the *projection* phase, the uncertainty region of each moving object is computed based on the uncertainty model used by the application (Fig. 6.12a shows the last recorded object

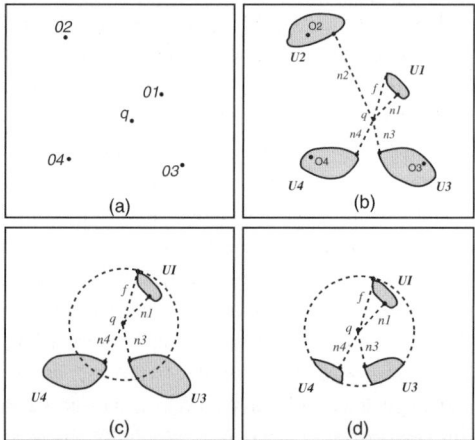

Fig. 6.12 An example of a PNNQ processing: (**a**) locations of objects, (**b**) uncertainty regions and distances from q, (**c**) bounding circle, and (**d**) bounded regions [13]

locations and Fig. 6.12b their uncertainty regions). During the *pruning* phase, the minimum f of the longest distances of the uncertainty regions from q is found and any object with shortest distance to q larger than f is eliminated (Fig. 6.12c shows how pruning removes objects that are irrelevant to q). During the *bounding* phase, a bounding circle C of radius f and center q is conceptually drawn and any object outside this circle is ignored. (This concept is depicted in Fig. 6.12d.) During the *evaluation* phase, for each resulted object o the probability that it is the nearest neighbor of q with distance r is calculated. This probability is given by the probability of o being at distance r from q times the probability that every other object is at a distance $\geq r$ from q.

An index structure, the so-called velocity-constrained index (VCI), has been proposed by Cheng et al. [13] and is particularly suited for handling uncertainty of free-moving objects. VCI is an R-tree-like index structure. Its difference from R-tree lies in the fact that each node is accompanied with an additional field, v_{max}, which is the maximum possible velocity of movement over all the objects that fall under that node. The only restriction imposed on the movement of objects is that they do not exceed a certain velocity. This velocity could potentially be adjusted if the object wants to move faster than its current maximum velocity. The index is built based on the locations of the objects at a given time point, t_0. However, it can also be used at a later time point t without being updated. The idea is that for a given VCI node, no object under this node can move faster than the maximum velocity stored in the node. Thus, if the MBB is expanded by $v_{max}(t - t_0)$, then the expanded region is guaranteed to contain all the points under this subtree.

The last subject related to the management of the location uncertainty in trajectory databases is the so-called *map-matching* problem, which deals with the problem of matching tracking data (via GPS or any other positioning method) in an underlying map containing, e.g., a road network. This problem occurs due to the fact that raw trajectory positions cannot be directly matched to the underlying infrastructure, and, they are mainly affected by two factors [9]: the measurement error introduced by, e.g., GPS, and the sampling error being up to the frequency with which position samples are taken, both contributing to the moving object's location uncertainty.

Related work in the subject of map matching includes, among others, [9, 16, 71, 76], which propose a variety of map-matching algorithms. Perhaps the most promising approach is the one presented in [9], where three map-matching algorithms are presented. These algorithms consider the trajectory nature of the data rather than simply the object's current position as it often happens in the typical map-matching case. The first one is an incremental algorithm, which matches consecutive portions of the trajectory to the road network, effectively trading accuracy for speed of computation. Specifically, Brakatsoulas et al. [9] first employ the similarity measure $s(p_i, c_j)$ [23] between a sampled position p_i and a network edge c_j, used in order to evaluate the likelihood of p_i to match each one of the candidate network edges c_j. Then, they propose an algorithm that looks ahead, and rather than calculating the similarity for just one sampled position, it takes into account the sum of the similarities between the l ahead positions against the local candidate path. The value of $l = 4$ is established empirically that is optimal in terms of matching quality and running time.

The other two algorithms compare the entire trajectory to candidate paths in the road network using the Fréchet and the weak Fréchet distances. The Fréchet and the weak Fréchet distances can be illustrated as follows: suppose a man is walking his dog and that he is constrained to walk on a curve and his dog on another curve. Both the man and the dog are allowed to control their speed independently, but are not allowed to go backward in the case of the simple Fréchet distance, while in the case of the weak Fréchet distance, they are. Then, the Fréchet and the weak Fréchet distances between the two curves is the minimal length of a leash that is necessary in each case. The proposed global map-matching algorithms find a curve in the road network that is as close as possible to the given trajectory. The underlying distance measure, i.e., the Fréchet distance and the weak Fréchet distance, also serves as a quality guarantee for the computed result.

Finally, the proposed algorithms are evaluated in terms of their running time and the quality of their matching result. Comparing the asymptotic running times, it is revealed that the incremental algorithm has a significant performance advantage over the global algorithms. On the other hand, the global algorithms were found to produce better matching results.

6.6 Handling Trajectory Compression

As addressed by [36], it is expected that all the ubiquitous positioning devices will eventually start to generate an unprecedented data stream of time-stamped positions. Sooner or later, such enormous volumes of data will lead to storage, transmission, computation, and display challenges. Hence, the need for compression techniques arises. However, existing work in this domain is relatively limited [10,36,50,51], and mainly guided by advances in the field of line simplification, cartographic generalization, and data series compression. According to [36], the objectives for trajectory data compression are:

- To obtain a lasting reduction in data size
- To obtain a data series that still allows various computations at acceptable (low) complexity
- To obtain a data series with known, small margins of error, which are preferably parametrically adjustable

As a consequence, our interest is with lossy compression techniques, which eliminate some redundant or unnecessary information under well-defined error bounds. Generally, the whole of the proposed compression algorithms that will be examined in this section deal with the compression of trajectory data in unrestricted spaces. To the best of our knowledge, the case of compression under network constraints has not been already examined in the research literature, and it will be consequently discussed in Sect. 6.7.3. Meratnia and By [36] exploit existing algorithms used in

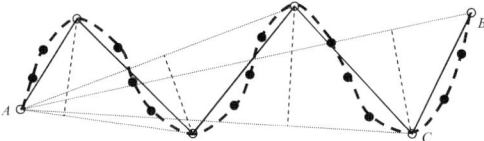

Fig. 6.13 Top–down Douglas–Peucker algorithm used for trajectory compression. Original trajectory is presented with *dotted lines* and compressed trajectory with *solid line* [36]

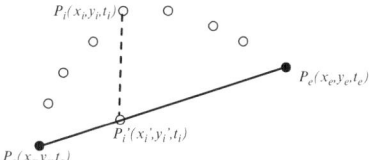

Fig. 6.14 The synchronous Euclidean distance (SED): the distance is calculated between the point under examination (P_i) and the point P'_i, which is determined as the point on the line (P_s, P_e) at the time instance t_i [36]

the line generalization field, presenting one top–down and one opening window algorithm, which can be directly applied to spatiotemporal trajectories. The top–down algorithm, named TD-TR, is based on the well-known Douglas–Peucker [18] algorithm (Fig. 6.13) introduced by geographers in cartography. This algorithm calculates the perpendicular distance of each internal point from the line connecting the first and the last point of the polyline (line AB in Fig. 6.13) and finds the point with the greatest perpendicular distance (point C). Then it creates lines AC and CB and, recursively, checks these new lines against the remaining points with the same method, and so on. When the distance of all remaining points from the currently examined line is less than a given threshold (e.g., all the points following C against line BC in Fig. 6.13), the algorithm stops and returns this line segment as part of the new, compressed, polyline. Being aware of the fact that trajectories are polylines evolving in time, the algorithm presented in [36] replaces the perpendicular distance used in the DP algorithm with the so-called *synchronous Euclidean distance* (SED), also discussed in [10, 51], which is the distance between the currently examined point (P_i in Fig. 6.14) and the point of the line (P_s, P_e) where the moving object would lie, supposed it was moving on this line, at time instance t_i determined by the point under examination (P'_i in Fig. 6.14)). The time complexity of such an algorithm is $O(N \log N)$.

Although the experimental study presented in [36] shows that the TD-TR algorithm is significantly better than the opening window in terms of both quality and compression (since it globally optimizes the compression process), it has the main disadvantage of not being an online algorithm and, therefore, it cannot be applied directly to trajectory segments at the time they are feeding a spatiotemporal database. Quite the opposite, it needs the a priori knowledge of the entire moving object trajectory.

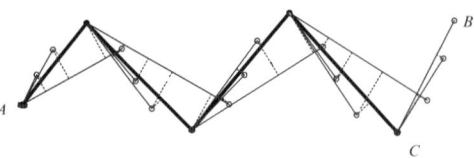

Fig. 6.15 Opening window algorithm used for trajectory compression. Original data points are represented by *closed circles* [36]

On the contrary, under the previously described conditions of online operation, the *opening window* (OW) class of algorithms can be easily applied. These algorithms start by anchoring the first trajectory point, and attempt to approximate the subsequent data points with one gradually longer segment (Fig. 6.15). As long as all distances of the subsequent data points from the segment are below the distance threshold, an attempt is made to move the segment's end point one position up in the data series. When the threshold is going to exceed, two strategies can be applied: either the point causing the violation (*normal opening window*, NOPW) or the point just before it (*before opening window*, BOPW) becomes the end point of the current segment, and also the anchor of the next segment. If the threshold is not exceeded, the float is moved one position up in the data series (i.e., the window opens further) and the algorithm caries on until the trajectory's last point; then the whole trajectory is transformed into a linear approximation. In the original OW class of algorithms, each distance is calculated from the point perpendicular to the segment under examination, while in the OPW-TR algorithm presented in [36], the SED distance is evaluated.

Although OW algorithms are computationally expensive – since their time complexity is $O(N^2)$ – they are very popular. This is because, they are online algorithms, and they can work reasonably well in presence of noise (but only for relatively short data series). Moreover, the time complexity is $O(N^2)$ regarding only the compression of the full data series; when dealing with each point update – that is in the online case – the complexity of determining whether each incoming point will be float or the next anchor is $O(N)$.

Recently, Potamias et al. [51] proposed several techniques based on uniform and spatiotemporal sampling to compress trajectory streams, under different memory availability settings: fixed memory, logarithmically or linearly increasing memory, or memory not known in advance. Their major contributions are two compression algorithms, namely, the *STTrace* and *Thresholds*. According to this work, there are two basic requirements when dealing with trajectory streams: the need for processing incoming points in high rates and the need for locally or globally constant allocated memory. To deal with the first requirement, they propose the Thresholds method with $O(1)$ time complexity. This method uses the current object's position, speed and direction in order to predict a *safe area*, where the next trajectory point will be located; when this area actually contains the next reported point, it can be approximated by the current moving point settings. The authors propose the calculation of the safe area using two methods: the first one, named *sample-based safe*

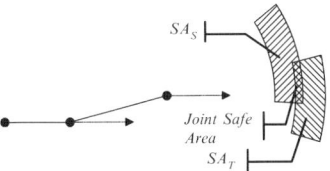

Fig. 6.16 Safe area used by the Thresholds algorithm

area, is calculated using each object's current position speed and direction in any case, despite of whether the object's current position was or was not eliminated by the heuristic. On the contrary, the second approach, named *trajectory based*, calculates the safe area using each object's last recorded position speed and direction. Because of several limitations that both approaches demonstrate, the safe area employed by the algorithm is calculated as the planar intersection of the sample-based and the trajectory-based one (SA_S and SA_T areas, respectively, in Fig. 6.16). The main advantage of the proposed algorithm compared with the opening window presented in [36] is its low-cost time complexity; however, although their results would possibly be comparable, they do not provide any experimental comparison between the two algorithms in terms of actual execution time, compression rate, and quality.

The second algorithm proposed in [51] is designed to fulfill the requirement of the preset amount of memory. The proposed algorithm, named *STTrace*, utilizes a constant for each trajectory amount of memory M. It starts by inserting in the allocated memory the first M recorded positions, along with each position's SED with respect to its predecessor and successor in the sample. As soon as the allocated memory gets exhausted and a new point is examined for possible insertion, the sample is searched for the item with the lowest SED, which represents the least possible loss of information in case it gets discarded. In the sequel, the algorithm checks whether the inserted point has SED larger than the minimum one found already in the sample and, if yes, the currently processed point is inserted into the sample at the expense of the point with the lowest SED. Finally, the SED attributes of the neighboring points of the removed one are recalculated, whereas a search is triggered in the sample for the new minimum SED. The proposed algorithm may be easily applied in the multiple trajectory case, by simply calculating a global minimum SED of all the trajectories stored inside the allocated memory.

6.7 Open Issues: Roadmap

6.7.1 Trajectory Indexing

Following the discussion of Sect. 6.3, we further present directions for future research on the subject of trajectory indexing. We sketch out this section by proposing that: (a) the need for trajectory preservation has to be traded back for the need of efficient coordinate-based query processing (and the opposite), (b) trajectory deletions call for support by existing indexes, and (c) new requirements arise when examining trajectory indexing under the prospect of trajectory compression. According to the survey presented earlier in this chapter, structures exploiting the network-constrained movement are much more efficient than those indexing objects in the unrestricted space; actually, the former usually outperform the later by orders of magnitude. However, none of the proposed index structures is designed to preserve trajectories: both FNR [20] and MON-tree [2] lack by definition a mechanism to retrieve trajectories and only care about the processing of coordinate-based queries. Moreover, SETI [11], which is the most efficient indexing scheme in unrestricted space regarding coordinate-based queries, suffers from the same drawback. However, as pointed earlier, the trajectory preservation is of great importance since it is required to process trajectory-based queries. As such, the first research direction arising on the subject of trajectory indexing is the development of indexing schemes efficiently supporting trajectory-based querying under both unrestricted and network-constrained space.

Trajectory indexing structures often ignore several real-world requirements. For example, deletions are often neglected when proposing indexing methods for moving object trajectories. The main argument is that deleting a three-dimensional line segment from an object's trajectory may sound meaningless. Although this might be assumed to be conceptually correct (transmitted positions are recorded, thus they do exist), deleting an entire object's trajectory is meaningful (trajectories of objects being no more useful could be deleted from the index). Therefore, the need for an efficient algorithm to support deletions of object trajectories arises. However, the support of such an operation would require the index to efficiently retrieve entire trajectories. As such, among the techniques surveyed in Sect. 6.3, it is only the TB-tree that would be capable to efficiently support deletion. However, its structure is not suitable for supporting deletion operations; a trajectory deletion would leave "holes" in the nodes. As for the rest index structures surveyed in Sect. 6.3 (such as the three-dimensional R-tree, the STR-tree and the SETI), since they do not have a mechanism to efficiently retrieve trajectories, they would have to answer a sequence of range queries, as described in [49] for the combined search of the three-dimensional R-tree and the STR-tree – a very expensive approach as shown in [49].

On the other hand, several experimental studies have been shown that the performance of the TB-tree decreases as the cardinality of moving objects increases, and other indexing structures usually outperform it [11, 21, 49]. Moreover, its spatial filtering capabilities are very limited, i.e., it is inefficient when the spatial extent of the spatiotemporal range is small since its insertion mechanism cares only about preserving trajectories and not about the spatial and temporal relations between objects. Then again, its efficiency in temporal filtering is based upon the assumption that data are inserted in chronological order. Therefore, it appears that the need for trajectory preservation has to be traded back to achieve efficiency in traditional coordinate-based queries (and the opposite) so as to find the best possible settings supporting both query types.

Another topic in the domain of spatiotemporal indexing is the utilization of compression techniques in existing trajectory databases, being indexed by a proper spatiotemporal index. On the basis of the experimental study presented in [36], it is shown that the TD-TR algorithm produces significantly better trajectories than the OPW-TR in terms of both quality rate. Then again, it has the disadvantage that it cannot be directly applied to trajectory segments as those are feeding the database system and it needs the a priori knowledge of the entire moving object trajectory. Therefore, only indexes preserving trajectories (e.g., the TB-tree) can exploit the TD-TR algorithm. As such, the TD-TR trajectory compression algorithm could be utilized over the TB-tree by reading each indexed trajectory one-by-one, compress it, and finally feed a new TB-tree with the compressed trajectory.

Moreover, many spatiotemporal indexes including TB-tree, SETI, and FNR-tree, assume that new entries are inserted in chronological order, placing them at the right "end" of the tree. However, a method for compressing existing indexes according to the previous discussion would place entire trajectories on this right side of the tree without considering their temporal ordering, leading thus to indexes with high temporal overlap decreasing their performance. To overcome this drawback, one would have to utilize intermediate steps processing all indexed trajectories, producing the new compressed ones, sorting them according to their temporal order and finally feed the new index. Nevertheless, such a technique would require processing the entire index in the main memory, or developing specialized algorithms to handle it efficiently. On the basis of the previous discussion, there are two basic requirements arising when dealing with optimal (e.g., using the TD-TR algorithm) compression of existing trajectory databases: the capability of *trajectory preservation* and *supporting of nonchronological insertions*.

6.7.2 Trajectory Query Processing

Although sufficient amount of research work exists in the context of trajectory query processing, still there are several issues to be handled. Outlining this section, we suggest that (a) nearest-neighbor search asks for more efficient support by the various indexing methods, (b) trajectory similarity search and derived information querying (involving speed, heading, etc.) need to be supported by general purpose indexes, (c) query optimization techniques must be further examined, and (d) trajectory querying under uncertainty needs further study.

6.7.2.1 Nearest-Neighbor Search

As previously mentioned, R-tree-like structures can efficiently support NN queries, while regarding the rest of the proposed spatiotemporal indexes, the corresponding papers do not consider NN search algorithms. However, for some of the proposals, NN querying can be easily supported. For example, since in the FNR-tree the underlying network is indexed by a conventional R-tree, the best-first algorithm described in [31] can be employed to find the spatial nearest neighbor; then, given that the network line segments (i.e., the spatial elements of the trajectory segments) are reported in incremental order of their distance from the query object, the algorithm would have to report such nearest segments until retrieving the first overlapping the query in the temporal dimension. A similar approach can be also employed in MON-tree, while SETI would have to search among all entries contained inside the corresponding cell in which the query point lies.

6.7.2.2 Trajectory Similarity Search

Although, as previously discussed, there is a sufficient number of research papers in the domain of similarity search, the majority of the proposals exploit dedicated index structures to prune the search space and efficiently support the *most similar trajectory* (MST) search. However, these dedicated structures require costly preprocessing steps and they do not conform to the requirement of online action, since there is no obvious way for them to be updated during the database operation. Therefore, future work needs to deal with the k-most similar trajectory search in MODs storing historical trajectory information, exploiting existing index structures, which can also be used to support other types of queries. Moreover, in order to use traditional index structures, future work should be based on novel metrics that follow triangle inequality, since the already proposed schemes [12, 70] typically use nonmetric measures that cannot be indexed with the majority of the proposed spatiotemporal indexes. One such proposal for the dissimilarity between trajectories completing the above requirements is the following [38]:

$$\text{DISSIM}(Q,T) = \int_{t_1}^{t_n} \text{Dist}_{Q,T}(t) dt, \tag{6.3}$$

where $\text{Dist}_{Q,T}(t)$ is the function of the Euclidean distance between trajectories Q and T with time. However, adopting the model where each trajectory is represented by a collection of discrete points with linear interpolation applied in between, the definition of dissimilarity can be transformed to [22]:

$$\text{DISSIM}(Q,T) = \sum_{k=1}^{n-1} \int_{t_k}^{t_{k+1}} \text{Dist}_{Q,T}(t) dt, \tag{6.4}$$

where t_k are the timestamps that objects T and Q recorded their position. Frentzos et al. [22] discuss exactly this problem, evaluating the above equation that provides a closed formula, nevertheless, expensive in terms of computational power. This formula is subsequently efficiently approximated using numerical analysis techniques. Moreover, Frentzos et al. [22] provide an efficient algorithm, based on several novel metrics and heuristics used for pruning purposes.

Moreover, additional algorithms to support other types of similarity search (i.e., directional, speed pattern, etc.), as already discussed in the previous chapter, have to be proposed. For example, since the evolution of the speed or heading of a given trajectory can be considered as an one-dimensional time series, metrics, and algorithms already utilized in the context of time series can be directly applied (such as edit distance, LCSS, and dynamic time wrapping).

6.7.2.3 Derived Information Queries

As discussed earlier, queries regarding derived information on trajectories have found limited interest. This category includes queries of the following types: *Find objects moving instantly with speed between V_{min} and V_{max}, inside a given time period (and/or given spatial extent)* or *Find objects moving with an average heading between dir_1 and dir_2*. Currently, such queries can be answered by employing a simple temporal index storing the speed or heading time series or through an exhaustive search over the trajectories stored inside the database.

On the other hand, assuming that trajectories are indexed by an R-tree-like spatiotemporal index, some preliminary information about each object's velocity vector can be found before accessing the leaf node containing its entries. For example, the dimensions of the bounding box can provide a first estimation of the average speed of the object during the temporal period covered by the node MBB. Moreover, by employing the TB-tree, which stores in each leaf node segments from the same trajectory, we can estimate some tighter bounds for the maximum and minimum average speed of a single trajectory without actually accessing leaf nodes, by using the fact that the length of the part of the trajectory being inside the leaf MBB can not be smaller than the MBB's diagonal and its lifetime is exactly determined by the MBB's temporal extension.

6.7.2.4 Spatiotemporal Query Optimization

Query optimization concerns about the development of selectivity estimation techniques and cost formulas for the execution of the several types of queries. On the subject of estimating the selectivity of several spatiotemporal queries against trajectory data sets, there are two equally significant directions. The first deals with the estimation of the *number of three-dimensional line segments* in the spatiotemporal data space retrieved by a given spatiotemporal range query, while the second deals with the actual *number of distinct trajectories* retrieved by the same query. The former is useful in the case of approximating the cost in the execution of a query, since all the proposed indexing schemes physically index three-dimensional line segments. Therefore, formulas of cost models already proposed for query optimization (as the one presented in 6.1 for the R*-tree [35]), involving the data set cardinality N (which is equal to the data set density in the unit space), would have to utilize the local density N' instead of N to produce a more accurate result, where N and N' refer to the number of line segments and not the number of distinct trajectories. On the basis of the above discussion, an extension of a simple spatial histogram, as the one presented in [1], in the spatiotemporal space, could be straightforwardly utilized to efficiently approximate the data set's local density.

On the other hand, regarding the later estimation (i.e., the number of distinct trajectories), it is not an easy task, since it involves the well-known *distinct-counting problem* [63]. The distinct-counting problem stands when an object samples its position in several timestamps inside a given query window, resulting to be counted multiple times in the query result. Tao et al. [63] provide a solution to the aforementioned problem by integrating spatiotemporal indexes with sketches, traditionally used for approximate query processing. However, their proposal reduces the space requirements only a few times (typically about the 40% of the original database size), while the corresponding index structure is maintained on the disk. Clearly, such an approach cannot be utilized instead of histograms (having a typical size of a few kilobytes [1]), since it introduces a sizeable overhead in terms of both memory space and processing time requirements.

In the same fashion, a spatiotemporal histogram concerning about the number of distinct trajectories would have to partition the space into several spatiotemporal buckets, counting the number of distinct trajectories inside each bucket. However, when trying to produce a selectivity estimation regarding a query window that contains more than one buckets, this estimation cannot be computed as the sum of the cardinality of two buckets since trajectories may be counted several times depending on the number of buckets they overlap. Figure 6.17 exemplifies this problem, illustrating four histogram buckets (B_1, B_2, B_3, B_4) along with their respective selectivity Sel(B_i): the selectivity of all four buckets Sel($\sum B_i$) = 3 is far from being the sum of $\sum$ Sel(B_i) = 7 because trajectories T_1, T_2, T_3 will be counted as many times, as the buckets each of them overlaps. Moreover, the same problem arises during the histogram construction following the methodology introduced in [1] for simple spatial histograms: the construction algorithm initially calculates the number of distinct objects inside each cell produced by a dense spatial grid, and then in each iteration it aggregates groups of cells to form more wide buckets, based on the *MinSkew* heuristic. However, during this aggregation, the number of trajectories inside each resulted bucket has to be calculated, clearly, not as the sum of the trajectories contained inside each fundamental cell.

Future work in the field of spatiotemporal query optimization includes the development of formulas for selectivity estimation on a variety of queries, such

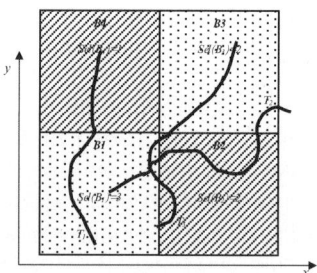

Fig. 6.17 The distinct-counting problem in trajectory histograms

as spatiotemporal joins, similarity-based search, derived information queries, etc. Moreover, since the work on spatiotemporal cost models is very limited, the need for cost models regarding the variety of the proposed spatiotemporal indexes arises. Finally, interesting directions arise when dealing with cost models under the presence of uncertainty. For example, while probabilistic queries have been well studied in the context of spatial and spatiotemporal databases, there is no work dealing with the estimation of the selectivity (or the cost of processing) of a probabilistic query of the form: *retrieve objects inside a given spatial (or spatiotemporal) interval with probability greater than x%*, neither in spatial nor in spatiotemporal databases.

6.7.2.5 Querying Imprecise Trajectories

There are several interesting issues for future work on query processing under location uncertainty. One important research direction is to lower the execution time of already proposed range queries and nearest-neighbor queries under uncertainty. An extension is to investigate other probabilistic queries, like k-nearest-neighbor queries and reverse nearest-neighbor queries. A critical issue in all these works is to define the quality of query answers over imprecise data, i.e., how reliable are the query results. In the same rationale, it is also important to bound the error associated with the provided answers.

Future work on trajectory querying under uncertainty must also include the determination of the error introduced in query results by measurement and sampling errors. For example, a very interesting MOD capability is to a priori provide the user with an estimation of the error introduced in query results. This approach can be considered as an alternative to the processing of probabilistic queries, already examined in the spatiotemporal domain. Possible research steps toward this direction are:

- To estimate the error (number of false negatives and false positives) introduced in queries over uniformly distributed spatial point data. The estimation of this error could be carried out by formulating the probability of a point to incur false hit regarding a single query and then integrate this probability over the entire space so as to produce its mean value.
- To extend this approach in the spatiotemporal domain assuming also uniform distribution of trajectories.
- Finally, to relax the uniformity assumption with the employment of spatiotemporal approximation techniques such as the one presented in [63] or spatiotemporal histograms as previously discussed.

Regarding the second step and time-slice queries, as illustrated in Fig. 6.18, the extension is straightforward since the temporal slice produces a set of (spatial) points along with their uncertainty area. In the general case, however, of range queries with nonzero temporal extent, this extension is not an easy task; nevertheless, we subsequently provide hints toward this direction. Consider for example Fig. 6.19 illustrating trajectories of three moving objects along with their uncertainty

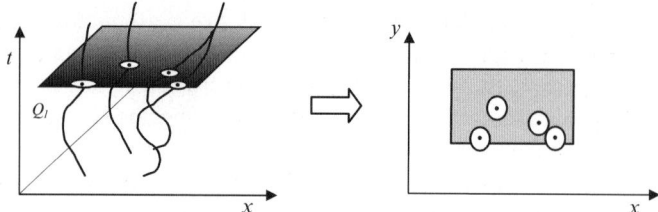

Fig. 6.18 The effect of uncertainty in time-slice queries over moving object trajectories is equivalent to the effect of the uncertainty in range queries over spatial point data

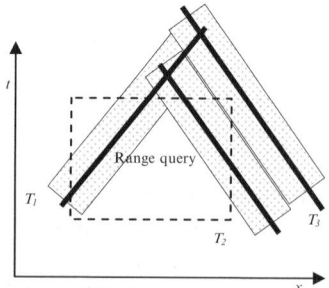

Fig. 6.19 The effect of uncertainty in general range queries

regions (e.g., the dotted areas) in the x–t space, along with a range query. Because of simplicity reasons, all trajectories are illustrated as line segments without loss of generality. Trajectories T_1 and T_2 cannot never encounter a false hit regarding the query window due to the fact that for at least one time instance their uncertainty region was entirely located inside it. On the other hand, trajectory T_3 may encounter a false hit because it is not inside the query window; nevertheless, its uncertainty region crosses it. Generalizing the above observation, we can state that only objects whose uncertainty area crosses the query window without being entirely inside it at any time instance may contribute to the number of false hits in the results of the query.

6.7.3 Trajectory Compression

Trajectory compression is a very promising field, since the amount of data collected by positioning devices will sooner or later expand with excessive rates. Therefore, research on trajectory compression has to be extended toward many directions. A first direction is based on the idea of the *STTrace* algorithm, where the amount of memory used to store each trajectory is a priori determined. Consider now the following task posed against a trajectory database: *compress a given trajectory T with n vertices to produce Q, which must contain only $j < n$ vertices*. Although the *STTrace* algorithm is suitable for such procedures applied on trajectory data streams, it concerns only local optimization. Therefore, in the above case where the trajectory is completely known in advance, it is possible to apply a different algorithm concerning *global optimization*. Such algorithms need to be examined in the future, providing also a bounded error.

A second direction also regarding the compression algorithms is the utilization of the insightful idea of amnesic approximation for streaming time series, which was initially introduced in [41], considering that the importance of each measurement generally decays with time. However, it is not possible to utilize this idea straightforwardly since the work presented in [41] only handles one-dimensional time series. Moreover, authors in [51] argue that apart from handling multidimensional points, the case for trajectories is different because of characteristics inherent in movement: not only spatial locations, but also speed and orientation should not be overlooked when approximating trajectories.

One first solution on this subject is presented with the *AmTree* introduced in [50]. The structure of an *AmTree* is illustrated in Fig. 6.20; each level i, except for the tree root, consists of two nodes R_i and L_i. Node R_0 at the lowest level accepts data in every timestamp that are reported. Each node at the ith level contains information about twice as many timestamps as a node at the $(i-1)$th level. Hence, a node at level i contains information about 2^i timestamps. *Amtree* is built in a bottom-up fashion. As new sampled positions are feeding it, they are added into position R_0, while the contents of node R_0 are shifted to L_0; with the addition of one more sampled position, the contents of L_0 and R_0 are combined using a simple function and propagated to the higher tree level. Potamias et al. [50] propose the employment of motion vectors (e.g. tuples of the form x, y, t_{start}, t_{end}, dx, dy) inside each tree node; then the combination function produces the summarization of the two vectors. A more sophisticated approach would include the treatment of the combination function with a similar algorithm as the one proposed in the previous section. Such

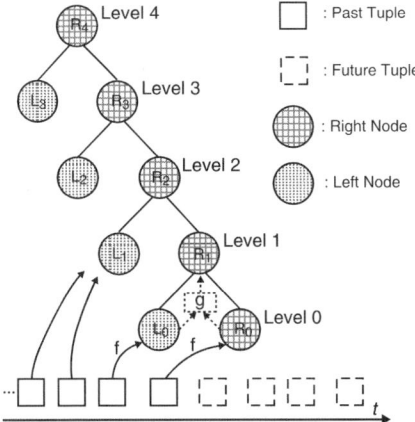

Fig. 6.20 The *AmTree* structure [50]

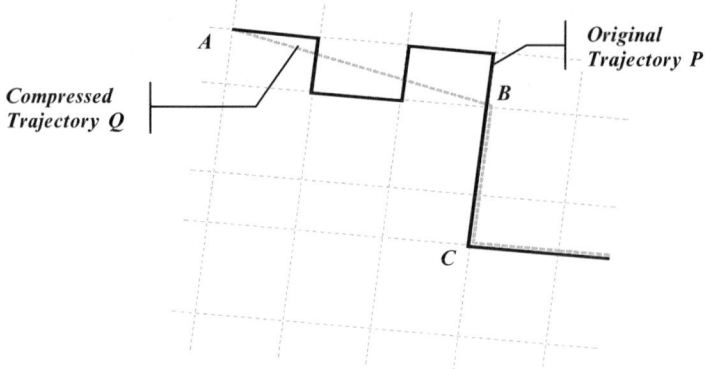

Fig. 6.21 Compressing trajectory data under network constraints

a function would compress a given partial trajectory T constituting of n vertices to produce Q, which must contain only $n/2$ vertices. This approach would lead to a compressed trajectory with higher compression rates – more abstract information – for aged data and lower compression rates for recent data, with the advantage that it incorporates at the same time the inherited requirements of trajectory data (e.g., considering together the spatial locations, the direction, and the speed of the objects).

A third research direction could consider network constraint compression of trajectory data. Consider for example Fig. 6.21 illustrating an uncompressed trajectory P constrained on an underlying road network and the respective compressed trajectory Q produced by one of the previously described algorithms; although the original trajectory is valid under the network constraint, the compressed one is not since Q directly connects nonneighboring network nodes (e.g., line segment AB in Fig. 6.21). A methodology towards the solution of this issue would be to initially compress the trajectory data with one of the existing algorithms and then use the underlying network to produce routes connecting the – under network constraints – unaffected by the compression and trajectory nodes. However, such an approach would be computationally inefficient since it would require performing many – possibly overlapping – routings through the network graph. Moreover, one has to examine whether the produced network route between two remained trajectory nodes still – after the routing – verifies the thresholds utilized by the compression algorithm.

The last issue on the subject of trajectory compression requires the determination of the error introduced in query results when trajectory data are compressed. For example, rather than determining the mean error of the trajectory approximation, as discussed in [36], it might be more meaningful for a system user to know the influence of the compression in the results returned by the database when executing a range or nearest-neighbor query against the trajectory data. Then, along with the query results, the database would inform the user about an estimation of the percentage of the false hits introduced in the query results. There are two approaches that may clarify this issue: either performing experiments under various compressed trajectory data sets or theoretically determining the effect of the compression by assuming uniformity of data and then using spatiotemporal histograms to overpass such an assumption.

6.8 Concluding Remarks

The domain of trajectory databases has attracted significant research effort during the last decade, developing trajectory database engines based on existing commercial systems, novel spatiotemporal indexes, queries, and query processing techniques. In this chapter, we conducted a brief presentation of the approaches in trajectory indexing, query processing, and query optimization, dealing also with the issues of location uncertainty and trajectory compression. Future work in this domain should include the development of all purpose indexes efficiently processing traditional (range/distance) and trajectory-based queries, novel query processing techniques about joining trajectories, performing similarity search, and derived information querying. Also, the domain of spatiotemporal query optimization is a very prominent field, since it has been concerned only at a relatively small extent.

Moreover, since the almost ubiquitous positioning devices will sooner or later produce a tremendous number of time-evolving positions, the need for trajectory compression arises. Existing techniques can be extended towards many directions, supporting compression under network constraints, compression of existing indexes, novel trajectory compression operators (such as *compress a trajectory so as to contain a predetermined number of vertices*), and the determination of the error introduced by the compression in query results.

References

1. S. Acharya, V. Poosala, and S. Ramaswamy. Selectivity estimation in spatial databases. In *Proceedings of the International Conference on Management of Data (SIGMOD '99)*, pp. 13–24, 1999.
2. V.T. Almeida and R.H. Güting. Indexing the trajectories of moving objects in networks. *GeoInformatica*, 9(1):33–60, 2005.
3. V.T. Almeida, R.H. Güting, and T. Behr. Querying moving objects in secondo. In *Proceedings of Seventh International Conference on Mobile Data Management (MDM '06)*, p. 47, 2006.
4. S. Arumugam and C. Jermaine. Closest-point-of-approach join for moving object histories. In *Proceedings of the 22th International Conference on Data Engineering (ICDE'06)*, p. 86, 2006.
5. P. Bakalov, M. Hadjieleftheriou, E.J. Keogh, and V.J. Tsotras. Efficient trajectory joins using symbolic representations. In *Proceedings Sixth International Conference on Mobile Data Management (MDM'05)*, pp. 86–93, 2005.
6. P. Bakalov, M. Hadjieleftheriou, and V.J. Tsotras. Time relaxed spatiotemporal trajectory joins. In *Proceedings of the 13th annual ACM international workshop on Geographic Information Systems (GIS'05)*, pp. 182–191, 2005.

7. R. Benetis, C.S. Jensen, G. Karciauskas, and S. Saltenis. Nearest neighbor and reverse nearest neighbor queries for moving objects. In *Proceedings of the International Symposium on Database Engineering and Applications (IDEAS'02)*, pp. 44–53, 2002.
8. D.J. Berndt and J. Clifford. Finding patterns in time series: A dynamic programming approach. In *Advances in Knowledge Discovery and Data Mining*, pp. 229–248. MIT Press, 1996.
9. S. Brakatsoulas, D. Pfoser, R. Salas, and C. Wenk. On map-matching vehicle tracking data. In *Proceeding on 31st International Conference on Very Large Data Bases (VLDB'05)*, pp. 853–864, 2005.
10. H. Cao, O. Wolfson, and G. Trajcevski. Spatio-temporal data reduction with deterministic error bounds. In *Proceedings of Discrete Algorithms and Methods for Mobile Computing and Communications-Principles of Mobile Computing (DIALM-POMC'03)*, pp. 33–42, 2003.
11. V.P. Chakka, A. Everspaugh, and J.M. Patel. Indexing large trajectory data sets with SETI. In *Proceedings of Conference on Innovative Data Systems Research (CIDR'03)*, 2003.
12. L. Chen, M.T. Özsu, and V. Oria. Robust and fast similarity search for moving object trajectories. In *Proceedings of the International Conference on Management of Data (SIGMOD'05)*, pp. 491–502, 2005.
13. R. Cheng, D.V. Kalashnikov, and S. Prabhakar. Evaluating probabilistic queries over imprecise data. In *Proceedings of the International Conference on Management of Data (SIGMOD'03)*, pp. 551–562, 2003.
14. K.L. Cheung and A.W.-C. Fu. Enhanced nearest neighbour search on the R-tree. *SIGMOD Record*, 27(3):16–21, 1998.
15. Y.-J. Choi and C.-W. Chung. Selectivity estimation for spatio-temporal queries to moving objects. In *Proceedings of the International Conference on Management of Data (SIGMOD'02)*, pp. 440–451, 2002.
16. A. Civilis, C.S. Jensen, J. Nenortaite, and S. Pakalnis. Efficient tracking of moving objects with precision guarantees. In *Proceedings of The Annual International Conference on Mobile and Ubiquitous Systems: Computing, Networking and Services (MobiQuitous'04)*, pp. 164–173, 2004.
17. S. Dieker and R.H. Güting. Plug and play with query algebras: SECONDO-a generic dbms development environment. In *Proceedings of the International Symposium on Database Engineering and Applications (IDEAS'00)*, pp. 380–392, 2000.
18. D. Douglas and T. Peucker. Algorithms for the reduction of the number of points required to represent a digitized line or its caricature. *The Canadian Cartographer*, 10(2):112–122, 1973.
19. L. Forlizzi, R.H. Güting, E. Nardelli, and M. Schneider. A data model and data structures for moving objects databases. In *Proceedings of the International Conference on Management of Data (SIGMOD'00)*, pp. 319–330, 2000.
20. E. Frentzos. Indexing objects moving on fixed networks. In *Proceedings of the 7th International Symposium on Spatial and Temporal Databases (SSTD'03)*, pp. 289–305, 2003.
21. E. Frentzos, K. Gratsias, N. Pelekis, and Y. Theodoridis. Algorithms for nearest neighbor search on moving object trajectories. *Geoinformatica*, 11(2):159–193, 2007.
22. E. Frentzos, K. Gratsias, and Y. Theodoridis. Index-based most similar search. In *Proceedings of the 23th International Conference on Data Engineering (ICDE'07)*, 2007.
23. J. Greenfeld. Matching gps observations to locations on a digital map. In *81th Annual Meeting of the Transportation Research Board*, pp. 164–173, 2004.
24. R.H. Güting and M. Schneider. Realm-based spatial data types: The rose algebra. *Very Large Data Bases Journal*, 4:243–286, 1995.
25. R.H. Güting, M.H. Böhlen, M. Erwig, C.S. Jensen, N.A. Lorentzos, M. Schneider, and M. Vazirgiannis. A foundation for representing and quering moving objects. *ACM Transactions Database Systems*, 25(1):1–42, 2000.
26. R. Güting, T. Behr, V. Almeida, Z. Ding, F. Hoffmann, and M. Spiekermann. SECONDO: An extensible dbms architecture and prototype. fernuniversitat hagen, informatik-report 313, 2004.
27. A. Guttman. R-trees: A dynamic index structure for spatial searching. In *Proceedings of the International Conference on Management of Data (SIGMOD'84)*, pp. 47–57, 1984.

28. M. Hadjieleftheriou, G. Kollios, V.J. Tsotras, and D. Gunopulos. Efficient indexing of spatiotemporal objects. In *Proceedings of Seventh International Conference on Extending Database Technology (EDBT'02)*, pp. 251–268, 2002.
29. M. Hadjieleftheriou, G. Kollios, and V.J. Tsotras. Performance evaluation of spatio-temporal selectivity estimation techniques. In *Proceedings of 15th International Conference on Scientific and Statistical Database Management (SSDBM'03)*, pp. 202–211, 2003.
30. M. Hadjieleftheriou, G. Kollios, V.J. Tsotras, and D. Gunopulos. Indexing spatio-temporal archives. *Very Large Data Bases Journal*, 15(2):143–164, 2006.
31. G.R. Hjaltason and H. Samet. Distance browsing in spatial databases. *ACM Transactions Database Systems*, 24(2):265–318, 1999.
32. Y.-W. Huang, N. Jing, and E.A. Rundensteiner. Spatial joins using R-trees: Breadth-first traversal with global optimizations. *The Very Large Data Bases Journal*, 396–405, 1997.
33. G.S. Iwerks, H. Samet, and K. Smith. Continuous k-nearest neighbor queries for continuously moving points with updates. In *Proceeding on 29th International Conference on Very Large Data Bases (VLDB'03)*, pp. 512–523, 2003.
34. J.A.C. Lema, L. Forlizzi, R.H. Güting, E. Nardelli, and M. Schneider. Algorithms for moving objects databases. *The Computer Journal*, 46(6):680–712, 2003.
35. Y. Manolopoulos, A. Nanopoulos, A. Papadopoulos, and Y. Theodoridis. *Rtrees: Theory and Applications*. Springer, Berlin Heidelberg New York, 2005.
36. N. Meratnia and R.A. de By. Spatiotemporal compression techniques for moving point objects. In *Nineth International Conference on Extending Database Technology (EDBT'04)*, pp. 765–782, 2004.
37. K. Mouratidis, D. Papadias, and M. Hadjieleftheriou. Conceptual partitioning: An efficient method for continuous nearest neighbor monitoring. In *Proceedings of the International Conference on Management of Data(SIGMOD'05)*, pp. 634–645, 2005.
38. M. Nanni. *Clustering Methods for Spatio-Temporal Data*. Ph.D. thesis, Computer Science Department, University of Pisa, 2002.
39. Oracle ®database documentation library, 10g release 1 (10.1), 2006.
40. J. O'Rourke. *Computational Geometry in C*. Cambridge University Press, Camridge (NY), 1998.
41. T. Palpanas, M. Vlachos, E.J. Keogh, D. Gunopulos, and W. Truppel. Online amnesic approximation of streaming time series. In *Proceedings of the 20th International Conference on Data Engineering*, pp. 338–349, 2004.
42. N. Pelekis. *STAU: A spatio-temporal extension to ORACLE DBMS, Ph.D., UMIST*. Ph.D. thesis, 2002.
43. N. Pelekis and Y. Theodoridis. Boosting location-based services with a moving object database engine. In *Proceedings of Workshop on Data Engineering for Wireless and Mobile Access (MobiDE'06)*, pp. 3–10, 2006.
44. N. Pelekis, Y. Theodoridis, S. Vosinakis, and T. Panayiotopoulos. Hermes – a framework for location-based data management. In *11th International Conference on Extending Database Technology (EDBT'06)*, pp. 1130–1134, 2006.
45. N. Pelekis, B. Theodoulidis, Y. Theodoridis, and I. Kopanakis. An Oracle data cartridge for moving objects. Technical report, University of Piraeus, 2005.
46. D. Pfoser and C.S. Jensen. Capturing the uncertainty of moving-object representations. In *Proceedings of Symposium on Advances in Spatial Databases (SSD'99)*, pp. 111–132, 1999.
47. D. Pfoser and C.S. Jensen. Querying the trajectories of on-line mobile objects. In *Proceedings of Workshop on Data Engineering for Wireless and Mobile Access (MobiDE'01)*, pp. 66–73, 2001.
48. D. Pfoser and C.S. Jensen. Indexing of network constrained moving objects. In *Proceedings of the 11th Annual ACM International Workshop on Geographic Information Systems (GIS'03)*, pp. 25–32, 2003.
49. D. Pfoser, C.S. Jensen, and Y. Theodoridis. Novel approaches in query processing for moving object trajectories. In *Proceeding on 26th International Conference on Very Large Data Bases (VLDB'00)*, pp. 395–406, 2000.

50. M. Potamias, K. Patroumpas, and T.K. Sellis. Amnesic online synopses for moving objects. In *Proceedings of Conference on Information and Knowledge Management (CIKM'06)*, 2006.
51. M. Potamias, K. Patroumpas, and T.K. Sellis. Sampling trajectory streams with spatiotemporal criteria. In *Proceedings of 18th International Conference on Scientific and Statistical Database Management (SSDBM'06)*, pp. 275–284, 2006.
52. C.M. Procopiuc, P.K. Agarwal, and S. Har-Peled. Star-tree: An efficient self-adjusting index for moving objects. In *Proceedings of the Fourth Workshop on Algorithm Engineering and Experiments (ALENEX'02)*, pp. 178–193, 2002.
53. S. Rasetic, J. Sander, J. Elding, and M.A. Nascimento. A trajectory splitting model for efficient spatio-temporal indexing. In *Proceeding on 31st International Conference on Very Large Data Bases (VLDB'05)*, pp. 934–945, 2005.
54. N. Roussopoulos, S. Kelley, and F. Vincent. Nearest neighbor queries. In *Proceedings of the International Conference on Management of Data (SIGMOD'95)*, pp. 71–79, 1995.
55. S. Saltenis and C.S. Jensen. Indexing of moving objects for location-based services. In *Proceedings of the 18th International Conference on Data Engineering*, pp. 463–472, 2002.
56. S. Saltenis, C.S. Jensen, S.T. Leutenegger, and M.A. Lopez. Indexing the positions of continuously moving objects. In *Proceedings of the International Conference on Management of Data (SIGMOD'00)*, pp. 331–342, 2000.
57. Z. Song and N. Roussopoulos. K-nearest neighbor search for moving query point. In *Proceedings of the Fourth International Symposium on Spatial and Temporal Databases*, pp. 79–96, 2001.
58. Y. Tao and D. Papadias. Time-parameterized queries in spatio-temporal databases. In *Proceedings of the International Conference on Management of Data (SIGMOD'02)*, pp. 334–345, 2002.
59. Y. Tao and D. Papadias. Performance analysis of R*-trees with arbitrary node extents. *IEEE Transactions on Knowledge and Data Engeneering*, 16(6):653–668, 2004.
60. Y. Tao, D. Papadias, and Q. Shen. Continuous nearest neighbor search. In *Proceeding on 28th International Conference on Very Large Data Bases (VLDB'02)*, pp. 287–298, 2002.
61. Y. Tao, J. Sun, and D. Papadias. Selectivity estimation for predictive spatio-temporal queries. In *Proceedings of the 19th International Conference on Data Engineering (ICDE '03)*, pp. 417–428, 2003.
62. Y. Tao, D. Papadias, and J. Sun. The TPR*-tree: An optimized spatio-temporal access method for predictive queries. In *Proceeding on 29th International Conference on Very Large Data Bases (VLDB'03)*, pp. 790–801, 2003.
63. Y. Tao, G. Kollios, J. Considine, F. Li, and D. Papadias. Spatio-temporal aggregation using sketches. In *Proceedings of the 20th International Conference on Data Engineering*, pp. 214–226, 2004.
64. Y. Theodoridis. Ten benchmark database queries for location-based services. *The Computer Journal*, 46(6):713–725, 2003.
65. Y. Theodoridis and T.K. Sellis. A model for the prediction of R-tree performance. In *Proceedings of Symposium on Principles of Database Systems (PODS'96)*, pp. 161–171, 1996.
66. Y. Theodoridis, M. Vazirgiannis, and T.K. Sellis. Spatio-temporal indexing for large multimedia applications. In *Proceedings of IEEE International Conference on Multimedia Computing and Systems (ICMCS'96)*, pp. 441–448, 1996.
67. Y. Theodoridis, E. Stefanakis, and T.K. Sellis. Cost models for join queries in spatial databases. In *Proceedings of the 14th International Conference on Data Engineering (ICDE '98)*, pp. 476–483, 1998.
68. G. Trajcevski. Probabilistic range queries in moving objects databases with uncertainty. In *Proceedings of Workshop on Data Engineering for Wireless and Mobile Access (MobiDE'03)*, pp. 39–45, 2003.
69. G. Trajcevski, O. Wolfson, K. Hinrichs, and S. Chamberlain. Managing uncertainty in moving objects databases. *ACM Transactions on Database System*, 29(3):463–507, 2004.
70. M. Vlachos, G. Kollios, and D. Gunopulos. Discovering similar multidimensional trajectories. In *Proceedings of the 18th International Conference on Data Engineering (ICDE '02)*, pp. 673–684, 2002.

71. C. Wenk, R. Salas, and D. Pfoser. Addressing the need for map-matching speed: Localizing globalb curve-matching algorithms. In *Proceedings of 18th International Conference on Scientific and Statistical Database Management (SSDBM'06)*, pp. 379–388, 2006.
72. O. Wolfson, A.P. Sistla, S. Chamberlain, and Y. Yesha. Updating and querying databases that track mobile units. *Distributed and Parallel Databases*, 7(3):257–387, 1999.
73. M. Worboys and M. Duckham. *GIS: A Computing Perspective*, 2nd edn. CRC Press, Florida, 2004.
74. Y. Xia and S. Prabhakar. Q+rtree: Efficient indexing for moving object database. In *Proceedings of The Eighth International Conference on Database Systems for Advanced Applications (DASFAA'03)*, pp. 175–182, 2003.
75. X. Xiong, M.F. Mokbel, and W.G. Aref. Sea-cnn: Scalable processing of continuous k-nearest neighbor queries in spatio-temporal databases. In *Proceedings of the 21th International Conference on Data Engineering (ICDE '05)*, pp. 643–654, 2005.
76. H. Yin and O. Wolfson. A weight-based map matching method in moving objects databases. In *Proceedings of 16th International Conference on Scientific and Statistical Database Management (SSDBM'04)*, pp. 437–438, 2004.
77. X. Yu, K.Q. Pu, and N. Koudas. Monitoring k-nearest neighbor queries over moving objects. In *Proceedings of the 21th International Conference on Data Engineering (ICDE '05)*, pp. 631–642, 2005.
78. H. Zhu, J. Su, and O.H. Ibarra. Trajectory queries and octagons in moving object databases. In *Proceedings of Conference on Information and Knowledge Management (CIKM'02)*, pp. 413–421, 2002.

Chapter 7
Towards Trajectory Data Warehouses

N. Pelekis, A. Raffaetà, M.-L. Damiani, C. Vangenot, G. Marketos, E. Frentzos, I. Ntoutsi, and Y. Theodoridis

7.1 Introduction

Data warehouses have received the attention of the database community as a technology for integrating all sorts of transactional data, dispersed within organisations whose applications utilise either legacy (non-relational) or advanced relational database systems. Data warehouses form a technological framework for supporting decision-making processes by providing informational data. A data warehouse is defined as a subject-oriented, integrated, time-variant, non-volatile collection of data in support of management of decision-making process [10].

In a data warehouse, data are organised and manipulated in accordance with the concepts and operators provided by a multi-dimensional data model that views data in the form of a data cube [1]. A data cube allows data to be modelled and viewed in multiple dimensions, where each dimension represents some business perspective, and is typically implemented by adopting a star (or snowflake) schema model. According to this model, the data warehouse consists of a fact table (schematically, at the centre of the star) surrounded by a set of dimensional tables related with the fact table, which contains keys to the dimensional tables and measures. A single entry in the fact table modelling the primitive analysis component is called *fact*.

Dimensions represent the analysis axes, while measures are the variables being analysed over the different dimensions. For example, in the marketing domain a kind of measure is the amount of sales and dimensions may be time, location and product. Under these assumptions, the data warehouse stores the amount of sales for a given product in a given region and over a given period of time. Each dimension is organised as a hierarchy (or even a set of hierarchies) of dimension levels, each level corresponding to a different granularity for the dimension. For example, year is one level of the time dimension, while the sequence <day, month, year>

N. Pelekis
Computer Technology Institute (CTI) and Department of Informatics, University of Piraeus, Greece, e-mail: npelekis@unipi.gr

defines a simple hierarchy of increasing granularity for the time dimension. Finally, the members of a certain dimension level (e.g. the different months for the time dimension) can be aggregated to constitute the members of the next higher level (e.g. the different years). The measures are also aggregated following this hierarchy by means of an aggregation function.

Data warehouses are optimised for online analytical processing (OLAP) operations. Typical OLAP operations include the aggregation or de-aggregation of information (called roll-up and drill-down, respectively) along a dimension, the selection of specific parts of a cube (slicing and dicing) and the re-orientation of the multi-dimensional view of the data on the screen (pivoting) [31]. Data warehouses following the paradigm of multi-dimensional data modelling have been widely investigated for conventional, non-spatial data. There is some initial research on spatial data warehousing where dimensions are categorised in three different types: descriptive (or thematic), temporal and spatial [2]. However, spatiotemporal data warehousing is still in its infancy. In this chapter, as a special case of the spatiotemporal domain, we focus on warehouses fed by time-dependent location data describing movements of objects (i.e. trajectories).

The motivation behind Trajectory Data Warehouses (TDWs) is to transform raw trajectories to valuable information that can be utilised for decision-making purposes in ubiquitous applications, such as location-based services, traffic control management, etc. Intuitively, the high volume of raw data produced by sensing and positioning technologies, the complex nature of data stored in trajectory databases, as well as the intricate and specialised query processing demands even for simple user queries make it a hard task to extract valuable information from them. As such, the idea is to extend traditional aggregation techniques so as to produce summarised trajectory information and provide OLAP style analysis.

The abundance of applications that would benefit from such a framework should be mentioned. Let us consider the application domain of supervision systems monitoring the road traffic (or else the movements of users) in a city and providing specialised location-aware services to their subscribers. Analysts, decision makers in the field as well as end users would have the advantage of prompt response to queries like 'what is the total number of users moving inside a district covered by a particular set of cells at a given temporal interval?' (here the important issue is to count the number of users rather than getting their ids), or 'which road has the highest traffic within a distance of 1 km from each hospital?', or given an emergency call, 'which is the nearest police station taking into account the current traffic condition?', or 'is there a substantial difference in the average speed of vehicles visiting downtown during weekends?'

Each of the above-mentioned queries can be answered by legacy systems; however, the computation cost and as such the response time is prohibitive for either real-time services or proper analysis of the application domain. Existing approaches covered in Sect. 7.2 try to provide working solutions to the problem by adopting ideas coming from the paradigm of spatial and spatiotemporal data warehouses. However, trajectory warehouses need a different approach for reasons that are described in Sect. 7.3. The requirements for modelling and constructing TDWs presented in that section implicitly disclose the next steps towards the development of efficient and robust TDWs, a road map for which is presented in Sect. 7.4 along with further issues that need to be addressed.

7.2 Preliminaries and Related Work

Research on extracting semantically rich information from raw space–time dependent data has focused on spatial and spatiotemporal data warehouses. As we would like to treat trajectory warehouses as a branch of spatiotemporal warehousing, the two subsequent sections present existing approaches in the area categorising the research efforts into, on the one hand, conceptual and logical modelling methodologies, and, on the other hand, implementation issues regarding aggregation techniques as the quintessence of the data warehousing concept.

7.2.1 Modelling

Research on spatial data warehouses (SDWs) is relatively recent. Since the pioneering work of Han et al. [8], several models have been proposed in the literature, aiming at extending the classical data warehouse models with spatial concepts and the OLAP tools with spatial operators (SOLAP). Despite the complexity of spatial data, current SDWs typically contain objects with simple geometric extents. Moreover, while a SDW model is assumed to consist of a set of representation concepts and an algebra of SOLAP operators for data navigation, aggregation and visualisation approaches proposed in literature often privilege either the concepts or the algebra; approaches which address both are rare.

Further, research in SDW modelling can be considered as addressing application requirements at either the logical or the conceptual data level. Mainstream solutions rely on the (logical level) relational data model [2, 25]. Relatively few developments focus on SDW conceptual aspects [3, 11, 14, 30]. The analysis presented in [22] asserts the moderate interest of the research community in conceptual multi-dimensional modelling. However, a significant percentage of data warehouses fail to meet their business objectives [22]. A major reason for failure is the poor or inappropriate design, mainly due to a lack of established DW design methods [23] and DW conceptual data models [23]. Similarly, the authors of [15] state that the proposed models either provide a graphical representation based on the E-R model or UML notations with few formal definitions, or provide formal definitions without any user-oriented graphical support.

Focusing on spatial modelling, it's a fact that existing approaches do not rely on standard data models for the representation of the spatial aspects. The spatiality of facts is commonly represented through a geometric element, instead of an open geospatial consortium (OGC) spatial feature, i.e. an object that has a semantic value in addition to its spatial characterisation [16].

Extending classical DW models to deal with spatial data requires allowing both dimensions and measures to hold spatial and topological characteristics. Indeed, dimensions and measures should be extended with spatiality in order to enrich the query formulation and the visualisation of the results. However, adding spatiality to both dimensions and measures is not enough. SDWs have further specific requirements that have been studied in the state of the art, such as different kinds of spatial dimensions and measures, multiple hierarchies in dimensions, partial containment relationships between dimensions levels, non-normalised hierarchies, many-to-many relationships between measures and dimensions and the modelling of measures as complex entities [2, 3, 11].

7.2.1.1 Spatial Dimensions

When adding spatiality to dimensions, most of the proposals follow the approaches by Stefanovic et al. [25] and Bédard et al. [2], which distinguish three types of dimension hierarchies based on the spatial references of the hierarchy members: non-geometric, geometric-to-non-geometric and fully geometric spatial dimensions. The non-geometric spatial dimension uses nominal spatial reference (e.g. name of cities and countries) and is treated as any other descriptive dimension [20, 21]. The two other types denote dimensions where the members of lower or all levels have an associated geometry. In the fully geometric spatial dimension, all members of all the levels are spatially referenced while in the geometric-to-non-geometric spatial dimension, members are spatially referenced up to a certain dimension level and then become non-geometric. More loosely, Malinowski et al. [14] extend this classification and consider that a dimension can be spatial even in the absence of several related spatial levels. In their proposal, a spatial level is defined as a level for which the application needs to keep its spatial characteristics, meaning its geometry, as this is represented by standard spatial data types (e.g. points, regions). This allows them to link the spatial levels of a dimension through topological relationships that exist between the spatial components of their members (contains, equals, overlaps, etc). Based on this, they define a spatial hierarchy as a hierarchy that includes at least one spatial level. In this connection, a spatial dimension is a dimension that includes at least one spatial hierarchy. As such, a spatial dimension is a dimension that contains at least one spatial level; otherwise it is a thematic dimension. An advantage of this modelling perspective is that different spatial data types are associated with the levels of a hierarchy. For example, assuming the hierarchy *user* $<$ *city* $<$ *county*, point type is associated to *user*, region to *city*, and set of regions to *county*.

Dimensions and their organisation into hierarchies are kept very simple in traditional and operational data warehouses. Levels of traditional non-spatial dimensions are usually organised into containment hierarchies such as *district* $<$ *city* $<$ *county* $<$ *country*. However, when dealing with spatial data, two spatial values may not only be either disjoint or one contained into the other, but they may also overlap. For

7 Towards Trajectory Data Warehouses

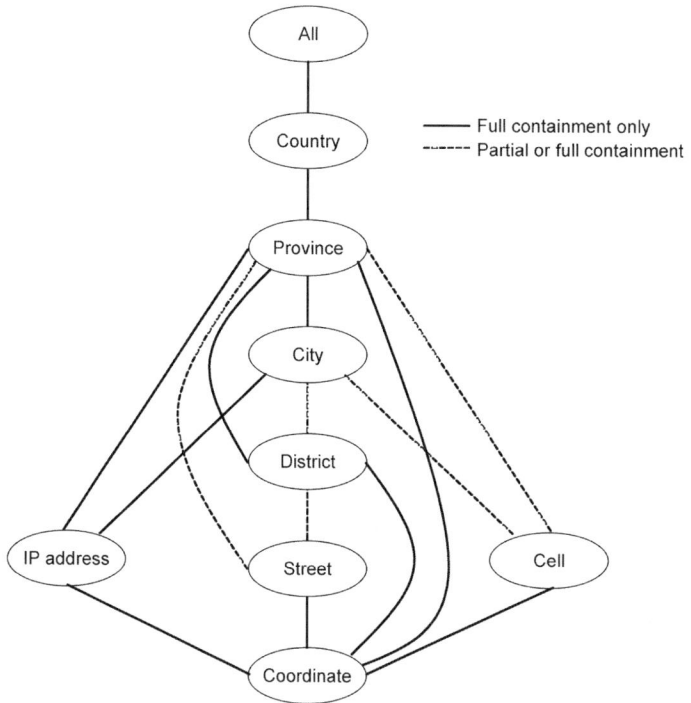

Fig. 7.1 Hierarchy with full and partial containment relationships (from [11])

instance, if we add the dimension level *cell* before the *district* level, a cell might overlap two districts. To address application requirements in a better way, a larger spectrum of possible hierarchies is being explored. Jensen et al. [11] propose a conceptual model that supports dimensions with full or partial containment relationships (see Fig. 7.1). The dimension hierarchies can contain levels that may be linked by full or partial containment relationships. For the members of a level linked by a partial containment relationship to members of another level, the degree of containment must be specified (e.g. 80% of this cell is contained in this district). Support for multiple hierarchies in a single dimension is also an important requirement proposed by the models of Jensen et al. [11] and Malinowski et al. [15]. It means that multiple aggregation paths are possible in a dimension (e.g. cells can be aggregated in districts or directly in counties). According to these models, multiple aggregation paths enable better handling of the imprecision in queries caused by partial containment relationships. Putting this idea into our example, they argue that the result of the aggregation of cells into county may give better results than aggregating cells into district, then into city and then into county. The models of Jensen et al. [11] and Malinowski et al. [15] support non-normalised hierarchy, i.e. hierarchies whose members may have more than one corresponding member at the higher level or no corresponding member (e.g. a cell may be related to two districts whereas a district may be related to no cells). Finally, in the model of

Malinowski et al. [15], simple hierarchies can be characterised as symmetrical (i.e. all levels of the hierarchy are mandatory), asymmetrical, generalised (i.e. including a generalisation/specialisation relationship between dimension members), non-strict (same as non-normalised) and non-covering (i.e. some levels of the hierarchy can be skipped when aggregating).

7.2.1.2 Spatial Measures

Similar to spatial dimensions, when adding spatiality to measures, most of the proposals distinguish two types of spatial measures [20, 21, 25]:

- Spatial measures represented by a geometry and associated with a geometric operator to aggregate it along the dimensions.
- A numerical value obtained using a topological or a metric operator.

When represented by a geometry, spatial measures consist of either a set of coordinates as in [3, 14, 18, 20, 21] or a set of pointers to geometric objects as in [25]. Finally, Bimonte et al. [3] and Malinowski et al. [14] advocate the definition of measures as complex entities. In [3], a measure is an object containing several attributes (spatial or not) and several aggregation functions (eventually ad hoc functions). In a similar way, Malinowski et al. [14] define measures as attributes of an n-ary fact relationship between dimensions. This fact relationship can be spatial, if it links at least two spatial dimensions, and be associated with a spatial constraint such as, for instance, spatial containment.

An important issue related to spatial measures concerns the level of detail they are described with. Indeed, spatial data are often available and described according to various levels of detail: for instance, the same spatial object can be defined as an area according to a precise level of detail and as a point according to a less detailed one. This is of particular importance with trajectories where the position of the objects is subject to imprecision. Damiani et al. [5] propose a model allowing to define spatial measures at different spatial granularities. This model, called MuSD, allows to represent spatial measures and dimensions in terms of OGC features. A spatial measure can represent the location of a fact at multiple levels of spatial granularity. Such multi-granular spatial measures can either be stored or they can be dynamically computed by applying a set of coarsening operators. An algebra of SOLAP operators including special operators that allow the scaling up of spatial measures to different granularities is proposed in [5].

7.2.1.3 Relationships Between Measures and Dimensions

Another requirement highlighted by Jensen et al. [11] and Bimonte et al. [3] concerns relationships between measures and dimensions. Indeed, while most of the models only propose the definition of one-to-one relationships between measures and dimensions, they advocate to define many-to-many relationships, which would allow the association of the same measure with several members of a dimension.

7.2.2 Aggregation Functions and Their Implementation

A related research issue that has recently gained increasing interest and is relevant for the development of comprehensive SDW data models concerns the specification and efficient implementation of the operators for spatial and spatiotemporal aggregation.

Spatial aggregation operations summarise the geometric properties of objects and as such they constitute the distinguishing aspect of SDW. Nevertheless, despite the relevance of the subject, a standard set of operators (like for example the SQL operators SUM, AVG, MIN) has not been defined yet. In fact, when defining spatial, temporal and spatiotemporal aggregates some additional problems have to be faced, which do not show up for traditional data. In particular, while for traditional databases only explicit attributes are of concern, the modelling of the spatial and temporal extent of an object makes use of interpreted attributes and the definition of aggregations is based on granularities.

A first comprehensive classification and formalisation of spatiotemporal aggregate functions is presented by Lopez et al. [12]. The operation of aggregation is defined as a function that is applied to a collection of tuples and returns a single value. To generate the collection of tuples to which the operation is applied, the authors distinguish three kinds of methods: group composition, partition composition and sliding window composition.

Recall that a (temporal/spatial) granularity creates a *discrete* image, in terms of *granules*, of the (temporal/spatial) domain. Given a spatial granularity G^S and a temporal granularity G^T, a *spatiotemporal group composition* forms groups of tuples sharing the same spatial and temporal value at granularity $G^S \times G^T$. An aggregate function can then be applied to each group. On the other hand, *spatiotemporal partition composition* is used when a finer level of aggregation is required and involves at least two granularities. The first one, which is the coarser, defines collections of tuples (the partitions). To each partition, a *sliding window composition* is performed. Instead of generating a single aggregate value for each partition, an aggregate value for every tuple in the collection at the finer granularity is computed. To slide through all tuples in the collection, a spatiotemporal sliding window is used.

In addition to the conceptual aspects of spatiotemporal aggregation, another major issue regards the development of methods for the efficient computation of this kind of operations to manage high volumes of spatiotemporal data. In particular, techniques are developed on the basis of combined use of specialised indexes, materialisation of aggregate measures and computational geometry algorithms, especially to support the aggregation of dynamically computed sets of spatial objects [17, 19, 27, 32]. Papadias et al. [17, 27] propose an approach based on two types of indexes: a *host index*, which manages the region extents and associates to these regions an aggregate information over all the timestamps in the base relation, and some *measure indexes* (one for each entry of the host index), which are aggregate temporal structures storing the values of measures during the history. For a set of static regions, the authors define the *aggregate R-B-tree* (aRB-tree), which

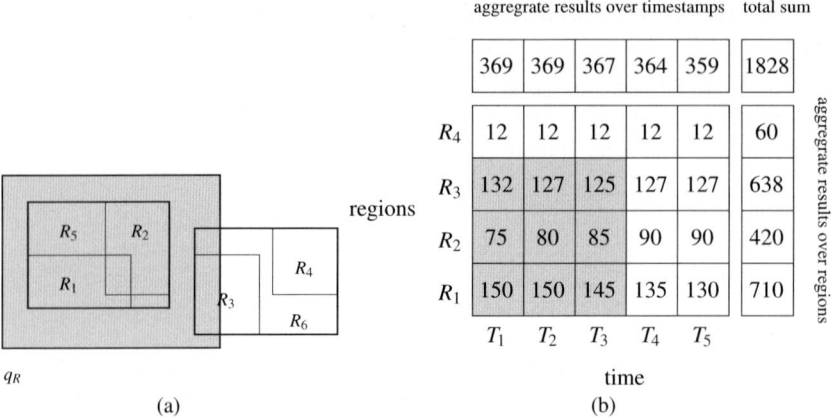

Fig. 7.2 (a) Regions of interest (b) A data cube example

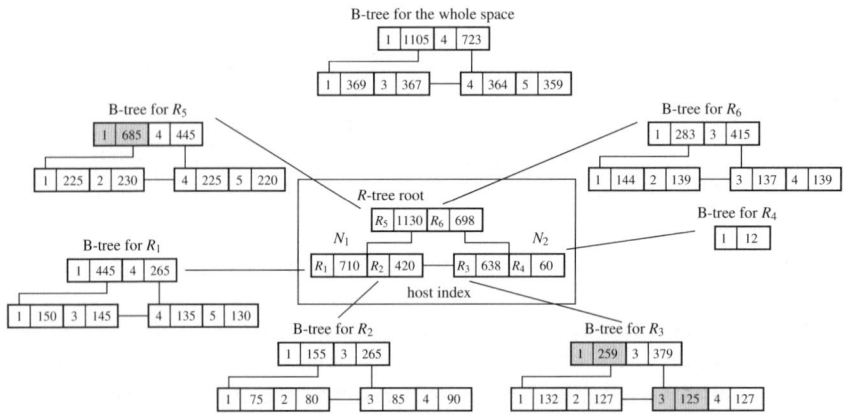

Fig. 7.3 The aRB-tree

adopts an R-tree with summarised information as host index, and a B-tree containing time-varying aggregate data as measure index.

To illustrate this concept, consider the regions R_1, R_2, R_3 and R_4 in Fig. 7.2a and suppose that the number of phone calls initiated in $[T_1, T_5]$ inside such regions is recorded as measure in the fact table depicted in Fig. 7.2b. Then Fig. 7.3 shows the corresponding aRB-tree.

This structure is well suited for the efficient processing of *window aggregate queries*, i.e. for the computation of the aggregated measure of the regions that intersect a given window. In fact, for nodes that are totally enclosed within the query window, the summarised measure is already available, thus avoiding to descend these nodes. As a consequence the aggregate processing is made faster.

For instance, let us compute the number of phone calls inside the shaded area in Fig. 7.2a during the time interval $[T_1, T_3]$. Since R_5 is completely included in the window query there is no need to further explore R_1 and R_2 once one accesses the B-tree for R_5. The first entry of the root of this B-tree contains the measure for the interval $[T_1, T_3]$, which is the value we are interested in. Instead, to obtain the sum of phone calls in the interval $[T_1, T_3]$ for R_3 one has to visit both an entry of the root of the B-tree for R_3 and also one leaf (the coloured nodes in Fig. 7.3).

Tao et al. [26] showed that the aRB-tree can suffer from the *distinct counting problem*, i.e. if an object remains in the query region for several timestamps during the query interval, it will be counted multiple times in the result. To cope with this problem, Tao et al. [26] proposed an approach that combines spatiotemporal indexes with sketches, a traditional approximate counting technique based on probabilistic counting [6]. The index structure is similar to the aRB-tree: an R-tree indexes the regions of interest, whereas the B-trees record the historical sketches of the corresponding region. However, this index differs from aRB-trees in the querying algorithms since one can exploit the pruning power of the sketches to define some heuristics allowing to reduce query time.

Finally, it is worth to mention the work by Shekhar et al. [24], who propose a traffic data warehouse model for the Twin Cities metropolitan area. Although building a warehouse for traffic management is easier than building a warehouse for trajectories (recall here that the main difficulty is that trajectories may extend to more than one cells), several interesting issues are analysed in this work. Of particular interest is the analysis regarding the aggregate functions. More specifically, following the work by Gray et al. [7], the authors classify aggregation functions into three categories:

- *Distributive*, whose values can be computed from the next lower dimension values.
- *Algebraic*, whose values can be computed from a set of aggregates of the next lower level data.
- *Holistic*, which need the base data to compute the results in all levels of dimensions.

For each category, the authors provide representative aggregation operations inspired from the GIS domain (Fig. 7.4), which also seem useful in our case.

Data Type	Aggregation Function		
	Distributive Function	Algebraic Function	Holistic Function
Set of numbers	Count, Min, Max, Sum	Average, MaxN, MinN, Standard Deviation	Median, Rank, MostFrequent
Set of points, lines, polygons	Minimal Orthogonal Bounding Box, Geometric Union, Geometric Intersection	Centroid, Center of mass, Center of gravity	Equi-partition, Nearest neighbour index

Fig. 7.4 Aggregate operations (from [24])

7.3 Requirements for Trajectory Data Warehouses

Extending traditional (i.e. non-spatial), spatial or spatiotemporal models to incorporate semantics driven by the nature of trajectories induces specific requirements as the goal is twofold: to support high-level OLAP analysis and to facilitate knowledge discovery from TDWs. Having in mind that the basic analysis constituents in a TDW (i.e. facts) are the trajectories themselves, in this section, we categorise the identified requirements into *modelling*, *analysis* and *management* requirements. The first considers logical and conceptual level challenges introduced by TDWs, the second goes over OLAP analysis requirements, while the third focuses on more technical aspects.

7.3.1 TDW Modelling Requirements

The following paragraphs investigate the prerequisites and the constraints imposed when describing the design of a TDW from a user perspective (i.e. conceptual model), as well as when describing the final application as a system in a platform-independent tool (i.e. logical model).

7.3.1.1 Thematic, Spatial, Temporal Measures

From a modelling point of view, a trajectory is a spatial object whose location varies in time (recall discussions on the nature of trajectories in Chaps. 1 and 5). At the same time, trajectories have thematic properties that usually are space- and time-dependent. This implies that different characteristics of trajectories need to be described in order to be analysed. As such, we distinguish (a) *numeric* characteristics, such as the average speed of the trajectory, its direction, its duration; (b) *spatial* characteristics, such as the geometric shape of the trajectory and (c) *temporal* characteristics, such as the timing of the movement. Additionally, as we pay particular attention to uncertainty and imprecision issues, a TDW model should include measures expressing the amount of uncertainty incorporated in the TDW due to raw data imprecision. Uncertainty should also be seen in granularities, while this implies that there are special aggregation operators propagating uncertainty to various levels.

In particular, depending on the application and user requirements, several numeric measures could be considered:

1. Number of trajectories found in the cell (or started/ended their path in the cell; or crossed/entered/left the cell).
2. Distance covered by trajectories in the cell.

Other measures could include motion characteristics of the trajectories, e.g. speed and change of speed (acceleration), direction and change of direction (turn), underlying spatial framework characteristics (e.g. network usage, frequency, density) and also the uncertainty associated with the locations of objects in the database. Handling uncertainty, the warehouse could even contain information regarding the quality of raw data (e.g. spatial/temporal tolerance of recordings).

As a final remark about measures, it is worth noticing that even restricting to numeric measures, the complexity of the computation can vary a lot. Some measures require little pre-computation and can be updated in the data warehouse while single observations of the various trajectories arrive, whereas others need a given amount of trajectory observations before updating. Braz et al. [4] propose the following classification of measures according to an increasing amount of pre-calculation effort:

(a) *No pre-computation*: The measure can be updated in the data warehouse by directly using each single observation.
(b) *Per trajectory local pre-computation*: The measure can be updated by exploiting a simple pre-computation, which involves only a few and close observations of the same trajectory.
(c) *Per trajectory global pre-computation*: The measure update requires a pre-computation that considers all the observations of a single trajectory.
(d) *Global pre-computation*: The measure requires a pre-computation that considers all the observations of all the trajectories.

For instance, the number of trajectories starting/ending their path in the cell can be of type (a) if the first/last point of the trajectories are marked; the distance covered by trajectories in the cell, the number of trajectories that entered and left the cell are of type (b); the number of trajectories that covered a total distance larger than a given value d is of type (c) and finally, the number of trajectories that intersect another trajectory only in the cell is of type (d).

The amount of pre-calculation associated with each type of measure has also a strong impact on the amount of memory required to buffer incoming trajectory observations. Note that, since observations may arrive in stream at different rates, and in an unpredictable and unbounded way, small processing time and small memory size are both important constraints.

Similar remarks can be found in [8] where Han et al. present three methods to compute spatial measures in spatial data cube construction. The first one consists of simply collecting and storing the corresponding spatial data but no pre-computation of spatial measures is performed. Hence such a method may require more computation on-the-fly. The second method pre-computes and stores some rough approximation/estimation of the spatial measures in a spatial data cube. For instance, if the measure is the merge of a set of spatial objects, one can store the minimum bounding rectangle (MBR) of the merge of the objects. Finally, one can selectively pre-compute some spatial measures. In this case the question is how to select a set of spatial measures for pre-computation. In [8] some criteria for materialisation of a cuboid are presented.

7.3.1.2 Thematic, Spatial, Temporal Dimensions

Regarding the supported dimensions, as starting point a TDW should support the classic spatial (e.g. coordinate, roadway, district, cell, city, province, country) and temporal (e.g. second, minute, hour, day, month, year) dimensions, describing the underlying spatiotemporal framework wherein trajectories are moving. Additionally, it is important to allow space–time related dimensions interact with thematic dimensions describing other sorts of information regarding trajectories like technographic (e.g. mobile device used) or demographic data (e.g. age and gender of users). This will allow an analyst not only to query the TDW for instance about the number of objects crossed an area of interest but also to be able to identify the objects in question. This is particularly important as in the first case we usually get quantitative information, while in the second case, the information is qualitative. Consequently, a flexible TDW design should include thematic, spatial, temporal and spatiotemporal dimensions.

- Temporal (time)
- Geographical (location)
- Demographics (e.g. gender, age, occupation, marital status, home postal code, work postal code, etc.)
- Technographics (e.g. mobile device, GPRS-enabled, subscriptions in special services, etc.)

Regarding the technographics and demographics dimensions, the idea behind them is to enhance the warehouse with semantic information.

An issue concerning the definition of dimensions is the considered level of detail for each dimension. Consider the spatial dimension: Since a trajectory is actually a set of sampled locations in time, for which the in-between positions are calculated through some kind of interpolation, the lowest-level information is that of spatial coordinates. This, however, implies a huge discretisation of the spatial dimension, thus more generic approaches should be followed. For example, instead of point positions cell positions could be used.

7.3.1.3 Hierarchies on Dimensions

Once having defined the dimensions, hierarchies on dimensions can be specified by users or generated automatically by data clustering or data analysis techniques. A general technique used to define hierarchies consists of discretising the values the dimension ranges over, resulting in a set-grouping hierarchy. A partial order can thus be established among these groups of values. Let us now analyse the different proposals and difficulties in creating hierarchies for the dimensions suggested in the previous subsection.

Defining hierarchies over the time dimension is straightforward, since typically there is an obvious ordering between the different levels of the hierarchy. For instance, a potential hierarchy could be *Year* > *Quarter* > *Month* > *Day* > *Hour* >

Minute > Second. Other hierarchies over the time dimension could concern seasons, time zones, traffic jam hours, and so on.

On the other hand, creating hierarchies over spatial data is more complicated as we have already discussed in Sect. 7.2. In fact, non-explicitly defined hierarchies might exist over the spatial data. For example, in the hierarchy *Country > City > District > Cell > Road*, it is not always the case that an inclusion relation holds between *District* and *Cell* and between *Cell* and *Road*. A *Road* value, for example, might cross more than one *Cell* value. To solve this problem, Jensen et al. [11] proposed a conceptual model, which supports dimensions with full or partial containment relationships. Thus, when a partial containment relationship exists between the different levels of a dimension, one should specify the degree of containment, e.g. 80% of this *Road* is covered by this *Cell*.

Besides the standard relation *City < Country*, further hierarchies could be defined over the spatial dimension depending on the application, e.g. a set-grouping hierarchies on districts according to the pollution.

Finally, as far as the demographic and technographic dimensions are concerned, the simplest solution is to create a hierarchy for each dimension. This solution, however, might cause complexity problems especially if the number of the dimensions considered is large. Another possibility is to combine attributes of these dimensions by creating groups of dimensions values and use these groups as the levels of abstraction. As an example of such a group consider the following one: {*gender = female, age = 25–35, marital status = single*}. The group definition could be performed by a domain expert or by carrying out some statistical pre-processing over the data. This approach reduces the number of dimensions, thus allowing for a simpler and more efficient data warehouse in terms of processing time and storage requirements.

As mentioned in Sect. 7.2, some recent approaches [11, 14] offer the support for creating multiple hierarchies for each dimension. This is an interesting topic to be investigated for the TDW case.

Collecting all these features in a single framework, we can depict an example of a general star schema for a TDW, having four dimensions (and their implied hierarchies), as shown in Fig. 7.5.

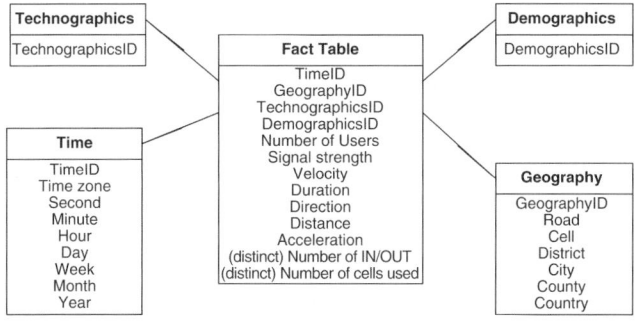

Fig. 7.5 An example of a star schema for TDW

7.3.2 OLAP Requirements

In traditional data warehouses, data analysis is performed interactively by applying a set of OLAP operators. In spatial data warehousing, particular OLAP operators have been defined to tackle the specificities of the domain [17]. Similarly, in our context, we expect an algebra of OLAP operators to be defined for trajectory data analysis. Such an algebra should include not only the traditional operators, such as roll-up, drill-down and selection properly tailored to trajectories, but also additional operators that account for the specificity of the spatiotemporal data type. Below we present these operators in more detail.

Roll-up. The roll-up operation allows us to navigate from a detailed level of abstraction to a more general level either by climbing up the concept hierarchy (e.g. from the level of 'city' to the level of 'country') or by some dimension reduction (e.g. by ignoring the 'time' dimension and performing aggregation only over the 'location' dimension).

As shown in [4], depending on the kind of analysed measures, the roll-up operation in TDWs can introduce some errors. For example, consider the TDW illustrated in Fig. 7.6, with the number of distinct trajectories in a spatiotemporal cell as measure. Assuming the object or trajectory identifier is not recorded, when summing up along the spatial and/or temporal dimension, one cannot obtain the *distinct* number of trajectories because there is only an aggregated information. This is a particular case of the already discussed distinct counting problem.

Another open issue concerns the application of the roll-up operation when uncertain data exist, which is the case for the trajectories. Indeed, two factors of uncertainty should be taken into account during the aggregation: the uncertainty in the values and the uncertainty in the relationships. The former refers to the uncertainty associated with the values of the dimensions and measures, which is propagated into the warehouse from the source data. The latter refers to the uncertainty imposed into the warehouse due to the non-explicitly defined concept hierarchies.

Fact Table

Xinterval	Yinterval	Tinterval	NTrajs
[0, 30)	[0, 30)	[0, 30)	200
[0, 30)	[0, 30)	[30, 60)	150
...	...	...	...

X	Y	T
Interval	Interval	Interval
[0, 30)	[0, 30)	[0, 30)
[30, 60)	[30, 60)	[30, 60)
...	...	...

Fig. 7.6 An example of a TDW

Drill-down. The drill-down operation is the reverse of the roll-up operation. It allows us to navigate from less detailed to more detailed data by either stepping down a concept hierarchy for a dimension (e.g. from the level of 'country' to the level of 'city') or by introducing additional dimensions (e.g. by considering not only the 'location' dimension but the 'time' dimension also). Similarly to the roll-up operation, drill-down is also 'sensitive' to the distinct counting problem and to the uncertainty associated with both values and relationships.

Slice, Dice. The slice operation performs a selection over one dimension (e.g. 'city = Athens'), whereas the dice operation involves selections over two or more dimensions (e.g. 'city = Athens and year = 2006'). The conditions can involve not only numeric values but also more complex criteria, like spatial or temporal query window. To support these operations, the selection criteria can be transformed into a query against the TDW and processed by adequate query processing methods.

In summary, traditional OLAP operations should be also supported by a TDW since they provide meaningful information. An open issue is whether other operations dedicated to trajectories exist. Examples include the following:

- Fold/unfold operators that dynamically modify the spatiotemporal granularity of measures representing trajectories.
- Operators, like medoid, which apply advanced aggregation methods, such as clustering of trajectories to extract representatives from a set of trajectories.
- Operators to propagate/aggregate uncertainty and imprecision present in the data of the TDW.

7.3.3 Management Requirements

The previous sections disclosed higher level requirements for TDWs as these can be captured by extended conceptual and logical data warehouse models. In this section we investigate the management requirements of a TDW from an implementation point of view, but still without restricting the discussion under a specific physical modelling framework.

7.3.3.1 ETL: Issues and Support for Continuous Data Streams

Having as main objective to build a data warehouse specialised for trajectories and considering the complexity and the vast volumes of trajectory data, we need to differentiate our architectural design from the one in traditional DWs. The situation is made even more complicated by the streaming nature of data sources such as logs from location-aware communication devices, which potentially come in continuous flows of unbounded size. Therefore, efficient and effective storage of the trajectories into the warehouse should be devised, capable of dealing with continuous incoming streams of raw log data, while the TDW itself must be equipped with suitable access methods to facilitate analysis and mining tasks. This poses extra challenges

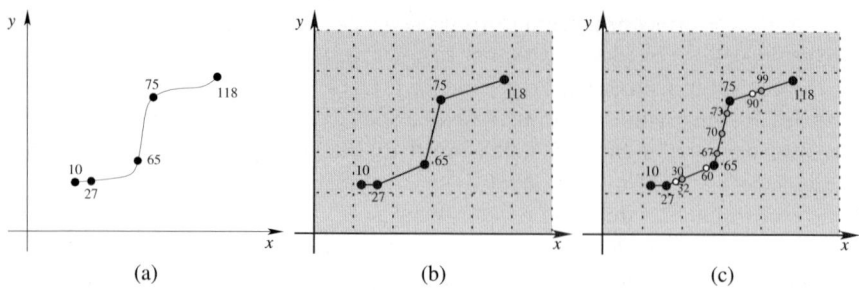

Fig. 7.7 (a) A 2D trajectory, with a sampling (b) Linear interpolation of the trajectory (c) The interpolated trajectory with the points matching the spatial and temporal minimum granularity

Table 7.1 A simple fact table for a trajectory warehouse

Time label	X interval	Y interval	T interval	N Trajs
10,27	[30,60)	[30,60)	[0,30)	2
65	[60,90)	[30,60)	[60,90)	1
75	[90,120)	[90,120)	[60,90)	1
118	[120,150)	[90,120)	[60,120)	1

to be solved as the ability of incrementally processing the data stream in an efficient and accurate way and the definition of adaptive strategies to make the hypercubes evolve with the data stream.

Also, because of the peculiarities of trajectories, some problems can arise in the loading phase of the fact table. To give an intuitive idea of these issues, consider a TDW where the facts are still the trajectories, but having only the spatial and temporal dimensions, discretised according to a regular grid, and as measure the number of distinct trajectories in the spatiotemporal cell, generated by the grid (see Fig. 7.6).

Moreover, assume that a trajectory is modelled as a finite set of *observations*, i.e. a finite subset of points taken from the actual continuous trajectory, later called *sampling*. For example, Fig. 7.7a shows a sampling of a trajectory.

The main issues are the following:

- The rough observations in a sampling cannot be directly used to compute the measures of interest in a correct way.
- These observations are not independent points; the fact that they belong to the same trajectory has to be exploited when computing some measures.

For instance, loading the fact table with the points in Fig. 7.7b results in Table 7.1. Notice that the first column of the table does not belong to the fact table, it is used to clarify which observations fall in the spatiotemporal cell.

It is evident that other cells might be crossed by the trajectory (e.g. the cell $[60, 90) \times [60, 90) \times [60, 90)$), meaning that some information can be missing. On the other hand, the same cell can contain more than one observation; the computed measure is not correct because it does not store the number of *distinct* trajectories (see the cell $[30,60) \times [30,60) \times [0,30)$).

To solve the first problem, Braz et al. [4] propose further intermediate points to be added by linearly interpolating the trajectory. The newly inserted points are the ones that intersect the borders of the spatiotemporal cell, considering all its three dimensions. Figure 7.7c shows the resulting interpolated points as white and gray circles. Note that the white interpolated points, associated with temporal labels 30, 60 and 90, have been added to match the granularity of the temporal dimension. In fact, they correspond to cross points of the temporal border of the 3D cell. On the other hand, the gray points, labeled with 32, 67, 70, 73 and 99, have been instead introduced to match the spatial dimensions. They correspond to the cross points of the spatial borders of some 3D cell or, equivalently, the cross points of the spatial 2D squares depicted in Fig. 7.7c.

The second problem concerning duplicates is more complex and an approach to cope with it is presented in Sect. 7.2. A thorough discussion about errors in the computation of different measures related to the described issues can be found in [4].

7.3.3.2 Support for Multiple Spatial Topologies

Certainly, a factor that characterises a TDW is the interrelationship between the development of the trajectories upon various possible spatial topologies represented by corresponding spatial dimensions. The base level partitioning of a spatial topology directly affects the multi-dimensional analysis of trajectories. Possible available topologies may be simple grids (e.g. artificial partitioning), complex polygonal amalgamations (e.g. suburbs of a city), real road networks and mobile cell networks. The first case is the simplest one as the space is divided in explicitly defined areas of a grid and thus it is easy to allocate trajectory points in specific areas. However, counting the number of objects that passed from an area may be proved hard for a TDW. This is because sampling frequency may not help in representing the actual trajectory [4]. Thus, it may be necessary to reconstruct the trajectory (as an ETL task) to add intermediate points between the sampling data (see Fig. 7.7c). The same problem stands for the general case of arbitrary polygons. In case of road networks, trajectories should be reconstructed so as to be network constrained, whereas managing cells is a more complex problem because the areas covered by cells may change from time to time depending on the signal strength of the base stations of the provider. Whatever the base of the spatial dimension relating with the trajectories, all spatial topologies are subject to the *distinct counting problem* [26] previously mentioned. Obviously, the reconstruction of the trajectories and the multiple counts of an object moving inside a region is straightforwardly dependent on the interpolation (e.g. linear, polynomial) used (if any) by the corresponding trajectory data model. The above discussion implies that an analyst has the ability firstly to analyse a bunch of trajectories according to a population thematic map and at a secondary level according to the road network of the most populated area.

7.4 Modelling and Uncertainty Issues

So far, we have addressed the modelling and operational requirements for building TDWs. In this section we present some further open issues that should be tackled, focusing on conceptual modelling and multiple trajectory representation issues as well as uncertainty handling.

7.4.1 Conceptual Modelling

While it is universally recognised that a data warehouse leans on a multi-dimensional model, little is said about how to carry out its conceptual design starting from the set of user requirements [23]. The domain of conceptual design for multi-dimensional modelling is still at a research stage. The analysis presented in [22] shows the little interest of the research community in conceptual multi-dimensional modelling. As stated by Malinowski et al. in [15] the proposed models either provide a graphical representation based on the E-R model or UML notations with few formal definitions or only provide formal definitions without any user-oriented graphical support. Considering spatial conceptual models for data warehouse, even fewer conceptual models have been proposed [3, 14, 21, 30] and as far as we know no conceptual models for TDW exists. Indeed as already presented in Sect. 7.2, we believe that TDWs can not be dealt with simply modelling trajectory data in a classical SDW: A TDW is more than a specific application. A conceptual model for a TDW entails particular requirements and among them, the following still comprise open issues:

- *Formal model.* A conceptual model for a TDW should be both spatial and temporal and it should rely on formal definitions [15].
- *Complex types.* Measures should be able to be defined as *complex types*. Indeed proposals for conceptual modelling of data warehouse very often remain very close to the star model where a fact table contains all the simple attributes to be analysed. However, in real world applications, properties are complex (compound, multi-valued). As by definition conceptual models are close to the way users perceive an application domain, the concepts they propose should reflect this. This need has been highlighted for traditional SDW by Bimonte et al. [3] and is particularly important for trajectories that are complex objects with many spatial, temporal and thematic characteristics to analyse.
- *Hierarchies.* Different kinds of hierarchies appear in real world applications and users should be able to describe them with adapted modelling concepts. Interesting works about complex hierarchies exist [11, 13] but yet no consensus has been reached.

7.4.2 Effects of Multiple Representation of Trajectories

While multiple representation has received a lot of attention in the spatiotemporal database community [9], multiple representation in trajectory modelling and particularly TDW is an open issue, as only few proposals exist tackling only spatial data [5]. Multiple representation means that we want to store and/or be able to retrieve several representations for the same trajectory. This may result from the description of the same trajectory according to different viewpoints but also, and more importantly, according to different spatial and temporal granularities (see Fig. 7.8). Granularity here refers to the notion that the world is perceived at different level of details, i.e. in the temporal aspect using more or less time steps and in the spatial aspect considering more or less location details. For instance, consider the trajectory of a person travelling from home in Lausanne to work in Geneva, some tasks might be only interested in analysing the trajectory from the starting point in Lausanne to the arrival point in Geneva and then use a coarse spatial and temporal granularity. On the contrary, another task might need a more detailed description: at 7.40 a.m. the person leaves home to walk to the bus station, then takes the bus to the train station where she/he waits 5 min, then she/he travels for 30 min and so on. In this example, the same trajectory is described at different levels of spatial and temporal granularity. The TDW has to provide a means to describe both of them as two representations of the same trajectory as well as the corresponding levels of granularity. Another case is when the same trajectory has only one representation but that includes different parts at different granularities: for instance, a detailed description of the trajectory between the person's home and the train station will be kept, while from Lausanne train station to Geneva train station no specific detail is necessary. Here the data warehouse has to be aware of the different granularities.

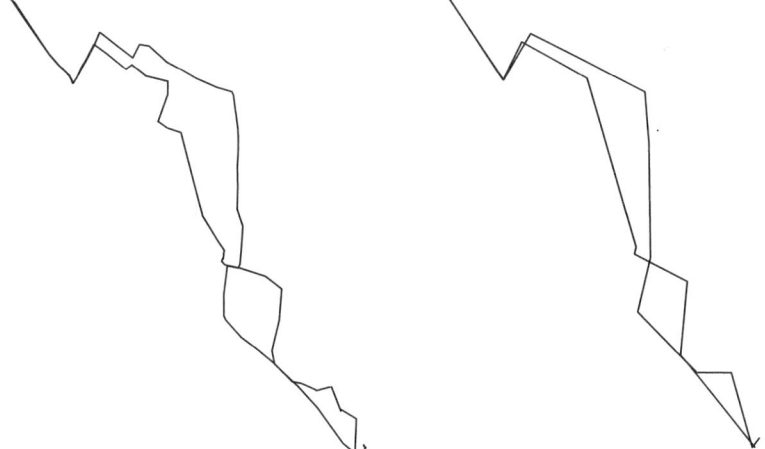

Fig. 7.8 A trajectory in different spatial and temporal granularities

The data warehouse has to provide for multiple representation concepts and conversion operations to shift between the multiple granularities.

7.4.3 Uncertainty Issues

As it has been already discussed in previous chapters, the recorded location of a moving object is rather imprecise due to several factors like GPS erroneous measurements and sampling errors. Since the data warehouse is built upon these data, it is obvious that information lying in the data warehouse is also subject to the uncertainty factor.

An interesting direction in the management of uncertainty in TDWs is to determine the way the locational uncertainty introduced in moving objects from the measurement and sampling errors propagates to the aggregate information stored in the data warehouse. Consider for example Fig. 7.9, illustrating a set of trajectories sliced across the temporal dimension so as to produce a set of points and a rectangular space partition used for aggregation. Adopting the uncertainty model introduced by Trajcevski et al. [29], this set of objects can be represented as a set of *points* along with the respective *uncertainty circles* inside which the actual point would be found. Then, considering the aggregation over the number of objects contained in each bucket (i.e. a cell of the data cube), along with this number of objects, the aggregation would also contain information about the percentage of objects reported inside, nevertheless being outside each bucket (and vice versa), applying a form of probabilistic query as the one discussed in [28]. For example, while bucket B_1 would be reported to contain three points, there is also the possibility to contain five points, since the uncertainty circles of p_1 and p_2 overlap B_1 – thus p_1 and p_2 might be possibly contained inside it.

Another preliminary conclusion gathered from Fig. 7.9 is that objects being spatially far from the bucket's boundary cannot contribute to the uncertainty introduced in the aggregation over the space partitioning. Consider for example point p_3 whose uncertainty circle does not intersect the bucket's boundary; then p_3 will contribute in the number of objects located in B_1 only. On the other hand, while points p_1 and p_2 are still counted in the aggregation of B_2, the possibility of them being

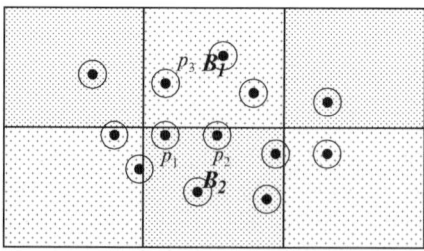

Fig. 7.9 Setting of uncertainty problem

located in B_1 exists. Such an observation leads to the conclusion that only points intersecting the *boundary* of the space partitioning may contribute to the overall aggregated uncertainty; therefore, the propagation from the actual to the aggregated data uncertainty depends on the size – e.g. length – of the boundary. As such, given that the actual data uncertainty can not be reduced, the uncertainty introduced in the aggregation can be minimised by minimising the *length* of the partitioning boundary. Concluding, the propagation of data uncertainty can be minimised by adopting a partitioning having minimum length for a given area that has to be covered; the above requirement is met in the *square* partitioning.

Another notion of uncertainty, namely the uncertainty in the relationships of spatial objects, has been addressed by Jensen et al. [11]. This type of uncertainty is due to the fact that non-investigated hierarchies might exist over data. In this work, the authors distinguish between two different types of binary relationships, namely full and partial relationships. For example, a coordinate is fully contained or not contained at all in a roadway (full relationship between coordinate and roadway entities), whereas a roadway may be fully or (only) partially contained in a district (partial relationship between roadway and district entities). When a partial containment relationship exists between the different levels of a dimension, the degree of containment should be specified, e.g 50% of the roadway is contained in the district. However, incorporating in the data warehouse the notion of weighting of the relationships is not so straightforward, mainly due to the presence of the measures. Consider for example the scenario where the average speed on a road is $50\,\mathrm{km\,h^{-1}}$ and this road is partially contained, with a weight of 10%, in a district. Consider also another road with average speed of $120\,\mathrm{km\,h^{-1}}$, partially contained, with a weight of 90%, in the same district. To find the average speed within the district, we have to take into account both roads. The simplest solution is to adopt a weighted technique; however, other approaches like fuzzy techniques could be also investigated. This procedure, though, might increase the uncertainty in the answers, thus the quality of the answers should be ensured by means of some pre-defined lower and upper bounds. Furthermore, the end user should be aware of the uncertainty accompanying its data at each level of aggregation. Also, capabilities that would allow the end user to query only data that fulfil some permissible uncertainty limit are useful.

7.5 Conclusions

In this chapter, we have discussed TDWs by means of transforming raw space–time dependent data in the form of trajectories to valuable information that can be used in decision-making processes. The starting point of our study was the investigation in the literature concerning spatial data warehousing, as it is the area that presents the highest commonalities with TDWs. This study has highlighted a set of modelling and management requirements arising from the specificities of trajectories that should be fulfilled in the development of efficient and semantically rich TDWs. Addressing these requirements forms a road map towards an effective construction of TDWs.

References

1. S. Agarwal, R. Agrawal, P. Deshpande, A. Gupta, J. Naughton, R. Ramakrishnan, and S. Sarawagi. On the computation of multidimensional aggregates. In *Proceeding on 22th International Conference on Very Large Data Bases (VLDB'96)*, pp. 506–521, 1996.
2. Y. Bédard, T. Merrett, and J. Han. Fundamentals of spatial data warehousing for geographic knowledge discovery. In *Geographic Data Mining and Knowledge Discovery*, pp. 53–73. Taylor & Francis, 2001.
3. S. Bimonte, A. Tchounikine, and M. Miquel. Towards a spatial multi-dimensional model. In *Proceedings of ACM 8th International Workshop on Data Warehousing and OLAP (DOLAP'05)*, pp. 39–46, 2005.
4. F. Braz, S. Orlando, R. Orsini, A. Raffaetà, A. Roncato, and C. Silvestri. Approximate aggregations in trajectory data warehouses. In *Proceedings of ICDE Workshop on Spatio-Temporal Data Mining*, pp. 536–545, 2007.
5. M.-L. Damiani and S. Spaccapietra. Spatial data warehouse modelling. In *Processing and Managing Complex Data for Decision Support*, pp. 12–27. Idea Group Publishing, 2006.
6. P. Flajolet and G. Martin. Probabilistic counting algorithms for data base applications. *Journal of Computer and System Sciences*, 31(2):182–209, 1985.
7. J. Gray, A. Bosworth, A. Layman, and H. Pirahesh. Data cube: A relational aggregation operator generalizing group-by, cross-tab, and sub-total. In *Proceedings of the 12th International Conference on Data Engineering (ICDE'96)*, pp. 152–159, 1996.
8. J. Han, N. Stefanovic, and K. Kopersky. Selective materialization: An efficient method for spatial data cube construction. In *Proceedings of Pacific-Asia Conference on Knowledge Discovery and Data Mining*, pp. 144–158, 1998.
9. K. Hornsby and M. Egenhofer. Modeling moving objects over multiple granularities. *Annals of Mathematics and Artificial Intelligence*, 36(1–2):177–194, 2002.
10. W. Inmon. *Building the Data Warehouse*, 2nd edn. Wiley, 1996.
11. C. Jensen, A. Kligys, T. Pedersen, C. Dyreson, and I. Timko. Multidimensional data modeling for location-based services. *The Very Large Data Bases Journal*, 13:1–21, 2004.
12. I. Lopez, R. Snodgrass, and B. Moon. Spatiotemporal aggregate computation: A survey. *IEEE Transactions o Knowledge Data Engeneering*, 2(17):271–286, 2005.
13. E. Malinowski and E. Zimányi. OLAP hierarchies: A conceptual perspective. In *Proceedings of the 16th International Conference on Advanced Information Systems Engineering (CAiSE'04)*, pp. 477–491, 2004.
14. E. Malinowski and E. Zimányi. Representing spatiality in a conceptual multidimensional model. In *Proceedings of the 12th annual International Workshop on Geographic Information Systems (GIS'04)*, pp. 12–21, 2004.
15. E. Malinowski and E. Zimányi. Hierarchies in a multidimensional model: From conceptual modeling to logical representation. *Data and Knowledge Engineering*, 59(2):348–377, 2006.
16. OpenGIS Consortium. Abstract Specification, Topic 1: Feature Geometry (ISO 19107 Spatial Schema), 2001. http://www.opengeospatial.org.
17. D. Papadias, Y. Tao, P. Kalnis, and J. Zhang. Indexing spatio-temporal data warehouses. In *Proceedings of the 18th International Conference on Data Engineering (ICDE'02)*, pp. 166–175, 2002.
18. T. Pedersen and N. Tryfona. Pre-aggregation in spatial data warehouses. In *Proceedings of the 5th International Symposium on Spatial and Temporal Databases (SSTD'01)*, vol. 2121 of *LNCS*, pp. 460–480, 2001.
19. F. Rao, L. Zhang, X. Yu, Y. Li, and Y. Chen. Spatial hierarchy and OLAP-favored search in spatial data warehouse. In *Proceedings of ACM 6th International Workshop on Data Warehousing and OLAP (DOLAP'03)*, pp. 48–55, 2003.
20. S. Rivest, Y. Bédard, and P. Marchand. Towards better support for spatial decision making: Defining the characteristics of spatial on-line analytical processing (SOLAP). *Geomatica*, 55(4):539–555, 2001.

21. S. Rivest, Y. Bédard, M. Proulx, M. Nadeau, F. Hubert, and J. Pastor. SOLAP: Merging business intelligence with geospatial technology for interactive spatio-temporal exploration and analysis of data. *Journal of International Society for Photogrammetry & Remote Sensing*, 60(1):17–33, 2005.
22. S. Rizzi. Open problems in data warehousing: Eight years later. In *Proceedings of the 5th Workshop on Design and Management of Data Warehouses (DMDW'03)*, 2003.
23. S. Rizzi and M. Golfarelli. Date warehouse design. In *Proceedings of International Conference on Enterprise Information Systems (ICEIS'00)*, pp. 39–42, 2000.
24. S. Shekhar, C. Lu, S. Chawla, and P. Zhang. Data mining and visualization of twin-cities traffic data. Technical Report, University of Minnesota, 2002.
25. N. Stefanovic, J. Han, and K. Koperski. Object-based selective materialization for efficient implementation of spatial data cubes. *IEEE Transactions on Knowledge and Data Engineering*, 12(6):938–958, 2000.
26. Y. Tao, G. Kollios, J. Considine, F. Li, and D. Papadias. Spatio-temporal aggregation using sketches. In *Proceedings of the 20th International Conference on Data Engineering (ICDE'04)*, pp. 214–225, 2004.
27. Y. Tao and D. Papadias. Historical spatio-temporal aggregation. *ACM Transactions on Information Systems*, 23:61–102, 2005.
28. G. Trajcevski, O. Wolfson, K. Hinrichs, and S. Chamberlain. Managing uncertainty in moving objects databases. *ACM Transactions on Database System*, 29(3):463–507, 2004.
29. G. Trajcevski, O. Wolfson, F. Zhang, and S. Chamberlain. The geometry of uncertainty in moving objects databases. In *Proceedings of 7th International Conference on Extending Database Technology (EDBT'02)*, pp. 233–250, 2002.
30. J. Trujillo, M. Palomar, J. Gómez, and I. Song. Designing data warehouses with OO conceptual models. *IEEE Computer, Special Issue on Data Warehouses*, 34(12):66–75, 2001.
31. P. Vassiliadis and T. Sellis. A survey of logical models for OLAP databases. *SIGMOD Record*, 28(4):64–69, 1999.
32. D. Zhang and V. Tsotras. Optimizing spatial Min/Max aggregations. *The Very Large Data Bases Journal*, 14:170–181, 2005.

Chapter 8
Privacy and Security in Spatiotemporal Data and Trajectories

V.S. Verykios, M.L. Damiani, and A. Gkoulalas-Divanis

8.1 Introduction

The European directive 2002/58/EC requires providers of public communication networks and electronic communication services to adopt techniques to ensure data security and privacy. This directive states, among others, that *"the provider of a publicly available electronic communication service must take appropriate technical and organizational measures to safeguard the security of its services having regard to the state of the art,"* and also that *"when location data relating to users can be processed, such data can only be processed when they are made anonymous or with the consent of the user."*

Data *security* and *privacy*, however, are concepts that although often used in conjunction represent two different facets of data protection for which various techniques have been developed. In particular, data security addresses the requirements of *confidentiality*, *integrity*, and *availability* [10]. Data confidentiality means to protect data against unauthorized disclosures; integrity means preventing unauthorized data modification; and availability means recovering from hardware and software errors, and malicious denials of data. Techniques that assure data confidentiality include *access control* and *authentication*; techniques for data integrity include *digital signatures* and *semantic integrity checking*; recovery mechanisms ensure *data availability*. On the contrary, privacy is defined as the right of individuals to determine for themselves *when*, *how*, and *to what extent* information about them is communicated to others [1]. Thus, privacy represents a specific form of data protection requiring flexible control over the disclosure of personal information, for which privacy-preserving data management techniques have been developed, including methods for the specification of privacy preferences by the individuals to whom the data refers, and anonymization techniques. It is important to mention here

---

V.S. Verykios
Department of Computer and Communication Engineering, University of Thessaly, Volos, Greece, e-mail: verykios@inf.uth.gr

that the latest developments propose to integrate privacy-preserving techniques with data security techniques, and in particular with access control methods. The rational behind this attempt is that privacy preservation is also ensured when appropriate data security is achieved. For instance, if an adversary cannot gain access to sensitive data stored in a data warehouse, due to an access control method that is in effect, then he or she cannot breach privacy. The EU directive, however, not only solicits the adoption of techniques to ensure security and privacy but also states that these techniques are to protect personal location data.

Personal location data is the spatiotemporal property that describes the identity along with the location of a user in time. In its most general form, personal location data is captured as tuples in the form of $<user\_id, loc>$, where $user\_id$ is the value of an attribute univocally identifying the individual, such as the social security number, and loc is a location that can either be a single position occupied at a given time or a set of positions temporally close to each other. Positions are described in terms of geometric objects (say for instance, a point or a region in a coordinated space) or semantically meaningful *spatial objects* and/or descriptions, such as a house, a park, a shop's name, or an address. Since geometry can be given one or more semantic descriptions and vice versa, we assume that the geometric and the semantic representations are interchangeable. Thus, saying that an individual is at point $P(x,y)$ or saying that he is located at "home" (supposing that "home" has some relation to P) makes no difference in our study. Position can also be represented at different spatial and semantic granularities, which allow control over the detail of the representation. Given a set of personal location data, the concern for *location privacy* arises when both the $user\_id$ and the location of an individual are disclosed. These privacy issues arise because wireless network providers and applications, like *location-based services* (LBS) and ubiquitous computing, are capable of collecting large amounts of personal location data that can then be stored, linked with external sources, and released to third parties without users' consent. Location data enables intrusive inferences that may reveal habits, social customs, religious and sexual preferences of individuals, and can be used for unauthorized advertisement, and user profiling. Thus, these applications pose challenging security requirements, including that of a controlled access to remote services and spatiotemporal data. It should be noticed, however, that the granularity of location, in its spatial and semantic dimension, is tightly related to location privacy, since the more precise the representation of location is, the higher is the privacy demand. As we are interested in techniques to safeguard privacy, in what follows, we assume locations to be available at fine granularity.

In this chapter, we survey the most popular state-of-the-art approaches for data security and privacy, and discuss the various privacy implications they aim to address. Moreover, we analyze the major context situations where such techniques have to be applied to ensure privacy preservation of individuals' identities. We conclude our presentation by providing a road map that highlights potential future directions that are expected to promote security and privacy preservation in the new era. This chapter is organized as follows. Section 8.2 provides an extensive presentation of the state-of-the-art methodologies proposed in the literature in the

8 Privacy and Security in STD and Trajectories

context of security and privacy for trajectories. Section 8.3 introduces an application scenario to demonstrate several open research issues and initiate discussion over potential future directions in this field of study. An architecture is provided as part of this section, which aims to address the special security and privacy requirements imposed in the spatiotemporal environment. Finally, Section 8.4 concludes our presentation by summarizing the most prevalent issues covered in the chapter.

8.2 State of the Art

In this section, we overview the state-of-the-art methodologies in the context of security and privacy in spatiotemporal data and trajectories. A *trajectory* consists of a set of positions that are temporally close to each other and pertain to the same individual. It captures the path that an entity followed in a specific time frame. We classify the existing approaches into four broad categories, namely *policy-based approaches*, *cryptography-based approaches for data access and release*, *K-anonymity approaches for personal location collection*, and *location privacy-based approaches for trajectories*. Each of these fields encompasses a distinct way of ensuring privacy preservation of individuals' sensitive data. Moreover, because of the variety of approaches depicted in some of these categories, we proceed to further partition the methodologies whenever necessary. Figure 8.1 classifies the different privacy and security approaches that appear in spatiotemporal data and trajectories. In this figure, we use capital letters P, S, and H to denote that the corresponding methodology belongs to the category of privacy, security, or hybrid approaches, respectively.

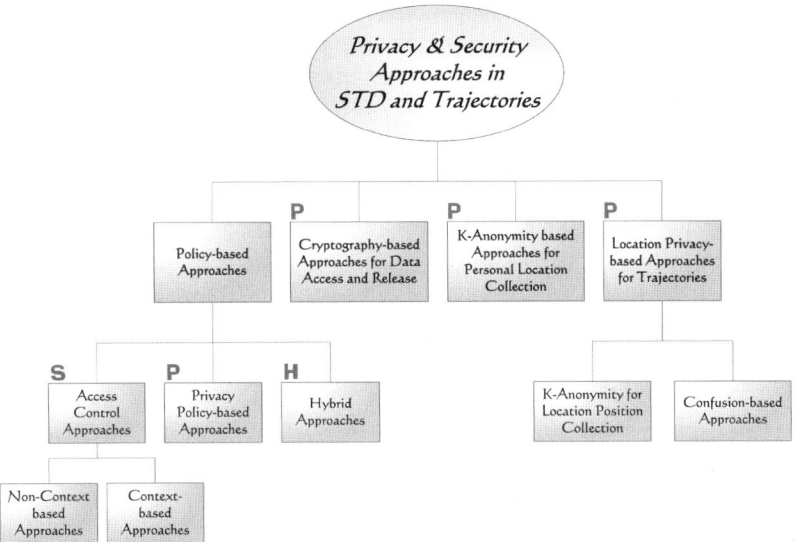

Fig. 8.1 Privacy and security approaches in spatiotemporal data and trajectories

8.2.1 Policy-Based Approaches

Policy-based approaches are based on the idea of regulating the access and usage of information through a set of rules that specify preferences and constraints. Such a set of rules constitutes a *policy*. Policy-based approaches are currently met both in the security and privacy fields.

Broadly speaking, data security aims at protecting information against security breaches, such as unauthorized data observation and modification. Unauthorized data observation results in disclosure of information to users not entitled to have access to such information. Incorrect data modification, either intentional or unintentional, results in the existence of inconsistent information, which may cause damage to users and/or organizations. Comprehensive solutions to data security are quite articulated and require the integration of different tools and techniques as well as specific organizational processes. Policy-based approaches, in the *security context*, are usually incorporated to allow system administrators to specify and apply an operational policy that will tie specific privileges to selected users. In particular, access control policies regulate the access to information resources, like files, database tables, and web services, so that only subjects who are explicitly authorized by a policy are allowed to use such resources.

On the other hand, policy-based approaches, in the *privacy context*, allow the user to define his or her own policy, based on which other users can gain access to his/her location information. The user is also capable to define the set of places from which, and the actual granularity in which, such information will be released to others. As an example, imagine a user who is willing to receive location-based advertisements on his/her cellular phone only from some merchants and under certain conditions. For instance, one may be interested to receive advertisements only from the restaurants in his/her neighborhood. Consequently, the data owner, who acquires the positions of users, cannot disclose such data to an arbitrary advertiser, but only to those who satisfy the conditions expressed by users. Individuals, thus, shall be able to dynamically specify their privacy preferences by defining a policy, which states a set of rules about who is allowed to access their personal data and for which purpose.

In the sections that follow, we present two basic directions of policy-based approaches, namely *access control* and *privacy policy-based* approaches. The first category of approaches is highly related to the security field, whereas the second one lies in the scope of privacy. To be more specific, the access control approaches provide the proper mechanism that guarantees the security of individuals' information after its collection and storage in the system. To achieve this goal, this type of approaches control the access to the sensitive resources and the type of operations that can be accomplished. On the other hand, privacy policy-based approaches protect the privacy of an individual's information before data collection. Based on the user's profile these approaches restrict the granularity of the returned location information regarding a user's current location, as a result of a location query. An in-between approach that shares some characteristics from both the first and the second categories is discussed in the final part of this section.

8.2.1.1 Access Control Approaches

Among the various tools, access control is fundamental to ensure confidentiality and integrity of information resources made available to multiple users. The component in charge of access control is the *access control system*. When a user tries to access an information resource, a component of the system, known as the *access control mechanism* (ACM), checks the rights of the requester against a set of authorizations, usually prespecified by a security administrator. An *authorization* is a regulation that states which user (or generally which subject) can perform which operation(s) on which object(s). It is commonly defined as a triple in the form of (**s**, **op**, **obj**), where **s**, **op**, and **obj** refer to the *subject*, the *operation*, and the *object*, respectively. The *subject* identifies the holder of the authorization, who can either be an individual or a group of individuals or programs. On the other hand, the *object* identifies the resource, whose nature and granularity are strongly dependent on the application domain. Finally, the *operation* defines what the subject is authorized to do to the specific object(s). The set of authorizations constitute what is known as the *access control policy*. Such a policy is usually described in a *policy specification language* that is enforced on the access control model. Because of its relevance to the context of solutions for information security, access control has been extensively investigated for Database Management Systems (DBMS), digital libraries, and multimedia applications. Yet, the importance of the spatial dimension in access control was only recently highlighted.

Two are the basic characteristics of spatial and spatiotemporal data that motivate the specification of spatially aware access control models: the richness and multiplicity of data representations, and the dynamic and mobile user population. Spatiotemporal data is characterized by multiple data representations (such as *attributive*, *vector-based*, and *topological*); additional representations are also often available, such as *raster images*. Spatiotemporal objects may also be complex, i.e., trajectories consisting of a set of subobjects. A suitable access control model for spatiotemporal data must thus (a) support the specification of authorizations against spatiotemporal objects, even at a fine level of granularity; (b) account for the various object dimensions and resolutions according to which certain objects can be viewed; and (c) support a variety of access rights that allow one to express authorizations in terms of the operations supported by the model according to which the data is represented. It is thus important that in addition to commonly met *access rights*, such as READ and WRITE (which provide user permission to read and modify data in the database, respectively), other access rights that correspond to meaningful operations on spatiotemporal objects are also supported. The dynamic and mobile user population is another interesting parameter of spatiotemporal data that needs to be addressed by access control systems. Many of the spatiotemporal applications, like LBS and ubiquitous computing, are characterized by a user population that constantly changes and moves. In many cases, users from different administrative domains or agencies may need to have access to data. Thus, a suitable access control model for spatiotemporal data must (a) support attribute-based and profile-based user specification because relying just on users' login names for authorization grant and revoke is a low-level approach; and (b) support authorizations that are dynamically enabled or disabled depending on the user's current location. An additional requirement that motivates the use of spatially aware access control models is related to the fact that organizations increasingly need to share spatial and spatiotemporal data even across national boundaries, for example, through spatial data infrastructures (SDIs) [23, 39].

Currently proposed access control approaches for spatiotemporal applications only partially succeed in addressing these requirements. We classify the existing approaches into two primary categories, namely *noncontext based* and *context based*. Noncontext based are the access control models proposed for accessing spatial data stored in databases, even available on the Web. In this case, access to resources depends exclusively on the permissions assigned to users, as specified in the access control policy. Conversely, in context-based models, access to resources depends also on the status of contextual variables, and in particular the actual position of the user when he or she initiates the request. Such models are usually applied to regulate access to LBS and pervasive-computing applications. Methodologies that belong in both categories of access control systems are discussed in the subsections that follow.

8.2.1.2 Noncontext-Based Access Control Systems

Noncontext-based access control models grant or revoke access to resources based solely on permissions that were preassigned to each user by the system administrator, without, however, taking into consideration the user's location at the time of the request. The first access control model for spatiotemporal databases based on their spatial extent, temporal duration, resolution of the images, and other spatiotemporal characteristics was presented in [2]. The *Geo-spatial data authorization model* (GSAM), as it is called, allows for the specification of access control based on the region covered by an image, the time of its capture, and the subject's credentials. These credentials correspond to the subject's qualifications and characteristics, and provide the basis for granting the authorizations. GSAM extends the conventional privilege modes of READ, INSERT, DELETE, and MODIFY to capture the special spatiotemporal properties. Thus, it supports several privilege modes, including VIEW, ZOOM-IN, DOWNLOAD, OVERLAY, IDENTIFY, ANIMATE, and FLY BY, that are all essential for providing controlled access to spatiotemporal data. For instance, the OVERLAY privilege allows users to integrate spatiotemporal data over a certain area by overlaying one image on top of another. An important property of GSAM is that the supported authorization specifications are capable of specifying access control policies based on spatiotemporal attributes in terms of not only the spatiotemporal object as a whole, but also on the area covered or contained within it. In addition, the authorization policies in GSAM can be specified based on subject's identities and credentials. Thus, they enable one to specify that a subject is allowed to view a specific image or region with a certain resolution, or is allowed to overlay a set of images that have a specific resolution. Credentials, in turn, are associated with spatial and temporal attributes that indicate the limits of their validity to a certain region and temporal interval, respectively. Each subject is assumed to hold credentials that are part of one or more *credential type hierarchies*. These hierarchies allow for a better structure of the permissions in the access control model. Spatiotemporal objects are also arranged in a hierarchy that defines type–subtype relationships.

An extension to the GSAM model is proposed by Atluri et al. in [3]. The extended model is used to control access to geo-referenced Earth images taken by a satellite at different levels of resolution. Similar to GSAM, the proposed scheme applies access control based on authorizations that define who are the users that are allowed to access which images and at which resolution (i.e., level of granularity). The authors, however, enhance the subject paradigm by associating subjects with a set of credentials and grouping them into classes, organized into a hierarchical structure. Each class is associated with a set of credentials and, based on the hierarchy, subjects within each class inherit privileges from its super classes. The notion of authorization (as presented in GSAM) is extended to include one more property, called *condition*, that is a logical expression over the *subject* and *object* attributes, which aims at enhancing the flexibility of the model. Moreover, the authors propose an implementation of authorizations that is no longer based on access control lists. The advantage of the new methodology is that an access request requires only one search in the proposed index structure instead of two (one for the authorizations and one in the images database). By employing a uniform index for both the image database and the authorization base, access requests are processed more efficiently than in the GSAM model. A final interesting novelty in this work is the capability of the model to capture both positive and *negative* authorizations, instead of only positive ones, as supported by GSAM. Negative authorizations correspond to denial of access of a set of users to specific system's resources.

The concept of limiting the scope of an authorization to a specific area or region was also investigated in the work of Bertino et al. [8]. In their paper, the authors discuss the central concept of a spatial authorization and propose a novel technique to control access to vector-based data on the Web. Since the application area is less complex than the one of geo-referenced data sets, the proposed model is simpler in nature than GSAM. However, it has two important advantages that need to be mentioned. First of all, it is based on existing standards for handling spatial data. On the contrary, GSAM deals with a very specialized type of data and fails to address any existing standards. Second, it has been implemented and its architecture is based on the modern Web-service paradigm, whereas GSAM has not been implemented and therefore its actual applicability has not been assessed. The authors define a *spatial authorization* as a tuple in the form of (**u**, **ft**, **p**, **w**), where **u** denotes the user (or role), **ft** is the object specified in terms of supported *spatial feature types* (such as "building," "house," etc.), **p** is the operation in the form of a Web service to be performed on spatial objects of the specified type, and **w** is the authorization window that indicates the geographical scope of the authorization, i.e., the portion of space in which the authorization is valid. This information is captured in the form of a polygon. The operations that users may execute on spatial data vary depending on both the user's identity and the object's position. This extension is important since it allows for authorizations to be *spatially constrained*; depending on the user's profile, the scope of the authorization may be limited to only one region. All potential users of a Web service are expected to be registered and classified to the system by subjects being assigned the role "administrator." This role is system-related and represents the top-level role that a subject can have. Administrators are given the entire set of privileges for the whole set of feature classes.

The notion of the authorization window was also integrated in the approach of Belussi et al. presented in [4], where an access control model was developed to regulate access to a multirepresentation spatial database. The proposed model is based on the *discretionary policy*, in which users are allowed to grant other users authorization to access certain data. The discretionary paradigm adds extra flexibility to the underlying model, thus it is adopted in several applications. The proposed access control model intends to regulate access of users to geographical maps. The authors assume that each map is composed of a set of features and each feature is represented in one or more maps by numerous spatial properties: *geometric properties*, that describe the shape, extension and location of the objects; and *topological properties*, that describe the topological relationships that exist among objects. The security administrator is allowed to define authorizations against map objects, at a very fine granularity level, taking into account the various spatial representations and the object's dimensions. The authorization mechanism was enhanced to support more capabilities. Such enhancements include the *authorization sign*, which defines whether the permission or privilege should be granted or denied; the *authorization type*, which specifies whether the authorization is weak (and can therefore be overridden) or strong; the *query*, which further restricts the set of objects over which the privilege is granted or denied; the *grantor*, which corresponds to the user who assigned the authorization to the subject (due to the discretionary policy); and the *grant option*, which, when true, denotes that the subject itself can subsequently grant the (positive) authorization to other users. An object hierarchy is also defined that permits inheritance and propagation of the authorizations, taking into account object dimension and type of spatial information.

Finally, a more recent approach, which integrates spatiotemporal and security standards to support controlled access to spatial information through geo-Web services, is presented in the work of Matheus [37]. In this approach, a *policy specification language*, GeoXACML, was defined as a spatiotemporal extension of the OASIS eXtensible Access Control Markup Language (XACML), and used for the declaration of the various access restrictions. GeoXACML provides the means for the declaration and enforcement of spatial access restrictions, and supports the specification of rules that enable or deny access to spatiotemporal objects based on several spatial criteria. The author provides the interface that allows one to formulate such rules by partitioning the various demands of geo-spatial data providers for access control restrictions along three distinct categories (a) *class-based* restrictions, which control the access to objects as they are instances of a particular class; (b) *object-based* restrictions, which are more specific in the sense that they restrict access to particular instances of information objects; and (c) *spatial* restrictions, which enable declaration of restrictions on objects on the basis of their spatial characteristics and geometry. To express such spatial restrictions, spatial attributes and functions are introduced that test topological relation of two geometries: one defining the boundary of the restricted area and the other being the geometry of a resource object.

8.2.1.3 Context-Based Access Control Systems

The aforementioned models were not conceived for use in a dynamic environment, where subjects constantly change position and access control needs to adapt to these changes. This instead is the main concern of context-aware access control models, which were initially proposed for pervasive-computing applications and then extended to support a variety of mobile applications. Since all the methodologies that are covered in this section are somewhat related to the basic *Role-Based Access Control* (RBAC) model, we first provide a small introduction to the most prevalent properties of this model. Having knowledge of these properties enables us to study the various extensions that authors proposed over the years to capture the special requirements of spatiotemporal data and the dynamic environment in which this data is accessed.

The RBAC model [42] is a reference model for RBAC, in which permissions are associated with roles and users are made members of appropriate roles. Roles are devised to bring together a set of users on one side and a set of permissions on the other side. Users can be reassigned from one role to another and roles can be granted new permissions as new applications and systems become available. The major purpose of RBAC is to facilitate *security administration* and *review*. Several sophisticated variations of RBAC have been proposed over the years that allow the establishment of relations among roles as well as between permissions and roles and between users and roles. Roles can also participate in inheritance relations, whereby one role inherits permissions assigned to another role, thus helping toward enforcing security policies. The basic RBAC model, also known as $RBAC_o$, sets the minimum requirements for any system that claims to support RBAC. It supports three sets of entities, called *users*, *roles*, and *permissions*. A user, in our context, is a human being. A role is a job function with some associated semantics regarding the authority and responsibility of someone who is a member of this role. A permission is an approval of a particular mode of access to one or more objects that reside in the system. Permissions can apply to single or to multiple objects. A user can be a member of many roles and a role can be assigned to multiple users. Similarly, a role can have many permissions and the same permission can be assigned to multiple roles. The key to the RBAC model lies behind these two relations of major significance.

The first extension to the basic RBAC model to capture the requirements of context-based access control systems was presented in [13]. The generalized RBAC model (GRBAC) introduces the concept of *environmental roles*, which are defined similarly to the basic subject roles of the RBAC model. Environmental roles allow for the establishment of a uniform access control framework that can be used to secure context-aware applications. They can be activated based on the value of the various conditions existing in the environment where the request was made. Environmental conditions, on the other hand, include time, location, and other contextual information relevant for access control. Depending on the environmental conditions, the user can be enabled or not to perform certain operations. Similar to the activation of roles in $RBAC_0$, environmental roles are defined by the system administrator who needs to specify the associated *environmental variables* and conditions that must hold in order for a role to be activated. However, unlike in $RBAC_0$, environmental roles are *dynamic* in nature since they control variables that constantly change value (e.g., temperature) and cannot be well-defined with respect to whether they are active or not in the processing of an access request. Testing every environmental role on every access control mediation would be prohibitively expensive. Thus, environmental roles are partitioned into role hierarchies, which provide an efficient means of role entry testing.

A similar approach is presented in [26], where the authors emphasize the fact that in mobile-computing environments, the availability of roles and permissions may depend heavily on users' location. To cope with this special requirement, they propose a *spatial role-based access control* (SRBAC) model that utilizes location information in security policy definitions. They introduce the notion of a *spatial role*, captured as a role that is automatically activated when the user is located within a given region. Location-sensing techniques are required to capture the actual position of the subject and estimate the location of mobile terminals. Several such techniques exist nowadays suitable both for indoor and outdoor location tracking [26]. In addition to obtaining the location information of the mobile terminal, the system must also be capable of identifying the authenticity of the spatial information obtained. In SRBAC, locations are represented using expressions in the form of certain location areas that are identifiable by the system. The spatial model that is incorporated is simple and targeted toward wireless network applications. It consists of a set of adjacent cells and the position of the user is depicted as either within a cell or as the aggregate of cells containing it. The model consists of five basic components: *users*, who are considered to be mobile units that can establish communication with system resources to perform certain activities; *roles*, which are sets of permissions to control access to system resources; *permissions*, which are approvals to execute some operation on one or more objects and highly depend on the role and the role owner's location; *sessions*, which are mappings of one user to possibly many roles, in a way that the user is capable of sequentially activating some subset of roles that he or she is a member of; and *locations*, which are represented by means of symbolic expressions that describe location domains identifiable by the system. The model also allows for hierarchies in which a role r_i inherits the permissions of another role r_j in locations L, if all the permissions of r_j in locations L are also permissions of r_i in the same locations. A constrained version of SRBAC is also presented, in which rules are partitioned in mutually exclusive sets and may not be executed simultaneously by a user.

GEO-RBAC [7] is an access control model also centered around the concept of a spatial role. In its context, a spatial role represents a geographically or spatially bounded organizational function, such as TAXI-DRIVER. In GEO-RBAC, positions can be captured as either *real* or *logical*. The real position corresponds to the position on the Earth of the user, obtained via a mobile terminal such as a Global Positioning System (GPS). On the other hand, the logical positions are presented in a way that is almost independent from the underlying positioning technology. They are modeled as *spatial features* and can be directly computed from real positions using specific mapping functions. Accordingly, a user can play a role only if he or she is located within the region denoted by the *role extent*. To clarify this, consider the example of a nurse working in a hospital. She may be authorized to access the record of patients through her PDA only when located within a specific hospital's department. Besides the concept of spatial role, GEO-RBAC introduces the concept of a *role schema* for the description of the invariant properties of a set of roles. A role schema defines some common properties of a set of spatially organizational functions in a similar fashion. A *role instance* is a role fulfilling the constraints defined at the schema level. Furthermore, the model allows the specification of *role hierarchies*, which can be defined in two levels: at the schema level, where given two role schemas r_{s_1} and r_{s_2}, if r_{s_2} inherits r_{s_1}, then r_{s_2} inherits all the permissions of r_{s_1}; and at the instance level, where given two role instances r_{1g_1} and r_{2g_2}, if r_{2g_2} inherits r_{1g_1}, then r_{2g_2} inherits the permissions of the role schema r_1 and the permissions that have been assigned specifically to the instance r_{1g_1}. Such a model has been recently extended to allow for privacy-preserving access to data. Finally, GEO-RBAC assumes that users can use a role only from within a particular location and that the role and its associated permissions are predefined for that location.

Fu and Xu in [19] present a framework to support a coordinated access control model enforcing both temporal and spatial constraints. To accomplish this, the authors propose the use of a Shared Resource Access Language (SRAL) for the specification of the access patterns and they proceed to extend the $RBAC_0$ model to specify and enforce spatiotemporal constraints. Relative to the spatial constraints, they define a Shared Resource Access Constraint language (SRAC) and extend $RBAC_0$ to incorporate SRAC as one of the constraint definition languages. To enforce these constraints, the authors proceed to introduce a new type of state that can be assigned to permissions, called *active*. Permissions are considered as *active* if, and only if, their associated role is assigned to the subject in a session and the related spatial constraints are satisfied. To meet the temporal requirements in accessing the time-sensitive resources, each permission is associated with a validity duration, which specifies the length of time period when the permission can be granted to a subject. Finally, a permission can be in one of the following three states: *inactive*, *active but invalid*, and *valid*.

8.2.1.4 Privacy Policy-Based Approaches

Privacy policy-based approaches attempt to ensure privacy by controlling access to individuals' location data based on their *location policy*. Location policies state *who* and *when* is allowed to get location information concerning a specific individual. In a location policy preservation system, an individual can access a person's location information only if he or she is permitted by that person's location policy. A location policy should thus often restrict the granularity of the returned location information, e.g., instead of the room in which the user resides, it should return the building. Moreover, location policies can potentially contain a set of locations and return location information only if the user who makes the request is at one of a set of predetermined locations. Finally, location policies can limit the time intervals in which access to a specific subject's information should be granted.

The first work that used privacy policies to protect *spatial privacy* was presented in [44]. The key concept in the proposed methodology is that of an observation of a located object. An observation typically includes the location where the object was observed, the time that the observation occurred, the identity of the located object, and the speed the located object was observed to have at the time of the observation. The authors model the observational accuracy using lattice structures, and describe the lattices corresponding to the degree of accuracy of location, time, identity, and speed. The key idea behind this methodology is that an individual should be able to adjust the accuracy at which observations are released to others, depending on parameters such as the intended use and the identity of the recipient. A policy scheme is thus applied to depict and model this information into rules that will later on be enforced by the system.

A complex privacy-aware system is presented in the work of Langheinrich in [36]. According to the proposed model, when a user enters an environment in which services are collecting data, a *privacy beacon* announces the privacy policies of each service. A user's privacy proxy checks these policies against the user's predefined privacy preferences. Privacy proxies are continuously running services that can be contacted and queried by data subjects anytime; they are configured using a preference language, such as the P3P Preference Exchange Language (APPEL) [14], proposed as part of the W3C's P3P specification [15]. An agreement is formulated as an XML document containing the data elements to be exchanged and the privacy policy that applies to them. Provided that the policies agree, the services are allowed to collect information and users can utilize them. A privacy framework that automates the privacy management decision-making process is presented in [38]. It allows users to apply general machine readable privacy policies and preferences in order to control the distribution of their information. Users subscribe to one or more location servers, registering their privacy requirements in each server. These requirements take the form of system components, called *validators*. Third-party applications seeking a user's location can issue queries to the location server that is responsible for the user. Validators check the acceptability of policies and determine whether the system should accept the request and release the information. For the specification of policies, the authors used the P3P Preference Exchange Language (APPEL).

Hengartner and Steenkiste were the first to propose the use of digital certificates in location privacy. In their work [28, 29], the authors introduce a privacy location mechanism that is based on *digital certificates*. A digital certificate is a signed data structure in which the signer states a decision concerning some other entity. The proposed implementation is based on SPKI/SDSI certificates [18]. When a user initiates a request concerning the location of an individual, the location service must first check whether the location policy grants access to the requester by building a chain of policy certificates from itself to the requester. If a location service receives a forwarded request, it must also check whether the location service from which it got the request is trusted. Similar to the previous step, the location service tries to build a chain of trust certificates from itself to the forwarding device. When these steps finish successfully, data is securely released to the user who initiated the corresponding request.

8.2.1.5 A Hybrid Approach

The work described in [16] can be considered as a hybrid approach, which shares several characteristics of both access control-based (and in particular of *context-based*) and privacy policy-based approaches. It presents an architecture, primarily grounded on the GEO-RBAC model [7], which incorporates a privacy-preserving access control system for mobile organizations. The system filters requests received from users and determines whether they must be accepted based on the user's *role* and *position*. It then proceeds to send an anonymous request together with a perturbed location to the application server, which provides a set of location-aware information services. Specifically, location is cloaked via a decrement of the spatial granularity of position (an approach known as *spatial generalization*). When a spatial generalization is performed over a position, the geometric space of the object is replaced by a coarser geometry. For example, the position in a building at the maximum granularity is represented as a point and at the coarsest detail is expressed by the whole building. The rational behind this approach is to specify the generalized location of the user in terms of a logical position, which is dynamically computed by applying the location mapping function. The mapping function is assigned the duty to implement the generalization criteria. Given a user's position p, the system determines whether a session role r is enabled in p and the corresponding logical position (which has been perturbed). Next, the access control system maps the actual position onto a location according to the specified policies; finally the logical location is forwarded to the application server. Another interesting characteristic is that the degree of spatial generalization can be dynamically specified at the user's preference. Since the role itself can be described at different granularities (because the set of roles constitutes a lattice), and each role is assigned generalization criteria, a user (having a given role) is allowed to specify the generalization associated to his role or to ancestor roles depicted in the lattice. For example, a user who is assigned the role TAXI-DRIVER can select the position to be communicated at the level of granularity associated to the role TAXI-DRIVER or to one of its ancestors, such as, i.e., the role GENERIC-DRIVER. Ultimately, the approach combines dynamic privacy preferences with spatial cloaking and access control. On the other hand, the generalization approach fails to ensure K-anonymity (a concept which is discussed later on in the chapter), and the privacy threat, related to the storage of trajectories, still remains an open issue.

8.2.2 Cryptography-Based Approaches for Data Access and Release

Several cryptography-based approaches have been developed over the years in order to support privacy preservation in the spatiotemporal domain. The key idea in each of these techniques is to make a cipher of the identity of the user prior to sending it to the service provider. In this way, the service provider has no way of knowing the real identity of the individual who initiated the request.

The first proposed architecture in this category of techniques was based on asymmetric key cryptography and is presented in [27]. The user acquires a *pseudonym* that uses as an asymmetric key, known only by the *Location Server* (LS) and the user. To disallow any other user from reading it, the pseudonym is encrypted with the public key of the LS. The key serves as a kind of reference with which potential subjects can address the user in location queries. To prevent an external attacker from matching queries of different subjects to the same target, the appearance of the cipher text of the pseudonym has to be different. For this reason, it is also encrypted with a unique piece of information, chosen individually for each subject.

A different privacy protocol is described in [35] and consists of three entities, namely *Users* (U), *Mobile Operators* (MO), and *Service Providers* (SP). It ensures trust (a) between users and mobile operators by generating a master secret key shared between the user and the MO, and (b) between mobile operators and service providers by the existence of a trusted *Public Key Infrastructure* (PKI); MO obtain a digital certificate and a private key and, similarly, SP obtain their digital certificate and private key. MO store the certificate of the SP (which contains the public key of the SP), and SP store the certificate of the MO (which contains the public key of the MO). Initially, the user detects his current location using a GPS. Then, he securely sends his location to the MO and requests for a list of the LBS available at that location. The MO replies with a list of services and take responsibility on behalf of the user to identify and authenticate the genuine SP. Then, the user selects a particular LBS from the provided list and (securely) sends the LBS parameters to the MO, who in turn send only the current location details, but not the identity of the user, to the SP. Finally, the SP cannot maintain the user's profile as they do not have knowledge to whom the service is being offered.

Two architectures that employ *hierarchical identity-based encryption* (HIBE) in order to ensure confidentiality of information are presented in [30]. The authors propose a proof-based schema and an encryption-based access control schema. In the proof-based schema, a client needs to assemble some access rights as a proof of access, which demonstrates to a service that the client is authorized to access the requested information. In an encryption-based schema, a service provides confidential information to any client but only in an encrypted form. Clients who are authorized to access the information have the corresponding decryption key that will allow them to decrypt it. Both schemas use HIBE [22], which is based on *Identity-Based Encryption* (IBE). In IBE the public key of an individual is an arbitrary string. The interested reader is encouraged to refer to the work in [11], which analyzes one of the first IBE schemes.

8.2.3 K-Anonymity Approaches for Personal Location Collection

Based on the Oxford English Dictionary [43], *anonymity* is defined as "being nameless" or "of unknown authorship." According to [40], *anonymity* is "the state of being not identifiable within a set of subjects, the anonymity set." The *anonymity set* was first introduced in [12] as "the set of participants who probably could have sent a certain request, as seen by a global observer." Given a certain area and time interval, an anonymity set can thus be defined as the set of users who were in that particular area, in that time interval, and could potentially initiate a request. Thus, in our context, anonymity means that location cannot be associated with a particular individual. As an example, consider an application for the monitoring of road traffic that collects data on vehicle movements. Suppose that cars can be univocally identified along with their position in time. For privacy purposes, however, car identifiers must not be disclosed. To address such a requirement, a naive approach is to remove the car identifiers from the collected data and replace them with pseudonyms. However, such an operation is insufficient to guarantee anonymity, since location represents a property that in some circumstances can lead to the identification of the individual. For example, if one is known to follow the same route almost every morning, it is very likely that the starting point is the home of the individual and the ending point is his or her working place. It is thus relatively easy for an observer to determine who the individual is. Because of this characteristic, location is said to represent a *quasi-identifier*. Quasi-identifiers are attributes, which, though not containing an explicit reference to the individuals' identity, can be easily linked with external data sources and reveal who the individual really is.

One of the most significant anonymity preserving methods is K-anonymity. It was introduced in the work of Samarati and Sweeney in [41]. The principle behind the notion of K-anonymity is that "each release of data must be such that every combination of values of private data can be indistinctly matched to at least K individuals." K-anonymity can be achieved by *generalizing* the value attributed to the quasi-identifiers. Generalization is performed by replacing the original attribute's value with a less specific, but semantically consistent, one.

The concept of location K-anonymity was first introduced in [24] as an extension of the K-anonymity model. It states that "a message sent from a user is K-*anonymous* with respect to spatial and temporal information when it is indistinguishable from the spatial and temporal information of at least $K - 1$ other messages sent from different users." The main idea of the proposed algorithm is to subdivide the area around the user's location and delay the request as long as the number of users in the specified area falls below the desired value of K. However, the proposed model in [24] has a basic drawback; it assumes a static K value for all messages, which hinders the quality of service for mobile nodes whose privacy requirements can be met using smaller values of K.

Instead of using the same K for all messages, the authors in [20, 21] describe a customizable K-anonymity model, which enables each message to specify an independent K-anonymity value and the maximum spatial and temporal tolerance resolutions it can tolerate based on its privacy requirements. The proposed algorithm tries to identify the smallest spatial area and time interval for each message, such that there exist at least $K - 1$ other messages, each from a different user with the same spatial and temporal dimensions.

Contrary to the previous models, the work described in [32–34] proposes a privacy system that takes into account only the spatial dimension. Following this approach, the area in which location anonymity is evaluated is divided into several regions and position data is delimited by the region. The desirable K-anonymity level is accomplished in two ways: the first, called *ubiquity*, assumes that K indicates the number of the regions where users stay. If the users satisfy K-anonymity, the number of regions where they stay must be at least K; the second, called *congestion*, assumes that K indicates the number of users in a region. If the users satisfy K-anonymity, then the number of users in that region is at least K. High ubiquity guarantees the location anonymity of every user, while high congestion guarantees location anonymity of local users in a specified region.

8.2.4 Location Privacy-Based Approaches for Trajectories

The approaches proposed in the literature in the context of location privacy for trajectories can be classified into two principal categories, namely *K-anonymity for location position collection* and *confusion-based techniques*. The techniques of the first category aim at preserving the anonymity of a user by obscuring his route, while techniques belonging to the second category aim at confusing an adversary by modifying true users' trajectories or by introducing "fake" ones.

8.2.4.1 K-Anonymity for Location Position Collection

The first published model for ensuring the anonymity of users' trajectories was presented in [5, 6]. It introduces the concept of *mixed zones* to enhance user privacy in LBS. To do so, it defines two zones: the *application zone*, which is the geographic space where LBS applications can trace users' movements, and the *mix zone*, which corresponds to the space where LBS applications cannot trace users' movements. When a user enters a mix zone, applications do not receive his or her real identity; instead what they receive is a pseudonym. Since applications do not receive any location information when users are in a mix zone, the identities are said to be "mixed." The pseudonym of any user changes whenever he or she enters a mix zone. In this way, LBS applications that see a user coming from the mix zone cannot distinguish that user from any others who were in the mix zone at the same time and cannot therefore link users entering the mix zone to others leaving it.

A similar classification of areas is discussed in [25] where the authors introduce the notion of the *sensitivity map*, which classifies locations as either "sensitive" or "insensitive," and proceed to describe three algorithms that hide users' positions in sensitive areas. The first algorithm, called Base, checks the sensitivity map for each location update and releases only locations belonging to areas classified as insensitive. The second algorithm, called Bounded-rate, reduces the amount of information released in insensitive areas to make it more difficult for an adversary to infer an individual's visit to sensitive areas. This is ensured by making sure that updates are not sent with a frequency higher than a predefined threshold. The last algorithm, called K-area, releases location updates only when these do not reveal which of at least K sensitive areas the user visited. The sensitivity map is partitioned into zones that include at least K distinct sensitive areas. A distinct sensitive area is an area that can be reached from at least one public area, and from which no other sensitive areas can be reached without traveling through public areas. Furthermore, if the individual visited a sensitive area in the previous zone, the algorithm suppresses the location updates to third-party applications.

Contrary to the notions of mixed zones and sensitivity maps, the approach introduced in [9] assumes that each location is depicted as a quasi-identifier. According to this approach, each user's trajectory is recorded as an ordered list of 3D points (x_i, y_i, t_i) that is known as the *personal history of locations* (PHL). Given a set R of requests issued to a *service provider* (SP), a subset R' of requests issued by the same user u is said to satisfy *historical K-anonymity* if, and only if, there exist $(K-1)$ PHLs: $P_1, P_2, ..., P_{K-1}$ for $(K-1)$ users different from u, such that each P_i contains one tuple with (x_i, y_i, t_i) identified by the user's request. A strategy is then applied to enforce privacy preservation: the exact position and time of a user is generalized when a request is forwarded to the SP. Generalization is performed by trying to preserve historical K-anonymity and is accomplished by enlarging the area and the time interval, thus increasing the uncertainty related to the user's real location at the time of the request.

8.2.4.2 Confusion-Based Techniques

Location is an attribute which is peculiar in not only that it is a quasi-identifier, but may also represent sensitive information. As an example consider the case of an individual that is transferred to a clinic. Knowledge of the target location may reveal private information concerning the actual health problems that he or she faces. For instance, if the user is captured in the cardiology's department, one can assume that he or she suffers from such kind of health diseases. It is thus necessary to provide techniques that achieve to hide the actual user's trajectory from any adversary that may potentially use it to expose sensitive information concerning user's activities. Trajectory *confusion* and *obfuscation* techniques are mechanisms that prevent an adversary from tracking a complete user trajectory. The principal idea behind these techniques is to modify users' trajectories or generate one or more "fake" trajectories in an attempt to confuse an adversary to learn the users' true routes. In this section, we survey some of the most prominent research directions in this area of study.

The first model, based on the technique of *path confusion*, was introduced in [31]. It presents an algorithm that aims at protecting users' privacy from adversaries who can use trajectory information to track paths. The work concentrates on a class of applications that continuously collect location samples from a large group of users. Every time two users' paths cross, there is a chance for the adversary to confuse the routes and follow the "wrong" user. For every user's trajectory, the algorithm generates one more trajectory. Then, it decides one point in space in which the trajectories will meet. This way spatial privacy is maximized, since the artificially generated trajectories can cause wrong trajectories being followed by the attacker; the attacker has no means of knowing which trajectory is the "correct" at the point they cross.

An idea based on the techniques of obfuscation and negotiation is presented in [17]. Obfuscation concerns the practice where an individual may degrade the quality of information about his or her location in order to protect his/her privacy. Negotiation, on the other hand, is used to provide the best quality of service, while revealing as little information as possible about the user's real location. In the proposed methodology, the space is represented as a graph where locations are modeled as a set of vertices V and the connectivity between pairs of locations is depicted by a set of edges E. A user's location is represented as a vertex $l \in V$. An obfuscation of a user's location is represented as a set O of vertices, such that $l \in O$. This way, the set O provides an imprecise representation of a user's location, since an adversary has no information about which one, among the various vertices depicted in O, corresponds to the user's actual location.

Similar to the first proposed technique is the work in [32–34]. The main idea behind this approach is the mixture of true users' trajectories with "fake" ones, called *dummies*, in an attempt to reduce the risk of exposing the true user's path to an adversary. For the generation of dummies, the authors propose three distinct methodologies, all of which are based on the assumption that space is formulated as a graph of locations V joined by edges E. The initial dummy's position (in this *connectivity graph*) is chosen at random within a specific *radius* around the *initial position* of the real user's trajectory. In the first methodology, called *random walk*, when a dummy reaches a node, the decision about the edge in which it will move is performed randomly. The drawback of this approach is that adversaries can easily distinguish true user's from dummies, since the dummy's generated trajectory does not appear natural. The second methodology, called *direction control*, aims at addressing this mishap. To do so, when a dummy reaches at a node, the decision about the new direction is made using a set of predefined likelihoods; these attempt to depict how possible it is for a true trajectory to move in each of the potential directions. For example, when a real user arrives at an intersection, he or she often goes straight, left, or right, but rarely goes back. In this method each dummy moves independently of the real trajectories. The third approach, called *collaborative direction control*, addresses this issue by having dummies move in a way that they and the true user frequently cross each other. This method extends the previous one by adding two conditions: (a) a dummy tries to *predict* the node that a true user will move to, when selecting its own direction, and (b) a dummy will *wait* on a node if it senses that the true user approaches this node.

8.3 Open Issues, Future Work, and Road Map

In this section we discuss open research issues related to the field of location privacy and security. To motivate our view, we focus on a challenging application area for security and privacy preservation techniques, namely *location-based* applications. In what follows, we first present an application scenario that depicts a wide variety of characteristics related to location privacy and security; then, we concentrate on the relevant open issues raised by this scenario, along with the guidelines for a possible reference architecture to address them.

8.3.1 An Application Scenario

Our application scenario concerns a road traffic information service for vehicles in a metropolitan area; from here on, we refer to it as *CityCruise*. Drivers subscribe to CityCruise in order to receive location-based traffic information, such as traffic density and accidents, on their on-board terminals. Registered vehicles are equipped with a high-resolution positioning system (such as a GPS), and communicate to the service through a high speed wireless network. In particular, vehicles transmit their position to a *location server* (LS) in predefined time intervals (known as, *sample rate*). Then, the LS communicates, in a privacy-aware manner, with a *traffic server* (TS), whose task is to dispatch traffic information to the vehicles based on their current contextual characteristics, i.e., position and time of request. The temporal dimension is crucial in our scenario, since certain functionalities are assumed to be delivered only at specific time periods. Moreover, we assume that traffic information can be delivered in two ways: either (a) upon *user's request* (known as the *pull mode*) or (b) *autonomously* by the LS (known as the *push mode*) at a given frequency.

CityCruise is provided by a business company; we name it "TrafficServiceInc." TrafficServiceInc offers personalized traffic information services based on user profiles. For that purpose, services are formulated into groups and provided to different categories of users on a payment basis. In our scenario, we consider the following user categories: PROFESSIONAL DRIVER (e.g., a taxi-driver), ROUTINE DRIVER (i.e., a regular user driving every day to reach his or her working place), and OCCASIONAL DRIVER (such as a tourist, visiting a country). Users, depending on their category, are expected to experience different requirements with respect to traffic information. For example, professional drivers usually require more detailed and geographically precise traffic information, possibly related to a wider area. On the contrary, information offered to occasional drivers can be provided at a coarser level of detail. Moreover, for marketing reasons, access to such a service may be subject to some spatial and/or temporal constraints; for instance, professional drivers may be enabled to receive traffic information during the whole day when they drive within any area in the city, while occasional drivers, such as tourists, may be enabled to receive traffic data only when located in specific areas of the city at certain time frames.

Notably, this service raises some important location privacy concerns since users' position, along with their identity, is constantly recorded by TrafficServiceInc. The recorded information may be used for various purposes, including accountability and log data analysis. It can be noticed, however, that users experience different sensitivity levels with respect to their privacy, depending on the category they belong in. For instance, professional drivers may be nearly inconsiderate with location privacy, since driving is part of their daily working activity; on the other hand, individuals driving for personal reasons or occasional drivers reasonably consider location privacy as a very important requirement. Moreover, there are also different privacy concerns about different service providers that cooperate with TrafficServiceInc to provide the CityCruise framework of services. Depending on the application context and, in particular, on whether TrafficServiceInc service providers' associates are trusted or not, we capture the following two scenarios, raising additional privacy issues that need to be properly addressed:

- *Untrusted service providers*. If the service providers are untrusted, then users fear that adversaries may gain access to their personal location data and use them for unauthorized purposes. As a consequence, users may be reluctant to subscribe to this service. It is thus to the interest of TrafficServiceInc to ensure that personal location data are not recorded at the level of detail that may endanger users' privacy. Even though this is not a major issue for professional categories of users, it can be so for users belonging in the other categories.
- *Trusted service providers*. A different situation occurs when the service providers are trusted. In such a case, location data are recorded comprehensively, bearing all details, including the *vehicle identifier* (ID), the *possible driver category*, the *coordinated position of the vehicle*, the *time of observation*, and possibly other additional parameters. However, as the movements of vehicles constitute an important data source for various tasks (like urban mobility planning), service providers may be eager to release such data sets to third parties. To comply to current legislation, service providers shall perform data anonymization and account for users' privacy preferences prior to releasing it.

8.3.2 Open Research Issues and Challenges

Based on the application scenario presented in Sect. 8.3.1, we proceed in discussing relevant research issues and challenges that remain open. In particular, we decide to focus on four issues of major importance:

1. In organizations and communities, users are often described not only by an identity but also by a *category*. The category of a user represents his or her *role* in the community. Role information is important for various reasons, including the customization of LBS and the assessment of privacy requirements. The introduction of the notion of a role raises a number of important issues, including (a) how to acquire role information about a user, and (b) how to customize services and, in particular, specify privacy requirements based on roles. As far as

we know, this notion of privacy related to the role performed by individuals has not been investigated yet. We refer to it as *role-based location privacy*; to protect it selected methods which may vary, depending on the context in which the location-based application is developed, need to be applied. In particular, on the basis of the above discussion on trusted and untrusted service providers, it seems important to distinguish privacy-preserving methods into two broad themes: (a) the ones that are applied *online*, during the interaction of the user with the system providing the service, and thus before data is recorded, and (b) the ones which are applied *offline*, to sanitize a database of location data prior to its release to third parties. We believe that comprehensive solutions should address both these aspects.
2. An important class of privacy-preserving methods are K-*anonymity* techniques and, in particular, those applied for the anonymization of location data. The notion of location K-anonymity needs to be extended in order to account for the fact that a user occupies not a single but a sequence of positions at different points in times. Thus, K-anonymity techniques have to be adapted to capture trajectory data. We refer to the latter issue as *trajectory K-anonymity*. Relevant questions to be addressed in this context include the following: (a) if a single location is anonymous, how can we make sure that a sequence of locations, occupied by the same individual, also remain anonymous? and (b) how can we measure and quantify trajectory K-anonymity? Especially in the role-based context, a relevant issue to be addressed regards the integration of the notion of a role with that of K-anonymity and, more specifically, how can role information be preserved while data remains K-anonymous. Finally, other issues to be addressed include, but are certainly not limited to, how to choose the value of K for the K-anonymity-based approaches, and if K is specified after the data has been collected, what is a "good" value for it? We believe that both of these questions relate to a broader issue that regards the interplay between law, policy, and the technology itself.
3. The definition of a framework for privacy-preserving location-based applications represents another important issue. Several approaches are currently proposed in the literature, but a comprehensive and flexible architectural framework is still lacking. Such an architecture should integrate different functionalities including, but not limited to, user authentication, service customization, privacy preservation, and system administration and control.
4. The issues of *efficiency* and *usability* regarding the algorithms for security and privacy preservation need also be addressed in future research work. To be more specific, with respect to efficiency, there are various methods that have been proposed over the years that are inefficient and difficult to implement in larger systems. For instance, the use of "dummy" information requires an increase in bandwidth, constituting this solution infeasible in the case of low-latency systems with a limited amount of resources. Another issue of great importance is the usability of the produced data. For instance, if the data is to be shared for secondary research purposes, the inclusion of dummy information, or confusion-based trajectories, can hinder the knowledge which can be mined from this data. For these reasons, and also due to the fact that such systems are usually aiming at

their use in the real-world environment that includes a very large area and population, we are confident that future algorithms must properly address both the issues of efficiency and usability in order to be useful.

8.3.3 A Reference Model System Architecture

The proposed reference architecture is of a location-based, temporal access control system; it aims at addressing most of the security and privacy implications discussed earlier in the context of spatiotemporal data environments. More specifically, our proposed reference architecture (a) provides role-based location privacy; (b) aims through its privacy-preserving components at identifying the frequent trajectories and providing trajectory K-anonymity; (c) is comprehensive and flexible to a high degree since it integrates different functionalities, such as user authentication, privacy preservation, and system administration and control; and (d) its design is that allows it to be efficient and usable since it is decentralized and has some mechanisms that enhance its functionality (such as the knowledge database (KDB) that, when applicable, alleviates the privacy-aware knowledge-sharing component from the time-consuming task of reproducing prior knowledge). Our proposed architecture supports both the *push* and the *pull* modes and adheres to the *client-server* paradigm, in which a set of servers offer certain services to a number of authorized clients. Some of its characteristics are borrowed from the work of Damiani et al. appearing in [16].

First of all, the system considers that LBS are offered based on roles that subscribers are assigned in the system by a system administrator. These roles primarily adhere to organizational and functional requirements and are triggered based on the user's actual position in space and the time of request. An *access control* component is the part of the system that regulates access to the spatiotemporal data warehouse, based on users' privileges, as indicated by their corresponding roles. Second, accessibility of services depends highly on the position of the user in the environment. Thus, the classical notion of ACM is extended to account for the mobility of users under the hypothesis of a bounded space and time frame. Third, privacy concerns are addressed to account for the capability of current technologies to collect, store, and disclose sensitive data regarding the location of individuals at a specific time period. The architecture of the model system is depicted in the figure below. When viewing the figure please bear in mind that both the *Req()* and the *Service()* operations may optionally carry a "data" field containing any applicable service-specific information that needs to be transmitted from the user to the system and vice versa.

The proposed architectural framework for a secure and privacy-preserving location-based application is thus centered around two main components: the ACM and the *Privacy-Preservation Components* (*online* and *offline* PPCs). The ACM filters the requests for service sent by mobile users and determines whether they should be accepted based on the role and contextual information (referenced in both space and time) of the user. We assume that the underlying access control model is based on GEO-RBAC [7].

8 Privacy and Security in STD and Trajectories

The *online* PPC is the component that regulates how data is stored in order to facilitate privacy preservation of location-based and trajectory data. It does so by addressing all users' privacy requirements as captured by their role in the system. We assume that the *online* PPC is part of the *request handler* (RH) component of the system. It operates at the actual time the request is received, where it needs to decide how this information tuple should be depicted in the log-file of the system in a privacy-aware manner.

On the other hand, the *offline* PPC component facilitates data sanitation techniques to ensure that the resulting data set can be safely released to third parties for mining purposes. To do so, it needs to interact with the SLC component in order to receive and sanitize data. This interaction is demonstrated through the dark curve that appears between two components, as shown in Fig. 8.2.

To demonstrate how the system operates, assume that a user (or client), recognized through a unique *ID*, initiates a request for a service (assigned a service ID, *SID*), when at location (x,y), at time t. The *request handler* (RH) receives this request and proceeds to query the ACM to ensure that the user has adequate permissions to request this service at the given place and time. ACM is a *trusted* (secure) component; it holds in the access policy a list of all registered users in the system, along with their user profiles, which determine their preferences, their role, and therefore the services that are allowed to request at given locations and time frames. It is used to perform the tasks of user authentication and services control. The response of ACM determines the next steps to be followed; if the user is not registered in the system or does not have sufficient privileges to request the service, ACM

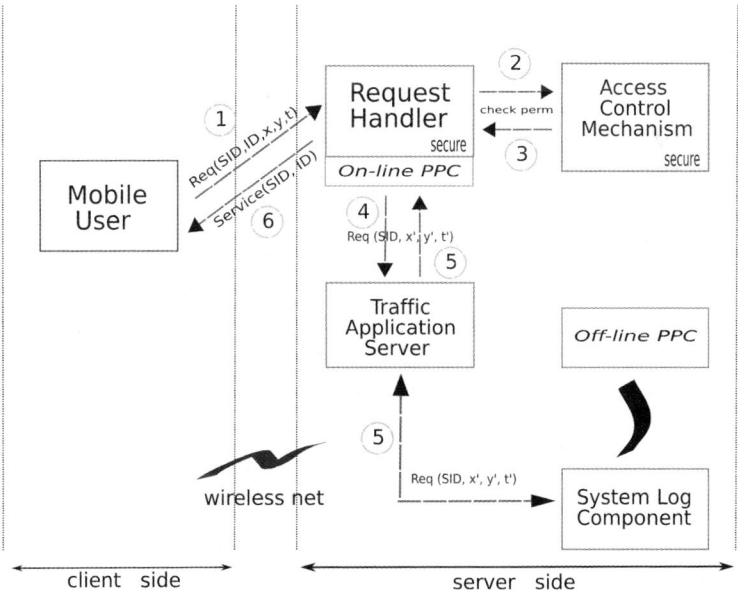

Fig. 8.2 The general architecture of the model system

sends a "denial of service" respond to the RH, who in turn terminates the processing of the request; otherwise, if the user's *ID* is among the recognized ones and his or her contextual information, corresponding to the requested service, are sufficient, ACM sends a "confirmation of validity" to the RH. Afterward, the request along with the role information and the additional information which the ACM may have provided is communicated to the *online* PPC. The *online* PPC accesses the role-based *privacy policy*. Such a policy may specify different privacy-preserving strategies. One of these is to cloak the contextual information by decreasing their spatiotemporal granularity. This decrement in granularity reduces the quality of the geometric and temporal representations; thus, it introduces a high level of uncertainty about the true position of the user at the time of request. Moreover, spatiotemporal generalization is both role and service dependent; the perturbation level is directly hinged on the user's role in the system and the tolerance constraints imposed by the requested service (to be still useful). After the *online* PPC finishes its essential processing, it forwards the privacy aware outcome to the *traffic application server* (TAS) to provide the requested service. Upon completion of the service, TAS forwards the request tuple to the *system log component* (SLC) in order to be stored. The SLC is assumed to communicate directly with a *session database*, which stores the users' sessions.

Figure 8.2 demonstrates the general system operation regarding the handling of location-based users' requests and the preservation of their location privacy. However, another important operation of the system regards the sharing of the induced knowledge that remains stored in the SLC component. In other words, after the data has been logged (in Step 5), the data is then protected by the server before it is shared to another party. Figure 8.3 depicts in detail how the sharing of knowledge is accomplished. The SLC component is depicted outside the regions of the knowledge-sharing subsystem since we envision it to enact as a bridge between the primary system operation and the operation accomplished in this subsystem. To explain how knowledge is shared by the system, assume that a mobile user wants

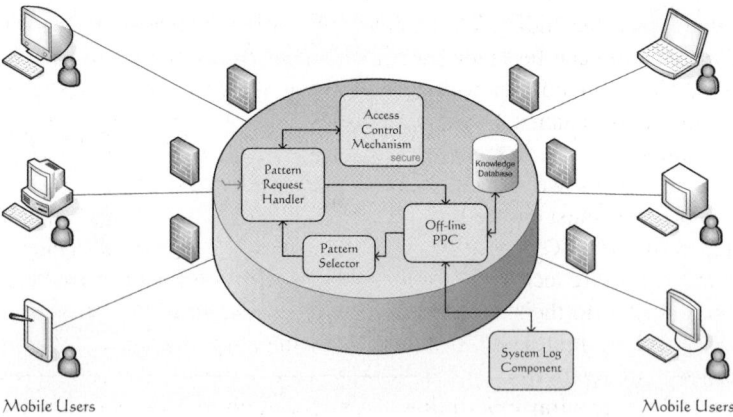

Fig. 8.3 Secure and privacy-aware sharing of knowledge

to retrieve knowledge regarding some specific pattern of behavior. To achieve this, he or she issues a request (similar to the ones presented earlier regarding location-based information) and sends this request to the system. This type of request has a specific *SID* that forces the RH component to forward it to the *pattern request handler* (PRH) component, presented in Fig. 8.3. This component works similarly to the RH; it first disambiguates the request based on its *SID* and then consults the ACM component to ensure that the user has the proper rights to receive the requested service. At this point we can either regard the ACM component of Fig. 8.3 as a different component than the one of Fig. 8.2 (as is demonstrated here) or we can think of them as the same component that lies outside the regions of the knowledge-sharing subsystem similarly to the SLC component. After the user's credentials are examined and access is granted, the PRH component updates the user's request to include the *pattern rights* associated to his or her profile and forwards the request to the offline PPC component. The *SID* of the service triggers the execution of the appropriate privacy-preserving data mining algorithm, existing within the regions of the offline PPC, and a set of knowledge patterns are produced. The produced knowledge is then stored to a KDB in order to be used for future reference and to relieve the offline PPC from the time-consuming task of reproducing it if needed in "recent" future requests. This means that if the same type of service is requested by some user within a small amount of time from the time it was produced by the offline PPC, then the system, instead of reproducing it, retrieves the appropriate patterns from the KDB. As a last step, after the knowledge has been produced, it is returned to the PRH after it passes through the *pattern selector* (PS). The PS acts as a filter whose purpose is to remove from the set of all identified patterns the ones that are off the region of the pattern rights associated to the user. The existence of pattern rights allows the system to enforce a strict policy that regulates who has access to which portion of knowledge, thus ensuring that knowledge is shared among the authorized users to the proper extent.

Concluding the presentation of the proposed framework, we devote a few lines to describe the envisioned privacy-preserving strategies, accomplished in the PPC parts of the system. The *offline* PPC encompasses a set of *data sanitization* techniques that allow the hiding of sensitive information depicted in the system's logfile. Information can be hidden at either *transactional* or *pattern* (i.e., knowledge) level. Moreover, information may be securely shared among interested parties, in a privacy-aware manner. Chapter 11 presents an extensive review of privacy-preserving methodologies that are suitable to be encompassed in the model system. Thus, in what follows, we concentrate only on the *online* PPC component.

The *online* PPC must ensure that location information is properly handled prior to its storage in the SLC subsystem. Thus, it is assigned the task of incorporating a set of privacy-aware techniques that protect data from revealing users' sensitive information related to their trajectories. To achieve this goal, we introduce to the proposed model the strategy of *K-anonymous spatiotemporal patterns*. In our context, the notion of a *pattern* is equivalent to the notion of a trajectory; what we want to secure is the user's trajectory information. Based on received location data, the RH component has knowledge regarding the trajectories of various subscribers. This trajectory data is communicated to the TAS whose role is to provide the requested service. The purpose of the applied *online* privacy-aware methodology is to identify *frequent trajectories* and incorporate techniques that guarantee K-anonymity preservation. There are several open issues to be addressed in this context, including (a) the identification of the most suitable granularity for the spatiotemporal representation, (b) the selection of the most applicable technique for frequent pattern mining in the spatiotemporal environment, and (c) the adoption of the most appropriate methodology for performing K-anonymization. As part of the online privacy-preservation approach, one needs to decide whether existing methodologies for generalization, suppression, and K-anonymization suffice to provide the required services or new techniques have to be devised.

8.4 Conclusion

In this chapter we provided a thorough examination of the state-of-the-art methodologies in the field of privacy and security in spatiotemporal data and trajectories. As we demonstrated, several attempts have been made over the years to address security and privacy implications in general purpose databases; however, very few approaches have been proposed to capture the special requirements raised by location data, and even fewer to address the requirements of trajectory data. New applications require the incorporation of novel techniques to handle the security and privacy implications, which are more complex and acrimonious in spatiotemporal settings. We investigated some new requirements originating from this novel type of application, and proposed a general purpose reference model system architecture that addresses most of these demands. Our study demonstrates that several considerations need to be made on both theoretical and technological grounds to ensure that security and privacy is readily provided and is fully guaranteed for all the people indiscriminatingly in future mobile services.

Acknowledgment

We thank the Ph.D. candidate Dimitrios Gougoulas for his comments and overall contribution in the state-of-the-art part of this chapter.

References

1. R. Agrawal, J. Kiernan, R. Srikant, and Y. Xu. Hippocratic databases. In *Proceedings of the 28th International Conference on Very Large Databases (VLDB'02)*, pp. 143–154, 2002.
2. V. Atluri and S.A. Chun. An authorization model for geospatial data. *IEEE Transactions Dependable Security Computing*, 1(4):238–254, 2004.
3. V. Atluri and P. Mazzoleni. Uniform indexing for geospatial data and authorizations. In *Proceedings of the 16th Conference on Database Security (DBSEC'02)*, pp. 207–218, 2002.

4. A. Belussi, E. Bertino, B. Catania, M.L. Damiani, and A. Nucita. An authorization model for geographical maps. In *Proceedings of the 12th International Workshop on Geographic Information Systems (GIS'04)*, pp. 82–91, 2004.
5. A.R. Beresford and F. Stajano. Location privacy in pervasive computing. *IEEE Pervasive Computing*, 2(1):46–55, 2003.
6. A.R. Beresford and F. Stajano. Mix zones: user privacy in location-aware services. In *Proceedings of the Second Conference on Pervasive Computing and Communications Workshops (PERCOM'04)*, pp. 127–131, 2004.
7. E. Bertino, B. Catania, M.L. Damiani, and P. Perlasca. GEO-RBAC: a spatially aware RBAC. In *Proceedings of the 10th Symposium on Access Control Models and Technologies (SACMAT'05)*, pp. 29–37, 2005.
8. E. Bertino, M.L. Damiani, and D. Momini. An access control system for a web map management service. In *Proceedings of the 14th International Workshop on Research Issues in Data Engineering (RIDE 2004)*, pp. 33–39, 2004.
9. C. Bettini, X.S. Wang, and S. Jajodia. Protecting privacy against location-based personal identification. In *Proceedings of the Second VLDB Workshop on Secure Data Management (SDM'05)*, pp. 185–199, 2005.
10. M. Bishop. *Introduction to Computer Security*. Addison-Wesley, Reading, MA, 2005.
11. D. Boneh and M.K. Franklin. Identity-based encryption from the weil pairing. In *Proceedings of the 21st Annual International Cryptology Conference (CRYPTO'01)*, pp. 213–229, 2001.
12. D. Chaum. The dining cryptographers problem: unconditional sender and recipient untraceability. *Journal of Cryptology*, 1(1):65–75, 1988.
13. M. Covington, W. Long, S. Srinivasan, A.K. Dev, M. Ahamad, and G.D. Abowd. Securing context-aware applications using environment roles. In *Proceedings of the 6th Symposium on Access Control Models and Technologies (SACMAT'01)*, pp. 10–20, 2001.
14. L. Cranor, M. Langheinrich, and M. Marchiori. A P3P preference exchange language 1.0 (APPEL 1.0), April 2002.
15. L. Cranor, M. Langheinrich, and M. Marchiori. The platform for privacy preferences 1.0 (P3P1.0), April 2002.
16. M. Damiani and E. Bertino. Access control and privacy in location-aware services for mobile organizations. In *Proceedings of the Seventh International Conference on Mobile Data Management (MDM'06)*, pp. 11–21, 2006.
17. M. Duckham and L. Kulik. A formal model of obfuscation and negotiation for location privacy. In *Proceedings of the Third International Conference on Pervasive Computing (Pervasive'05)*, pp. 152–170, 2005.
18. C. Ellison, B. Frank, B. Lamson, R. Rivest, B. Thomas, and T. Ylonen. *SPKI Cerificates Theory. RFC 2693*, September 1999.
19. S. Fu and C.-Z. Xu. A coordinated spatio-temporal access control model for mobile computing in coalition environments. In *Proceedings of the 9th International Parallel and Distributed Processing Symposium(IPDPS'05)*, 2005.
20. B.G. Gedik and L. Liu. A customizable K-anonymity model for protecting location privacy. Technical Report GIT-CERCS-04-15, Georgia Institute of Technology, April 2004.
21. B. Gedik and L. Liu. Location privacy in mobile systems: a personalized anonymization model. In *Proceedings of the 25th International Conference on Distributed Computing Systems (ICDCS'05)*, pp. 620–629, 2005
22. C. Gentry and A. Silverberg. Hierarchical ID-based cryptography. In *Proceedings of the 8th International Conference on the Theory and Application of Cryptology and Information Security (ASIACRYPT'02)*, pp. 548–566, 2002.
23. Global spatial data infrastructure association (gsdi). http://www.gsdi.org/.
24. M. Gruteser and D. Grunwald. Anonymous usage of location-based services through spatial and temporal cloaking. In *Proceedings of the First International Conference on Mobile Systems, Applications, and Services (MobiSys'03)*, 2003.
25. M. Gruteser and X. Liu. Protecting privacy in continuous location-tracking applications. *IEEE Security & Privacy Magazine*, 2(2):28–34, 2004.

26. F. Hansen and V. Oleshchuk. Spatial role-based access control model for wireless networks. In *Proceedings of the Vehicular Technology Conference (VTC'03)*, pp. 2093–2097, 2003.
27. C. Hauser and M. Kabatnik. Towards privacy support in a global location service. In *Proceedings of the IFIP Workshop on IP and ATM Traffic Management*, pp. 81–89, 2001.
28. U. Hengartner and P. Steenkiste. Protecting access to people location information. In *Proceedings of the First International Conference of Security in Pervasive Computing (SPC'03)*, pp. 25–38, 2003.
29. U. Hengartner and P. Steenkiste. Implementing access control to people location information. In *Proceedings of the 9th Symposium on Access Control Models and Technologies (SACMAT'04)*, pp. 11–20, 2004.
30. U. Hengartner and P. Steenkiste. Exploiting hierarchical identity-based encryption for access control to pervasive computing information. In *Proceedings of the First IEEE/CreateNet International Conference on Security and Privacy for Emerging Areas in Communication Networks (SecureComm'05)*, pp. 384–393, 2005.
31. B. Hoh and M. Gruteser. Protecting location privacy through path confusion. In *Proceedings of the IFIP Workshop on IP and ATM Traffic Management*, pp. 194, 205, 2005.
32. H. Kido. Location anonymization for protecting user privacy in location-based services. Master's thesis, Graduate School of Information Science and Technology, Osaka University, February 2006.
33. H. Kido, Y. Yanagisawa, and T. Satoh. An anonymous communication technique using dummies for location-based services. In *Proceedings of the Third International Conference on Pervasive Computing (Pervasive'05)*, pp. 88–97, 2005.
34. H. Kido, Y. Yanagisawa, and T. Satoh. Protection of location privacy using dummies for location-based services. In *Proceedings of the 21st International Conference on Data Engineering (ICDE'05)*, pp. 118–122, 2005.
35. D. Konidala, C.Y. Yeun, and K. Kim. A secure and privacy enhanced protocol for location-based services in ubiquitous society. In *Proceedings of GLOBECOMM'04*, pp. 931–936, 2004.
36. M. Langheinrich. A privacy awareness system for ubiquitous computing environments. In *Proceedings of the 4th International Conference on Ubiquitous Computing (UbiComp'02)*, pp. 237–245, 2002.
37. A. Matheus. Declaration and enforcement of fine-grained access restrictions for a service-based geospatial data infrastructure. In *Proceedings of the 10th Symposium on Access Control Models and Technologies (SACMAT'05)*, pp. 21–28, 2005.
38. G. Myles, A. Frifay, and N. Davies. Preserving privacy in environments with location-based applications. *IEEE Pervasive Computing*, 2(1):56–64, 2003.
39. National spatial data infrastructure (nsdi). http://www.fgdc.gov/nsdi/nsdi.html.
40. A. Pfitzmann and M. Köhntopp. Anonymity, unobservability, and pseudonymity – a proposal for terminology. In *Proceedings of the International Workshop on Design Issues in Anonymity and Unobservability*, pp. 1–9, 2000.
41. P. Samarati and L. Sweeney. Protecting privacy when disclosing information: k-anonymity and its enforcement through generalization and suppresion. In *Proceedings of the Symposium on Research in Security and Privacy*, pp. 384–393, 1998.
42. R.S. Sandhu, E.J. Coyne, H.L. Feinstein, and C.E. Youman. Role-based access control models. *IEEE Computer*, 29(2):38–47, 1996.
43. J.A. Simpson and E.S.C. Weiner. *Oxford English Dictionary*, 2nd edn. Clarendon Press, Oxford, 1989.
44. E. Snekkenes. Concepts for personal location privacy policies. In *Proceedings of the 3rd Conference on Electronic Commerce (EC-'01)*, pp. 48–57, 2001.

Part III
Mining Spatiotemporal and Trajectory Data

Chapter 9
Knowledge Discovery from Geographical Data

S. Rinzivillo, F. Turini, V. Bogorny, C. Körner, B. Kuijpers, and M. May

9.1 Introduction

During the last decade, data miners became aware of geographical data. Today, knowledge discovery from geographic data is still an open research field but promises to be a solid starting point for developing solutions for mining spatiotemporal patterns in a knowledge-rich territory. As many concepts of geographic feature extraction and data mining are not commonly known within the data mining community, but need to be understood before advancing to spatiotemporal data mining, this chapter provides an introduction to basic concepts of knowledge discovery from geographical data.

In performing knowledge discovery in a spatial data set, the first important question is how to use the spatial dimension in the discovery process. At least two viewpoints can be considered: either spatial relationships are made explicit prior to data mining or specialised algorithms are directly applied to spatial and non-spatial data. The first approach claims that the spatial dimension is somewhat more basic than the other features, and, then, it can be used to prepare the data set for a successive knowledge extraction step. The exploitation of the spatial dimension for selecting the values of attributes to be used in the mining step can be quite complex, and it may depend both on the structure of the domain and on the kind of model one is looking for. This approach offers the advantage of allowing the reuse of standard data mining technology on data extracted according to the spatial dimension. The second approach aims at exploiting the spatial features dynamically during the discovery process. This implies a complete reinvention of the data mining technology, but it allows a more flexible use of spatial knowledge.

Mining geographic data poses additional challenges, which include the exploitation of background knowledge as well as the handling of spatial autocorrelation

---

S. Rinzivillo
KDD Laboratory, Dipartimento di Informatica, Università di Pisa, Italy, e-mail: rinziv@di.unipi.it

and highly erroneous data. Although many data mining algorithms extend over multi-dimensional feature spaces and are thus inherently spatial, they are not necessarily adequate to model geographic space. The first specialised algorithms for geo-referenced data were introduced by Koperski and Han [25] and Ester et al. [13].

This chapter provides an overview of knowledge discovery from geographic data. In Sect. 9.2 we revise basic spatial concepts and the representation of geographic data. Section 9.3 introduces Geographic Information Systems (GIS) and first approaches to enrich these systems with data mining capabilities. Section 9.4 focuses on the extraction of implicit features and relationships from geographic data. Algorithms for mining geo-referenced data are discussed in Sect. 9.5, and in the subsequent section we provide an example that connects all presented aspects of the knowledge discovery process. In Sect. 9.7 we construct a road map that views approaches to geographic knowledge discovery in the light of spatiotemporal data, and we conclude the chapter with a short summary.

9.2 Geographic Data Representation and Modelling

9.2.1 Conceptual Models of Space

Conceptual models are an abstract representation of reality, reflecting the main characteristics of objects and events from a user's point of view. In the spatial domain, they depict measurements or observations (of objects) referenced in space. Conceptual models of space are independent of any restrictions imposed by a subsequent representation in information systems. On the conceptual level, two major approaches can be distinguished [19,27]. The first model regards the spatial domain as a continuous surface, each point of which can be mapped to one and only one value of some attribute. This paradigm is called *field* model and represents a function of location in two or three-dimensional space. Typical applications of field models are measurements of mineral and pollutant concentrations or temperature in soil and air. The second conceptual model is based on discrete *objects*. An object may be a point, line or polygon and may represent a tree, street or city, respectively. In contrast to fields, the world of object models is empty except for places that are occupied by objects.

9.2.2 Representation of Spatial Data

The continuous geographic space must be digitalised before it can be stored in a computer. Two main data structures, tessellation and vector, have been developed to represent geographic data in a discrete way. Although apparently related, both forms can be used to represent the concept of fields or objects.

9 Knowledge Discovery from Geographical Data

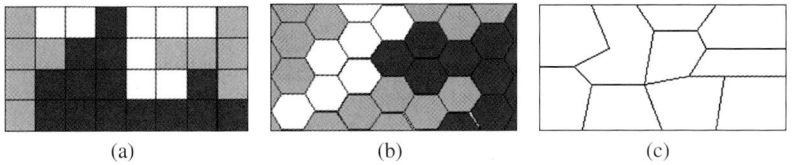

Fig. 9.1 Regular and irregular tessellations: (**a**) square cells, (**b**) hexagonal cells, (**c**) irregular cells

Fig. 9.2 Vectors: points, polylines and polygons

9.2.2.1 Tessellation

Tessellation models partition the space into a number of cells, and each stores a value of the associated attribute [7, 37]. The grid can be regularly spaced, in this case it is also called a raster, or irregularly spaced. Figure 9.1 shows a regular grid of square and hexagonal cells as well as an irregular tessellation. The intensity of colour indicates different attribute values. All variation within a cell is lost. Thus, the size of the cell defines the level of resolution. Regular tessellation models possess very efficient indexing structures (run-length encoding, quadtrees) and are well suited to model continuous change. Their disadvantages include the memory space and computational costs involved to manage high resolutions. Regular tessellation models are commonly applied to satellite and environmental data. Irregular tessellations are used, for example, to represent administrative units.

9.2.2.2 Vector

The vector model is most commonly used in current GIS [26, 37]. In the vector model, infinite sets of points in space are represented as finite geometric structures. More precisely, a vector datum consists of a tuple of the form (*geometry, attributes*), where a geometry can be a point, polyline or polygon. A point is typically given by its rational coordinates. A polyline is represented by a sequence of points and a polygon takes the form of a closed polyline. Examples of vector data are shown in Fig. 9.2. The advantage of the vector model is the concise representation of objects. However, it involves complex data structures, and the computation of spatial operations, such as intersection and overlay, may take considerable time and resources [7, 37].

9.3 Geographic Information Systems

9.3.1 Definition

Geographic Information Systems (GIS) have been defined in many ways. Today, it is no longer easy to give a clear definition of GIS. During the past 20 or more years GIS have evolved from Systems to Science, a complex interaction of theory, technology and systems. But the main concern of GIS remains to handle geographic information about places on the Earth's surface and to deal with knowledge about *what is where when* (recently, time has also taken its place in GIS [35]). A popular definition [10, 21] says that a GIS is 'a system of hardware, software and procedures designed to support the capture, management, manipulation, analysis, modelling and display of spatially-referenced data for solving complex planning and management problems'.

On its technological side, GIS rely on techniques like Global Positioning Systems (GPS) and Remote Sensing. In the past, GIS have cynically been called 'maps with a database behind it', but the data models allow a complex representation of the real world that can support querying, analysis and decision support.

The data stored in a GIS is typically divided over thematic layers. Grosso modo, we can say that each of these layers is modelled in the tessellation or vector model as described above.

9.3.2 Thematic Layers

Real world data contains many different aspects. In the description of a city or a region we can, for instance, distinguish between the road network, cadastral information about parcels and houses, hydrographic information, topography (terrain elevation), etc. Following this thematic division, data in a GIS is typically organised by *layers*, which correspond to themes in the application. For instance, one layer could contain information about the road network, whereas another could contain information about the rivers and lakes and yet another could contain information on elevation. Although data is divided over thematic layers, there is a way of integrating different layers, namely using explicit location on the Earth's surface. Using the geographic location as an organizing principal between layers, they can be overlayed or spatially joined.

Each layer represents a common feature and therefore the information in one layer is of a similar type, whereas information in different layers may be of quite different nature. Layers are described by two types of data: spatial data that describes the location of objects and thematic or attribute data that specifies the characteristics of the data in a traditional alpha-numeric way (these data are usually stored in a relational database). The spatial part of the information within one layer can be stored in any physical representation, depending on the need of the data and the application.

9.3.3 Integration of Knowledge Discovery and GIS

Most commercially available GIS provide extended functionality to store, manipulate and visualise geo-referenced data, but rely on the user's ability for exploratory data analysis. This approach is not feasible with regard to the large amount and high dimensionality of geographic data, and demands for integrated data mining technology.

The integration of data mining methods and GIS functionality does not only facilitate data analysis, but also allows for an efficient implementation of algorithms. One prospective area is spatial feature extraction. In general, the application of spatial operations for feature extraction is computationally expensive. When the feature extraction and data mining step are interweaved, a dynamic selection of objects, for which some spatial relationship must be computed, can take place.

To our best knowledge, there are only a few software systems that join the power of data mining techniques and GIS, namely GeoMiner, SPIN! and INGENS. GeoMiner [20] has been among the first approaches to mine geographic data from large spatial databases and focuses on the discovery of spatial association rules (SARs). SPIN! [32, 42] is a spatial data mining platform that integrates several algorithms for spatial data mining, which include multi-relational subgroup discovery, rule induction and spatial cluster analysis. It pays special attention to the scalability of algorithms allowing for a tight coupling with the database, and it provides an extensive interface for visual data exploration. INGENS [29] (INductive GEographic iNformation System) is a prototypical GIS that possesses an inductive learning capability. It can generate first-order logic descriptions for geographic objects, and it includes a training facility that allows the interactive selection of examples and counter examples of geographic concepts.

9.4 Spatial Feature Extraction

A spatial feature describes some characteristic of a geographic object. We use the term *feature* in compliance with the definition commonly used in data mining, and not according to the Open GIS Consortium terminology where it corresponds to a real world or abstract entity [34].

A geographic object is characterised by a spatial component, e.g. a geometric object that represents its position in the geographic space, and a set of attributes that describe the non-spatial dimensions of the object, e.g. the type of a road or the construction year of a building. While the non-spatial information can be queried in traditional ways, the information of spatial relationships is implicitly encoded and must be extracted prior to data mining. Spatial feature extraction poses the challenge to reveal meaningful information of geographic objects, with a particular interest in their relationships.

This section describes relation-based and aggregation-based spatial features. It gives an overview of the state-of-the-art of spatial feature extraction. We conclude with the enhancement of feature extraction using background knowledge.

9.4.1 Relational Features

Information about spatial objects can be derived from single objects or from the relationship between two or more objects. The former are called *unary* features (such as length, area and perimeter), the latter *relational* features. Probably the most prominent relational feature is the *distance* between two objects, which can be measured, for example, using Euclidean distance. In this section we will give an introduction to two further relational features of spatial objects, namely *topological* and *directional* relations.

9.4.1.1 Topological Relations

The topological relations are invariant under homomorphisms, i.e. they are preserved if the considered objects are rotated, scaled or moved. The formal definition of such relations is based on the *point-set topology* theory. Each geometry G is considered as composed of three parts: the *interior* (denoted by $G°$), i.e. the set of all the points that are inside the geometry; the *boundary* (denoted by δG), i.e. the limit of the geometry; and the *exterior* (denoted by G^-), i.e. all the remaining points that do not belong to the object.

The *9-intersection model*, proposed in [12], gives a formal description of the topological relation between two objects. The model is based on the evaluation of all the possible intersections among the interiors, the boundaries and the exteriors of two objects. In particular, given two geometries A and B, the nine possible intersections define the relation between the two geometries, and it is represented by means of the *9-intersection matrix*:

$$R(A,B) = \begin{pmatrix} \delta A \cap \delta B & \delta A \cap B° & \delta A \cap B^- \\ A° \cap \delta B & A° \cap B° & A° \cap B^- \\ A^- \cap \delta B & A^- \cap B° & A^- \cap B^- \end{pmatrix}.$$

Each of these intersections is tested if it is empty or not, which results in a total of 2^9 combinations. However, many of these cases can be discarded due to geometric properties of the considered objects. For example, if we consider 2 two-dimensional objects, say A and B, there are only eight possible relations between A and B (shown in Fig. 9.3), i.e. there are eight possible distinct configurations of the matrix.

9.4.1.2 Directional Relations

Directional relations are defined over a reference system determined by two orthogonal axes, x and y. Based on relationships between point objects, the definition

9 Knowledge Discovery from Geographical Data

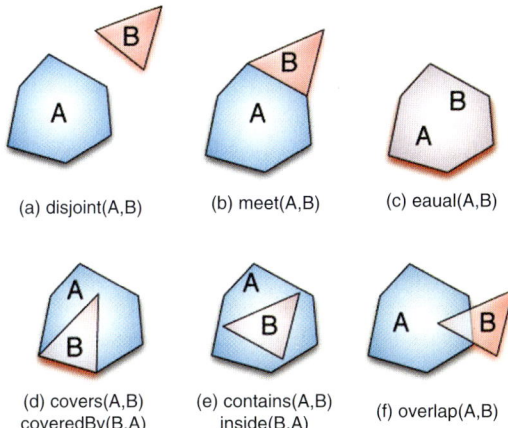

Fig. 9.3 Topological relations

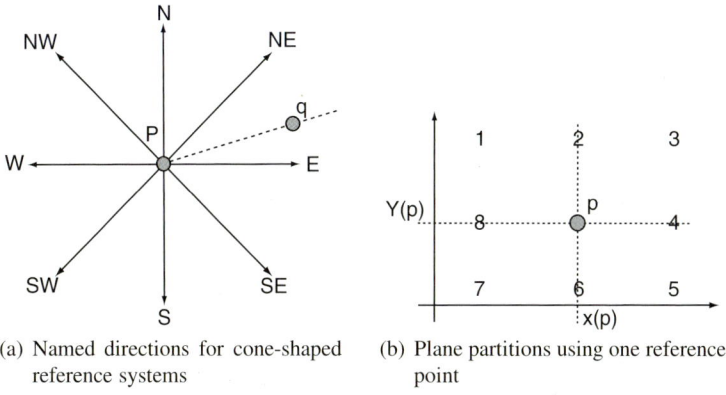

(a) Named directions for cone-shaped reference systems

(b) Plane partitions using one reference point

Fig. 9.4 Examples of directional relations

of directional relations can be extended to objects of arbitrary shape [36]. The approaches used for the formal definition of directional relations are mainly based on two methods [17]: *cone-shaped areas* and *projections*.

The *cone-shaped areas* method relates the cardinal direction between two points p and q by considering the angular direction with reference to some fixed direction in space. For example, the directional symbols for the system presented in Fig. 9.4a are $V_9 = \{N, NE, E, SE, S, SW, W, NW, 0\}$. The direction 0 (zero) represents the situations when two points are not distinct. The direction through the two points p and q is closer to the E direction; therefore, the symbol E is assigned to the direction pq.

The *projection-based* method uses projection lines to determine the direction between two points in space. Let us consider a reference point p. If we draw two orthogonal projections from point p we obtain nine partitions of the space (four open line segments, four open regions and the intersection point, see Fig. 9.4b). The position of a second point q inside one of these regions determines the direction.

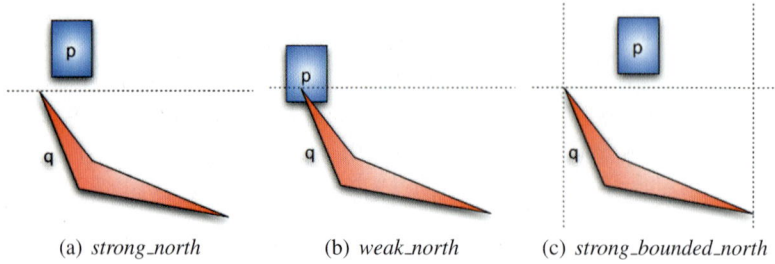

(a) *strong_north*   (b) *weak_north*   (c) *strong_bounded_north*

Fig. 9.5 Directional relations for extended spatial objects

Directional relations can be generalised for two objects of arbitrary shape using the above definitions. Given two spatial objects p and q, we denote with p_i (q_i) a generic point of object p (q). The relation *strong_north* $\equiv \forall p_i \forall q_i north(p_i, q_i)$ denotes that all the points of p are north of all points of q (Fig. 9.5a). The relation *weak_north* holds when some points of p are north of all points of q, for each point of p there exist some points of q such that p_i is north of q_i, and some points of p are south of some points of q (Fig. 9.5b).

Directional relations can be defined by using the minimum bounding box (MBB) approximation. A MBB is the axis-parallel rectangle that is spanned by the two coordinates $c_1 = (\min(x_p), \min(y_p))$ and $c_2 = (\max(x_p), \max(y_p))$. We denote with MBB($p$) the bounding box of object p. The relation

$$\text{strong\_bounded\_north}(p, q) \equiv \text{strong\_north}(\text{MBB}(p), \text{MBB}(q))$$

holds when all the points of p are bounded by the horizontal line that passes through the northernmost point of q and by the two vertical lines that also bound q (Fig. 9.5c). Similarly, the relations for other cardinal directions can be defined.

9.4.2 Spatial Aggregation

Aggregation of data is commonly applied to summarise information and to derive features that cannot be measured at a single point. Within the spatial domain, aggregation is also used to attach information about the local environment to some entity. For example, to compare birth rates of European countries the number of live births and inhabitants must be summarised for each country. The birth rate itself is a variable that cannot be measured at a single location but must be derived for some areal unit. For urban planning, a smaller areal unit may be chosen. For example, a city council might evaluate locations for a new kindergarten based on socio-demographic data of the respective municipal districts.

This example shows that spatial information can be aggregated at several levels of resolution. The choice of resolution is not always obvious, which gives rise to the

Fig. 9.6 Aggregations within Frankfurt, Germany: (**a**) 1,500 m buffer; (**b**) 4 min drive time zone

modifiable areal unit problem. The modifiable areal unit problem [15] comprises two parts, the scale effect and the zoning effect. The scale effect may lead to different statistical results if information is grouped at different levels of spatial resolution. The zoning effect refers to the variability of statistical results if the borders of spatial units are differently chosen at a given scale of resolution. Both effects need to be carefully considered when aggregating spatial data.

In the example above, administrative borders were chosen to define spatial units. This paragraph presents two techniques to aggregate spatial data based on distance and time. Distance-based units, also called buffers, contain all objects that lie within a predefined distance to the object in question. Continuing the example from above, the city council could count the number of households with young children within a 2 km distance to each potential location. Yet, distance alone may not always yield the desired result. Imagine, one location is situated close to a river and no bridge is nearby. In this case, it may be more important *how long* it takes for a parent to reach the kindergarten. Units that contain all objects reachable within a given amount of time are called drive time zones. They are calculated using Dijkstra's algorithm to determine the shortest path between two points in a graph. The graph is formed by the underlying street network with edges weighted according to the average or maximal speed allowed. Figure 9.6 contrasts a 1,500 m buffer and a 4-min drive time zone in the middle of Frankfurt, Germany.

9.4.3 State-of-the-Art Feature Extraction

The extraction of spatial features from geographic data, such as topological and distance, is the most effort- and time-consuming step in the whole discovery process [41], but only little attention has been devoted to this problem. On the one side the user must choose the appropriate spatial and non-spatial features. On the other side the extraction process itself requires high computational costs. Spatial features can be extracted from geographic data by functionalities provided by GIS and geographic database management systems. Several approaches to extract spatial features for data mining and knowledge discovery have been proposed. Spatial features can be extracted either in the data preprocessing or during the data mining task.

Most approaches extract spatial features in data preprocessing, where any spatial relation may be computed and geographic objects may have any geometric representation (e.g. point, line). In [25] a top-down progressive refinement method is proposed and spatial approximations are calculated in a first step. In a second step, more precise spatial relationships are computed to the outcome of the first step. This method has been implemented in the GeoMiner system [20]. Ester et al. [14] proposed new operations such as graphs and paths to compute spatial neighbourhoods. However, these operations are not implemented by most GIS, and to compute all spatial relationships between all geographic objects, in order to obtain the graphs and paths, is computationally expensive for real applications. In [28] all spatial relationships are computed and converted to a first-order logic database. This process is computationally expensive for real problems and many spatial relationships might be unnecessarily extracted. A feature extraction module named Featex has been implemented in the ARES system [3], where the user can choose the geographic object types and non-spatial attributes. An approach that uses geo-ontologies as prior knowledge to filter spatial features has been proposed by Bogorny et al. [5]. In this approach the semantics of geographic objects is considered, and geo-ontologies are used to compute only topological features semantically consistent.

The approach of [22] deals with geographic coordinates directly and extracts spatial features during the mining task, yet it considers only distance features. Another drawback is the input restriction to point primitives. For geographic objects represented by *n*-dimensional primitives, their centroid is extracted. This process may lose significant information and may generate imprecise patterns. For example, the Mississippi River intersects many states considering its real geometry, but it will be far from the same states if only the centroid is considered.

9.4.4 Improvement of Feature Extraction Using Background Knowledge

In geographic space, many features represent natural geographic dependences in which some objects are *always* related to other objects. For instance, islands are naturally *within* water bodies, ports are naturally *adjacent* to water bodies and bus stops *intersect* roads.

A large amount of natural geographic dependences is well-known by geographers and geographic database designers. These dependences are normally explicitly represented in geographic database schemas through one-to-one and one-to-many cardinality constraints [41] in order to warrant the spatial integrity of geographic data [40]. Since natural dependences are intrinsic to geographic data they are also

9 Knowledge Discovery from Geographical Data

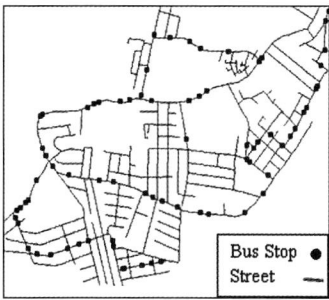

Fig. 9.7 Region of the Porto Alegre city

represented in geographic ontologies [5]. Geo-ontologies and geographic database schemas are rich knowledge repositories that can be used as background knowledge to accelerate the spatial feature extraction.

Well-known spatial features require computational time to be extracted and generate patterns which are in most cases non-novel and non-interesting for data mining [4]. But which spatial features can be considered either interesting or well-known? Figure 9.7 shows a geographic map in which it is possible to visualise that all bus stops intersect streets. It is well-known that bus stops only exist on streets, but their spatial relationship is normally not explicitly stored in geographic databases, and needs to be extracted with functionalities provided by GIS. Considering this example, the topological feature between both geographic objects could be retrieved from the knowledge base instead of performing a spatial join operation. Since both geographic objects have a mandatory topological relationship, the distance between these objects is zero, and no spatial operation is required to extract either topological or distance features.

Background knowledge can be used to improve the discovery process from many different perspectives, but only a few approaches have used prior knowledge in geographic data mining. In [3] prior knowledge is defined by the data mining user and is used to reduce the number of well-known patterns. In [4] background knowledge is extracted from geographic database schemas to reduce spatial joins in feature extraction and to reduce well-known patterns. In [5] background knowledge is extracted from geo-ontologies and is used to improve topological feature extraction. In this approach only topological relationships that are semantically consistent are computed.

9.5 Spatial Data Mining

This section presents an overview of spatial data mining techniques that are applied to geographic data. It is important to notice the difference between the term *spatial* and *geographic* [33]: "'Spatial' concerns any phenomena where the data objects can be embedded within some formal space that generates implicit relationships among the objects. [...] 'Geographic' refers to the specific case where the data objects are georeferenced and the embedding space relates (at least conceptually) to locations on or near the Earth's surface'. Spatial data mining thus includes geographic data mining as a special case.

In Sect. 9.5.1 we describe challenges of spatial data mining that arise due to the nature of geographic data. The remaining sections present recent approaches to clustering, classification, regression, association rule mining and subgroup discovery using geographic data.

9.5.1 Challenges for Mining Geographic Data

Geographic data often violate assumptions that are essential to traditional data mining techniques. The most predominant characteristic of geographic data is known as *Tobler's Law* [43], which states that '[...] everything is related to everything else, but near things are more related than distant things'. It means that attribute values of spatial objects are the stronger correlated the closer two objects are in location. Usually, geographic objects exhibit strong positive autocorrelation and show similar values within their local neighbourhood. This behaviour directly contrasts the often made assumption of independent, identical distributions in classical data mining and causes poor performance of algorithms that ignore autocorrelation [8]. A second characteristic of geographic data is its variation across several scales of resolution. Dependencies on a small scale turn into random variation when analysed using broader units of measurement. Thus, discovered patterns depend on the choice of resolution and are subject to random variation. A third challenge for mining geographic data pose the implicitly defined relationships between spatial objects. They can be extracted, as described in Sect. 9.4 either previously to the application of algorithms or dynamically [2].

In general there are two alternatives how algorithms treat geographic data. The first approach uses traditional algorithms and includes spatial attributes either as ordinary variables or requires feature extraction during pre-processing. The second approach relies on specialised algorithms that incorporate feature extraction or are able to handle geographic dependencies directly. In the remaining section, we present several algorithms that are applied to geographic data and emphasise their strategy for feature extraction and their ability to handle autocorrelation.

9.5.2 Clustering

Clustering divides a given set of objects into non-overlapping groups, such that similar objects are within the same group and objects of different groups are most heterogeneous. As clustering relies on the distance between objects, it is inherently spatial. Yet, the assumption of convex clusters (e.g. *k*-means) is inappropriate for many geographical data sets (see Fig. 9.8). Ester et al. [13] developed a density based algorithm for point data that finds clusters of arbitrary shape. The idea of this

9 Knowledge Discovery from Geographical Data

Fig. 9.8 Spatial clusters: (**a**) convex shape; (**b**) arbitrary shape

approach is that a cluster can be recognised by a high density of points within, while only few points are found in the surrounding environment. It requires the definition of a *neighbourhood*, which is used to iteratively join points, and a *density* which is used to delineate the borders of a cluster. In [39] this approach is extended to cluster vector data (e.g. polygons).

9.5.3 Classification and Regression

In classification and regression, the unknown target value of some object is predicted given a set of training instances. If the target variable is discrete, the learning task is called classification. If it is continuous it is referred to as regression. We start with the well-known k-nearest neighbour method, which can be applied to both classification and regression tasks. The second part presents spatial model trees, geographically weighted regression, and we conclude this section with Kriging. Kriging is a popular regression method in geostatistics and takes explicitly advantage of autocorrelation.

9.5.3.1 k-Nearest Neighbour

The k-nearest neighbour algorithm (kNN) is an instance-based learning method that classifies unknown instances according to the target value of the k most similar training examples. It assumes that objects with similar characteristics also possess similar class values. In case of classification, the most frequent target value among the neighbours will be assigned to the instance. In case of regression, a (weighted) mean is calculated. To determine the similarity between two objects, kNN requires a distance measure for each attribute. As geographic coordinates can be used to determine the distance between two locations, they can be directly included in the algorithm. Thus, kNN relies on objects that are within the geographic neighbourhood and exploits positive autocorrelation of the target variable.

9.5.3.2 Model Trees

Model trees [45] operate similar to decision trees, but possess leaves that are associated with (linear) functions instead of fixed values. While internal nodes of the tree partition the sample space, leave nodes construct local models for each part of the sample space. Malerba et al. [30] developed a spatial model tree that is able to model local as well as global effects. Their induction method, Multi-relational Spatial Stepwise Model Tree Induction (Mrs-SMOTI), places regression nodes also within inner nodes of the tree and passes these regression parameters to all child nodes. Mrs-SMOTI exploits spatial relationships over several layers and possesses a tight database integration to extract spatial relations during the induction phase.

9.5.3.3 Geographically Weighted Regression

Geographically weighted regression (GWR) [16] extends the traditional regression framework such that all parameters are estimated within a local context. The model for some variable z at location i then takes the following form:

$$z_i = \beta_0(x_{ix}, x_{iy}) + \sum_k \beta_k(x_{ix}, x_{iy}) x_{ik} + \varepsilon_i.$$

In the equation above, (x_{ix}, x_{iy}) denotes the pair of coordinates at location i, $\beta_k(x_{ix}, x_{iy})$ is the localised parameter for attribute k, x_{ik} is the value of attribute k at location i and ε_i denotes random noise. The GWR model assumes that all parameters are spatially consistent. Therefore, parameters at location i are estimated from measurements close to i. This is realised by the introduction of a diagonal weight matrix W_i, which states the influence of each measurement for the estimation of regression parameters at i:

$$\widehat{\beta}(x_{ix}, x_{iy}) = (X^T W_i X)^{-1} X^T W_i \, z.$$

The weight matrix can be built according to several weighting schemes, such as a Gaussian or bi-square function. GWR is a local regression method that takes advantage of positive autocorrelation between neighbouring points in space.

9.5.3.4 Kriging

Kriging [9, 11, 44] is an optimal linear interpolation method to estimate unknown values in geographic field data. Let x denote a location in an index set $D \subset \mathbb{R}^n$ in n-dimensional space and $Z(x)$ a random variable of interest at location x. Generally, each variable $Z(x)$ can be decomposed into three terms: a structural component representing a mean or constant trend, a random but spatially correlated component that denotes autocorrelation, and random noise expressing measurement errors or variation inherent to the variable of interest [7].

A technique most widely used is Ordinary Kriging, which assumes intrinsic stationarity with an unknown, but constant mean of the random target variable $Z(x)$. Given a set of measurements, Kriging estimates unknown values as weighted sum of neighbouring sample data (Fig. 9.9a) and uses the *variogram* to determine optimal

9 Knowledge Discovery from Geographical Data

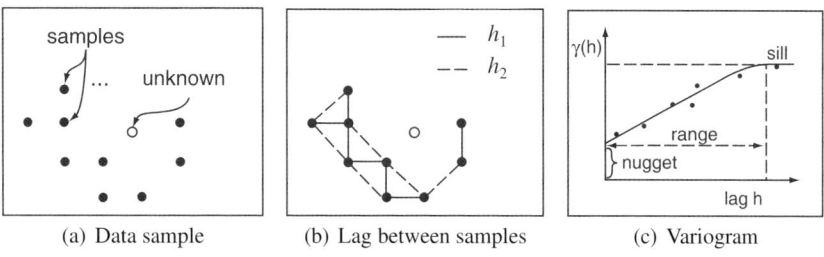

Fig. 9.9 Variance of sample increments

weights (Fig. 9.9c). Variograms model spatial dependency between locations and are a function of distance for any pair of sites:

$$\gamma(h) = \frac{1}{2}\text{Var}[Z(x+h) - Z(x)].$$

A variogram of the data can be obtained in two steps. First, the experimental variogram is calculated from the sample by calculating the variance between samples for all increments h. Figure 9.9b shows all pairs of sample points with a lag h_1 (solid lines) and a second lag of h_2 (dashed lines). In a second step, the experimental variogram serves to fit a theoretical variogram which is used in Ordinary Kriging. Depending on the data, different model types may be appropriate for the theoretical variogram. Often, a spherical model is used and its parameters are adapted to reflect the experimental variogram. Each variogram is characterised by three parameters: nugget, range and sill as depicted in Fig. 9.9c. The nugget effect represents random noise, as by definition $\gamma(0) = 0$. Within the range, the variance of increments increases gradually with distance in this example. It directly shows the spatial dependency. The closer two points are the more likely are their values similar. Finally, the curve levels are off at the sill. The variance has reached its maximum value and is independent of distance.

Ordinary Kriging estimates the unknown value at a location x_0 as weighted sum of neighbouring sample points x_i ($i = 1\ldots n$):

$$Z^*(x_0) = \sum_{i=1}^{n} w_i\, Z(x_i).$$

The weights w_i are determined in conformance with two restrictions. First, $Z^*(x_0)$ must be an unbiased estimate of the true value $Z(x_0)$, which means that on average the prediction error for location x_0 is 0. Because the model assumes a constant mean $m = E[Z(x_i)]$ ($i = 0..n$), this claim bounds the sum of weights to 1.

$$0 = E[Z^*(x_0) - Z(x_0)] = E\left[\sum_{i=1}^{n} w_i\, Z(x_i) - Z(x_0)\right]$$

$$= m\left(\sum_{i=1}^{n} w_i - 1\right) \quad \Rightarrow \quad \sum_{i=1}^{n} w_i = 1.$$

Second, we require an optimal estimate which minimises the error variance σ_E^2 of the estimate. The second equation expresses the variance in terms of the variogram.

$$\sigma_E^2 = \text{Var}\left(Z^*(x_0) - Z(x_0)\right) = E\left[(Z^*(x_0) - Z(x_0))^2\right]$$

$$= \sum_{i=1}^{n}\sum_{j=1}^{n} w_i w_j \gamma(x_i - x_j) - 2\sum_{i=1}^{n} w_i \gamma(x_i - x_0) + \gamma(x_0 - x_0).$$

The derivatives of the error variance with respect to $w_i (i = 1..n)$ yield a linear system of n equations. In combination with the restriction on the weights, a Lagrange parameter ϕ is introduced and a total of $n + 1$ equations is obtained. For each location x_0, the optimal weights w_i are estimated using the following system of equations, given in matrix form:

$$\begin{pmatrix} \gamma(x_1 - x_1) & \cdots & \gamma(x_1 - x_n) & 1 \\ \vdots & \ddots & \vdots & 1 \\ \gamma(x_n - x_1) & \cdots & \gamma(x_n - x_n) & 1 \\ 1 & \cdots & 1 & 0 \end{pmatrix} \begin{pmatrix} w_1 \\ \vdots \\ w_n \\ \phi \end{pmatrix} = \begin{pmatrix} \gamma(x_1 - x_0) \\ \vdots \\ \gamma(x_n - x_0) \\ 1 \end{pmatrix}.$$

Note that Ordinary Kriging is an exact interpolator. If the value of a location in the data sample is estimated, it will be identical with the measured value. Several variants of Kriging have been developed, which extend interpolation to data that contains a trend (Universal Kriging [9, 11]), involves uncertainty (Bayesian Kriging [9]) or contains temporal relations (Spatiotemporal Kriging).

9.5.4 Association Rules

Association rules consist of an implication of the form $X \rightarrow Y$, where X and Y are sets of items co-occurring in a given tuple of the data set ψ [1]. The support s of an itemset X is the percentage of rows in which the itemset X occurs as a subset. The support of the rule $X \rightarrow Y$ is given as $s(X \cup Y)$. The rule $X \rightarrow Y$ is satisfied in ψ with confidence factor $0 \leq c \leq 1$, if at least $c\%$ of the instances in ψ that satisfy X also satisfy Y. The confidence factor is given as $s(X \cup Y)/s(X)$.

Spatial association rules (SARs) consist of an implication of the form $X \rightarrow Y$, where X and Y are sets of predicates, and at least one element in X or Y is a spatial predicate [25]. The problem of mining SARs is decomposed in at least three main steps, where the first is usually performed as a data preprocessing method because of the high computational cost:

- spatial predicate computation: the spatial predicate is a spatial relationship (e.g. distance) between two geographic objects (e.g. closeToRiver),
- find all frequent predicate sets: a set of predicates is frequent if its support is at least equal to a minimum support *minsup*,
- generate strong association rules: a rule is strong if it reaches minimum support and the confidence is at least equal to the threshold *minconf*.

Existing SAR mining algorithms are Apriori-like approaches, since the computational cost relies on spatial feature extraction and not on the candidate generation as in transactional rule mining [41]. SAR mining algorithms can be classified into two main approaches. The first is based on quantitative reasoning, which mainly computes distance relationships during the frequent set generation. These approaches [22] deal with geographic data (coordinates x, y) directly. Although they have the advantage of not requiring the definition of a reference object, they have some general drawbacks: usually they deal only with points, consider only quantitative relationships, and they normally do not consider non-spatial attributes of geographic data.

The second approach [3, 6, 25, 28, 38] is based on qualitative reasoning, which usually considers different spatial relationships and features between a reference geographic object type and a set of relevant object types represented by any geometric primitive (e.g. points, lines). Spatial features are normally extracted in a first step, in data *preprocessing* tasks, as explained in Sect. 9.4, while frequent sets are generated in another step.

The main problem in both spatial and non-spatial association rule mining is the generation of huge amounts of rules. Both qualitative and quantitative reasoning approaches have proposed different methods for mining and filtering SAR. Koperski and Han [25] presented an approach which exploits taxonomies of both geographic object types and spatial relationships for mining SARs at different granularity levels. Only minimum support is used to prune frequent sets and association rules. In [28] both frequent sets and association rules are pruned a posteriori.

9.5.5 Subgroup Discovery

Subgroup discovery analyses dependencies between a target variable and several explanatory variables. It detects groups of objects that show a significant deviation in their target value with respect to the whole data set. For example, given a discrete target attribute, a subgroup displays an over-proportionally high or low share of a specific target value. More precisely, the quality q of a subgroup h accounts for the difference of target share between the subgroup p and the whole data set p_0, as well as the size n of the subgroup [23]:

$$q(h) = \frac{|p - p_0|}{\sqrt{p_0(1 - p_0)}} \sqrt{n}.$$

Subgroups are usually defined by simple conjugation of attribute values, which are then applied to the data set in question. Spatial subgroups are formed if the subgroup definition involves operations on spatial components of the objects. For example, a spatial subgroup could consist of all city districts that are intersected by a river [24]. However, spatial operations are expensive and, due to early pruning, it may not be necessary to compute all relations in advance. Klösgen and May [24] developed a spatial subgroup mining system, which integrates spatial

9.6 Example: Frequency Prediction of Inner-City Traffic

Research within the transportation domain as outlined in Sect. 2.6.1 does not only contribute to improved traffic management, but leads also to fruitful applications in other domains. The *Fachverband Außenwerbung* (FAW) is the governing organisation of German outdoor advertisement. Among the development of advertising media and other responsibilities, FAW regulates prices of poster sites. The value of each site is characterised by a quantitative measure, the number of passing pedestrians, vehicles and public transports, and a qualitative measure which specifies the average notice of passers-by. Therefore, to calculate poster prices it is vital for FAW to know inner-city traffic frequencies. However, the large number of streets within Germany prohibits empirical measurements for all locations. Within the FAW project, Fraunhofer IAIS developed a method to predict traffic frequencies using spatial data mining [18].

The input data comprises several sources of different quality and resolution. The primary objects of interest are street segments, which generally denote a part of street between two intersections. Each segment possesses a geometry object and has attached information about the type of street, direction, speed class, etc. For a small sample of segments one or more frequency measurements are available. In addition, demographic and socio-economic data about the vicinity as well as nearby points of interest (POI) are known. Demographic and socio-economic data usually exist for official districts like post code areas and are directly assigned to all contained streets. In contrast, POI simply mark attractive places like railway stations or restaurants. Clearly, areas with a high density of restaurants will be more frequented than quiet residential areas. To utilise POI, the data must first be aggregated. As described in Sect. 9.4.2 buffers were created around each street segment to calculate the number of relevant POI within the neighbourhood.

To infer reliable frequencies for all remaining street segments, a k-nearest neighbour algorithm (kNN, see Sect. 9.5.3) has been applied [31]. It possesses the advantage to incorporate spatial and non-spatial information based on the definition of appropriate distance functions. The frequency of a street segment is calculated as weighted sum of frequencies from the most similar k segments in the data sample. The kNN algorithm is known to use extensive resources as the distance between each street segment and available measurement must be calculated. For a city like Frankfurt this amounts to 43 million calculations (about 21,500 segments and 2,000 measurements). While differences in numerical attributes can be determined very fast, the distance between line segments is computationally expensive. Fraunhofer IAIS implemented the algorithm to perform a dynamic and selective calculation of distance from each street segment to the various measurement locations. First, at any time only distances to the top k neighbours are stored, replacing them dynamically during the iteration over measurement sites. Second, a step-wise calculation of distance is applied. If the summarised distance of all non-spatial attributes already exceeds the maximum total distance of the current k neighbours, the candidate neighbour can be safely discarded and no spatial calculation is necessary. For the city of Frankfurt this integrated approach sped up calculations from nearly 1 day to about 2 h. In addition, the dynamic calculations reduced the required disc space substantially.

9.7 Roadmap to Knowledge Discovery from Spatiotemporal Data

Spatial data has proved to be a rich source of information about our environment, taken at a fixed moment in time or aggregated over some period of time. However, spatial patterns do not only develop in space, they also extend in (and possibly change over) time. A great challenge therefore lies in the knowledge discovery from spatiotemporal data. In this section we will look at feature extraction, usage of background knowledge and data mining from a spatiotemporal point of view.

9.7.1 Feature Extraction

The main actors for knowledge discovery from spatiotemporal data are the environment and the objects under consideration. The temporal dimension influences both of them by having an environment that changes along with time, and in parallel a group of individuals that change their position. Depending on the type of pattern that the analyst investigates, different approaches for feature extraction can be taken according to which entity is evolving during time. The methods and the techniques discussed in the chapter focus their attention on the relations among objects in the space. Note that also various methods for feature extraction from imagery data sets (e.g. satellite images, field bitmaps, etc.) exist. However, because of lack of space we decided not to include these methods in the chapter.

The feature extraction process is mainly based on the exploration of relations among the objects in the data set. But, how are these relations influenced by the temporal dimension? Some relations, lets call them *time invariant*, do not change during time. For example, the *Leaning Tower of Pisa* has a *contained* relation with the *Piazza dei Miracoli*. And this relation will continue to hold for a long time, at least as long as the tower is still leaning. In contrast, the environment can change over time. Consider, for example, a holiday at the sea with the water coming and going during tide and ebb.

When an object moves in time it modifies its relations with the environment. Actually, it changes only its position: the new location determines new relations with the new neighbourhood. For the feature extraction approaches presented so far, an object located at the same position at different time instants has the same relations with the environment. So, changes in the relations are determined by the modified position alone. However, this is a simplification. Consider, for example, the location of an employee in the morning and evening of a working day. Probably, the employee will travel along the same road from home to work, but the *status* of the same object is different.

It is already challenging to find valid methods to extract meaningful and useful features from geographic data. The temporal dimension adds a new level of difficulty to this task. The example above, where time is used as a pre-condition to determine if a relation holds or not, represents just a starting point of investigation. The role of time is limited to enabling or disabling a certain feature. The real challenge is one step forward: the definition of feature extraction approaches that explore also the temporal dimension. For example, consider the analysis of pollution traffic density. Here, time is *embedded* in the description of the various scenarios: 'clouds' of moving areas moving along together, objects moving far away from each other. The evolution of the whole scene depends both on the positions of each object, and also on the evolution of the status of each observed area: its composition, its density and all the other properties that characterise it during time.

9.7.2 Background Knowledge

Background knowledge comprises valuable information about an object of interest and originates from explicit domain knowledge of some expert or additional data sources. It fulfils several tasks during knowledge discovery, which include feature extraction and data mining. During feature extraction, background knowledge can be used to distinguish interesting and non-interesting relations, and thus to speed up the feature extraction process. In addition, it advances data mining techniques by restricting the hypothesis space. However, the integration of geographic background knowledge is still a field for exploration.

When we add time to the geographic setting, the integration of background knowledge becomes even trickier. It then does not suffice to treat static information, but necessitates the inclusion of dynamic knowledge. For example, attractive points of interest at daylight differ from those at night time. Shopping centres or museums become desolate places after closing hours, while night clubs just start their business at that time. Also, weekly, monthly or long-term fluctuations need to be considered.

9.7.3 Data Mining Algorithms

Geographic references form an inherent part of spatiotemporal data. Therefore, insights gained in geographic data mining should be applied for spatiotemporal data mining. Yet, how can we incorporate time? Given a trajectory of a moving object, a simple approach might flatten time by reducing the trajectory to its pure spatial dimension. Obviously, this results in a great loss of information. Temporal anomalies as traffic jams, locations of interest to a person (home, work, shops) or the means of transportation (by car, on foot) cannot be inferred without a temporal reference. A second approach might consider a sequence of time slices, where spatial patterns are discovered independently within each time slice and are later on combined. Basically, this approach performs spatial and temporal mining in a sequential order. It is clearly limited as it relies on synchronous observations and cannot exploit space–time dimensions concurrently. Obviously, both approaches are not optimal to make extensive use of spatiotemporal structures. Again, we will need specialised algorithms, which will be discussed in detail in Chap. 10.

9.8 Summary

Knowledge discovery from geographic data is not a trivial process and cannot be solved by classical data mining approaches. On the contrary, it requires an understanding of fundamental geographic concepts, sophisticated feature extraction and specialised algorithms. In this chapter, we presented geographic data models and the role of GIS to manage geographic data. We described several methods to detect hidden relationships between geographic objects and reviewed the state-of-the-art of feature extraction. The section on geographic data mining motivates the use of specialised algorithms. It emphasises the need for dynamic feature extraction and the tight integration of spatial databases in the mining process. The various aspects of knowledge discovery are illustrated by an example from the traffic domain. Finally, we pose a number of open research questions when extending geographic knowledge discovery to the dimension of time.

References

1. R. Agrawal and R. Srikant. Fast algorithms for mining association rules in large databases. In *Proceedings of 20th International Conference on Very Large Data Bases (VLDB'94)*, pp. 487–499. Morgan Kaufmann, 1994.
2. G. Andrienko, D. Malerba, M. May, and M. Teisseire. Mining spatio-temporal data. *Journal of Intelligent Information Systems*, 27(3):187–190, 2006.
3. A. Appice, M. Berardi, M. Ceci, and D. Malerba. Mining and filtering multi-level spatial association rules with ARES. In *Proceedings of the 15th International Symposium on the Foundations of Intelligent Systems (ISMIS'05)*, pp. 342–353. Springer, 2005.
4. V. Bogorny, S. Camargo, P. Engel, and L.O. Alvares. Mining frequent geographic patterns with knowledge constraints. In *Proceedings of the 14th Annual International Workshop on Geographic Information Systems (GIS'06)*, pp. 139–146. ACM, 2006.
5. V. Bogorny, P. Engel, and L.O. Alvares. Enhancing the process of knowledge discovery in geographic databases using geo-ontologies. In H.O. Nigro, S.G. Cizaro, and D. Xodo (eds.), *Data Mining with Ontologies: Implementations, Findings and Frameworks*. Idea Group, 2007.

6. V. Bogorny, J. Valiati, S. Camargo, P. Engel, B. Kuijpers, and L.O. Alvares. Mining maximal generalized frequent geographic patterns with knowledge constraints. In *Proceedings of the 6th International Conference on Data Mining (ICDM'06)*, pp. 813–817. IEEE Computer Society, 2006.
7. P.A. Burrough and R.A. McDonnell. *Principles of Geographical Information Systems*. Oxford University Press, 2000.
8. S. Chawla, S. Shekhar, W. Wu, and U. Ozesmi. Modelling spatial dependencies for mining geospatial data. In H.J. Miller and J. Han (eds.), *Geographic Data Mining and Knowledge Discovery*, Chap. 6. Taylor & Francis, 2001.
9. J.-P. Chilés and P. Delfiner. *Geostatistics – Modeling Spatial Uncertainty*. Wiley, 1999.
10. D.J. Cowen. GIS versus CAD versus DBMS: what are the differences? *Journal of Photogrammetric Engineering and Remote Sensing*, 54:1551–1555, 1988.
11. N.A.C. Cressie. *Statistics for Spatial Data*. Wiley, 1993.
12. M. Egenhofer. Reasoning about binary topological relations. In *Proceedings of the 2nd International Symposium on Advances in Spatial Databases (SSD'91)*, pp. 143–160. Springer, 1991.
13. M. Ester, J. Sander, H.-P. Kriegel, and X. Xu. A density-based algorithm for discovering clusters in large spatial databases with noise. In *Proceedings of the 2nd International Conference on Knowledge Discovery and Data Mining (KDD'96)*, pp. 226–231. AAAI Press, 1996.
14. M. Ester, A. Frommelt, H.-P. Kriegel, and J. Sander. Spatial data mining: database primitives, algorithms and efficient DBMS support. *Journal of Data Mining and Knowledge Discovery*, 4(2–3):193–216, 2000.
15. A.S. Fotheringham and P.A. Rogerson. GIS and spatial analytical problems. *International Journal of Geographical Information Systems*, 7(1):3–19, 1993.
16. A.S. Fotheringham, C. Brunsdon, and M. Charlton. *Geographically Weighted Regression*. Wiley, 2002.
17. A.U. Frank. Qualitative spatial reasoning: cardinal directions as an example. *International Journal of Geographical Information Systems*, 10(3):269–290, 1996.
18. Fraunhofer Institut Intelligente Analyse- und Informationssysteme (IAIS). http://www.iais.fraunhofer.de, 2007.
19. R. Haining. *Spatial Data Analysis: Theory and Practice*. Cambridge University Press, 2003.
20. J. Han, K. Koperski, and N. Stefanovic. GeoMiner: a system prototype for spatial data mining. In *Proceedings of the International Conference on Management of Data (SIGMOD'97)*, pp. 553–556. ACM, 1997.
21. D.A. Hastings. *Geographic Information Systems: A Tool for Geoscience Analysis and Interpretation*. 1992.
22. Y. Huang, S. Shekhar, and H. Xiong. Discovering colocation patterns from spatial data sets: a general approach. *IEEE Transactions on Knowledge and Data Engineering*, 16(12):1472–1485, 2004.
23. W. Klösgen. Subgroup discovery. In W. Klösgen and J. Zytkow (eds.), *Handbook of Data Mining and Knowledge Discovery*, Chap. 16.3. Oxford University Press, 2002.
24. W. Klösgen and M. May. Spatial subgroup mining integrated in an object-relational spatial database. In *Proceedings of the 6th European Conference on Principles and Practice of Knowledge Discovery in Databases (PKDD'02)*, pp. 275–286. Springer, 2002.
25. K. Koperski and J. Han. Discovery of spatial association rules in geographic information databases. In *Proceedings of the 4th International Symposium on Advances in Spatial Databases (SSD'95)*, pp. 47–66. Springer, 1995.
26. R. Laurini and D. Thompson. *Fundamentals of Spatial Information Systems*. Vol. 37. APIC Series. Academic Press, 1992.
27. P.A. Longley, M.F. Goodchild, D.J. Maguire, and D.W. Rhind. *Geographic Information Systems and Science*, Chap. 3. Wiley, 2001.
28. D. Malerba and F.A. Lisi. An ILP method for spatial association rule mining. In *Proceedings of Workshop on Multi-Relational Data Mining (MRDM'01)*, pp. 18–29, 2001.

29. D. Malerba, F. Esposito, A. Lanza, F.A. Lisi, and A. Appice. Empowering a GIS with inductive learning capabilities: the case of INGENS. *Journal of Computers, Environment and Urban Systems*, 27(3):265–281, 2003.
30. D. Malerba, M. Ceci, and A. Appice. Mining model trees from spatial data. In *Proceedings of the 9th European Conference on Principles and Practice of Knowledge Discovery in Databases (PKDD'05)*, pp. 169–180. Springer, 2005.
31. M. May. Data mining cup, presentation, 2006. http://www.data-mining-cup.de/2006/Fachkonferenz/Programm/.
32. M. May and S. Savinov. SPIN! – an enterprise architecture for spatial data mining. In *Proceedings of the 7th International Conference on Knowledge-Based Intelligent Information & Engineering Systems (KES'03)*, pp. 510–517. Springer, 2003.
33. H.J. Miller. Geographic data mining and knowledge discovery. In J.P. Wilson and A.S. Fotheringham (eds.), *Handbook of Geographic Information Science*. Blackwell, 2006.
34. Open GIS Consortium. OpenGIS abstract specification, 1999. http://www.opengeospatial.org/standards/as.
35. T. Ott and F. Swiaczny. *Time-integrative Geographic Information Systems – Management and Analysis of Spatio-Temporal Data*. Springer, 2001.
36. D. Papadias and Y. Theodoridis. Spatial relations, minimum bounding rectangles, and spatial data structures. *International Journal of Geographical Information Science*, 11(2):111–138, 1997.
37. P. Rigaux, M. Scholl, and A. Voisard. *Spatial Databases. With Application to GIS*. Morgan Kaufmann, 2001.
38. S. Rinzivillo and F. Turini. Extracting spatial association rules from spatial transactions. In *Proceedings of the 13th Annual International Workshop on Geographic Information Systems (GIS'05)*, pp. 79–86. ACM, 2005.
39. J. Sander, M. Ester, H.-P. Kriegel, and X. Xu. Density-based clustering in spatial databases: the algorithm GDBSCAN and its applications. *Journal of Data Mining and Knowledge Discovery*, 2(2):169–196, 1998.
40. S. Servigne, T. Ubeda, A. Puricelli, and R. Laurini. A methodology for spatial consistency improvement of geographic databases. *Geoinformatica*, 4(1):7–34, 2000.
41. S. Shekhar and S. Chawla. *Spatial Databases: A Tour*. Prentice Hall, 2002.
42. SPIN! Spatial mining for public data of interest, 2007. http://www.ais.fraunhofer.de/KD/SPIN/.
43. W. Tobler. A computer movie simulating urban growth in the Detroit region. *Journal of Economic Geography*, 46(2):234–240, 1970.
44. H. Wackernagel. *Multivariate Geostatistics*. Springer, 1998.
45. Y. Wang and I. Witten. Inducing model trees for continuous classes. In *Proceedings of the 9th European Conference on Machine Learning (ECML'97), Poster Papers*, pp. 128–137, 1997.

Chapter 10
Spatiotemporal Data Mining

M. Nanni, B. Kuijpers, C. Körner, M. May, and D. Pedreschi

10.1 Introduction

After the introduction and development of the relational database model between 1970 and the 1980s, this model proved to be insufficiently expressive for specific applications dealing with, for instance, temporal data, spatial data and multi-media data. From the mid-1980s, this has led to the development of domain-specific database systems, the first being temporal databases, later followed by spatial database systems.

In the area of data mining, we have seen a similar development. Many data mining techniques – such as frequent set and association rule mining, classification, prediction and clustering – were first developed for typical alpha-numerical business data. From the second half of the 1990s, these techniques were studied for temporal and spatial data and sometimes specific, previously well studied, techniques such as time-series analysis were introduced in the data mining field. For an overview of mining techniques in spatial and geographic data, we refer to Chap. 9.

For spatiotemporal data, this development has only just started. This field is no longer in an embryonic state; now, in 2007, we can say that with the organization of a few workshops, this field has just been born. In this chapter, we give an overview of what has been done in spatiotemporal data mining, with a focus on mining trajectories of moving objects, and we mainly emphasize the challenges that this field faces.

This chapter is organized as follows. In Sect. 10.2, we outline, by means of examples, challenging tasks for spatiotemporal mining. In Sects. 10.3 and 10.4, we discuss, respectively, spatiotemporal clustering and patterns. Spatiotemporal prediction and classification, including time series, are discussed in Sect. 10.5. In Sect. 10.6, the role played by uncertainty in spatiotemporal data mining is briefly described. Finally, in Sect. 10.7, we summarize the main problems and issues

---

M. Nanni
KDD Laboratory, ISTI-CNR, Pisa, Italy, e-mail: mirco.nanni@isti.cnr.it

discussed in this chapter, and propose a taxonomy of spatiotemporal data mining tasks based on the variation from exact presence to complete absence of time in the mined patterns.

10.2 Challenges for Spatiotemporal Data Mining

During the last five years, attempts have been made to extend many techniques for knowledge discovery in classical relational or transactional data, such as association rule mining, frequent pattern discovery, clustering, classification, prediction and time-series analysis, to knowledge discovery in the context of data with a spatial annotation [17, 33, 34]. Much of this research discusses some simple classes of patterns and focusses mainly on algorithmic aspects, often involving some approximation techniques. The research in this field has not yet produced a theoretical framework for spatial data mining.

This makes research in data mining in the context of moving objects more challenging. And the objectives in this area are manifold. First, we have to discover the relevant patterns to mine for. Second, a taxonomy of these patterns will make it clear for which mining tasks new techniques will have to be developed. Third, suitable algorithmic solutions will have to be proposed to implement these mining tasks. Finally, this new research field could benefit from a clean unified theoretical framework. These objectives are rather ambitious and we will not try to approach and tackle them for general moving or changing objects, but rather in the more restricted setting of moving object data or trajectory data. Still, these tasks go beyond the research in spatial mining, because we will always assume spatial information as background data for the moving object data and it will be involved and appear in the mined patterns. We will also not touch on algorithmic problems in this chapter.

When we think of moving object or trajectory data that represent traffic situated in some city or province, obvious tasks we would like to perform concerning everyday phenomena include detecting traffic jams, predicting traffic jams and discovering relations between traffic jams. A typical example is

$$\text{Find all traffic jams in Pisa between 7 and 9 AM.} \quad (10.1)$$

Traffic jams can be defined in terms of the density and speed of the traffic and there is an obvious relation to *clustering*. It is also clear that detecting traffic jams is typically done on certain fragments of the data. Since there are many ways in which distances or similarity measures between trajectories can be defined, many variations of clustering are possible. Typical for moving objects is that they have speed, and clustering can be directed to detect similarly fast-moving objects. The following example asks for clustering of the cars, bicycles and pedestrians:

$$\text{Find three clusters of objects that have similar speed (slow, medium and fast).}$$
$$(10.2)$$

10 Spatiotemporal Data Mining

Physical quantities of trajectories, like speed, acceleration and length, can be expected to play a role in much of the knowledge to be discovered about moving objects.

In many cases, different traffic jams are temporarily related. Relations between spatiotemporal phenomena can be expressed using *association rules*, as for example

$$\text{traffic\_jam}(\text{Pisa}, 7.30\text{ AM}) \Rightarrow \text{traffic\_jam}(\text{Lucca}, 8.30\text{ AM}), \tag{10.3}$$

meaning that whenever the first event (a traffic jam in Pisa at 7.30 AM) occurs, usually it is followed by the second one (a traffic jam in Lucca at 8.30 AM). A more general version of this rule could be

$$\text{traffic\_jam}(\text{Pisa}, t) \Rightarrow \text{traffic\_jam}(\text{Lucca}, t+1\text{ h}), \tag{10.4}$$

in which time appears as a parameter. Rules could also be discovered after even further abstracting time, as the following generalization of (10.3) and (10.4) shows

$$\text{traffic\_jam}(\text{Pisa}) \Rightarrow \text{traffic\_jam}(\text{Lucca}). \tag{10.5}$$

In the same style of these examples, *frequent patterns* can be discovered in the trajectory data.

Finding examples of *classification* concerning trajectory data appears to be more difficult. Problem (10.2) can be seen as the task of classifying the trajectories into three groups that are defined in terms of the length, speed and other particularities of the trajectories. Other classification tasks can involve the recognition of a situation, like distinguishing traffic jams from normal traffic, or the aim of the individuals, such as whether he/she is going to work, shopping, taking a walk, etc.

There are far more opportunities for discovering *sequential patterns* in trajectory data. Suppose we associate some events or features to trajectories like passing at location $A, B, C, \ldots$, then a pattern

$$A \to B \to C \tag{10.6}$$

(possibly associated with a support) indicates that A, B and C appear in that temporal order. Also, we can have the pattern

$$A \to_3 B \to_7 C, \tag{10.7}$$

where $\to_i$ indicates temporal delay of i minutes. Another promising class of spatiotemporal patterns is *spatiotemporal trends*. An example of a trend is

$$\text{The speed of objects increases as they move away from Pisa.} \tag{10.8}$$

Some patterns are also largely pre-defined and can be seen as a query. The following example, which addresses a typical aspect of human behaviour, namely periodicity, can be seen as an illustration of this:

$$\text{Find all periodic patterns (for a given period).} \tag{10.9}$$

A pattern could be defined to be periodic if enough and the same objects repeat it with some fixed interval of time. Also other behavioural patterns, like traffic jams and flocking, belong to this category.

A last category of tasks that we want to mention concerns *extrapolation* of trajectory data. An example is the question

$$\text{How many trajectories will cross Pisa tomorrow at time 5 PM?} \qquad (10.10)$$

The above-discussed categories of mining problems will be discussed in detail in the following sections of this chapter, where an overview of the existing spatiotemporal mining approaches will also be included. This overview does not have the intention to be exhaustive.

We argue that there is a need of a theory for spatiotemporal data mining and that a theoretical foundation in this field can contribute to a better understanding of the expressive power needed in the mining patterns, on the one hand, and efficient techniques for the implementation of these tasks, on the other hand. In the conclusion of this chapter, we will discuss a preliminary taxonomy of transformation groups of time and classify mining tasks or patterns according to the groups under which they are invariant. We will also address spatiotemporal transformation groups that leave patterns invariant.

It will be clear to the reader that, underlying this chapter, there are motivating questions, such as "What is spatiotemporal data mining?", "What is a spatiotemporal pattern?" and "What is the input to a spatiotemporal data mining process?". It is the objective of this chapter to get closer to the answers of these basic questions.

10.3 Clustering

A common need in analysing large quantities of data is to divide the data set into logically distinct groups, such that the objects in each group share some property that does not hold (or holds much less) for other objects. As such, clustering searches a global model of data, usually with the main focus on associating each object with a group (i.e. a cluster), even though in some cases we are interested (also) in understanding *where* clusters are located in the data space.

In this section, we focus on the context of moving objects and, thus, on the trajectories that describe their movement. In this setting, clustering consists essentially in trying to outline groups of individuals that show similar behaviours.

As for other forms of complex data, we can have two main approaches to the problem: (1) applying generic notions of clustering and generic clustering algorithms by defining some distance function between trajectories, which will be the only information on trajectories known by algorithms (with some slight exceptions, as described later in this section) and (2) defining ad hoc notions and algorithms tailored around the specific data type. In particular, in the first case, the semantics of trajectory data is completely encapsulated in the distance function, while in the second case, it is exploited in the whole clustering algorithm. In what follows, we will discuss possible solutions and existing proposals to trajectory clustering, dividing the treatment for the two approaches outlined above, and emphasizing, in both cases, which notions of clustering have been chosen. Note that each definition of clustering adopted also induces a corresponding symmetric problem, not explicitly treated in this chapter, *outlier detection*, i.e. discovering those elements that do not fit well into any cluster found.

10.3.1 Distance-Based Trajectory Clustering

Defining a distance between objects implicitly determines, to a large extent, which objects should be part of the same cluster and then which kind of clusters we are going to discover. How exactly each cluster is composed, then, is decided by the choice of the (generic) clustering algorithm adopted. For instance, centre-based algorithms like *k-means* will yield a flat set of spherical, tendentially compact clusters; hierarchical methods will organize clusters in a multi-level structure of clusters and sub-clusters and density-based methods will form maximal, crowded (i.e. dense) groups of objects, thus not limiting the group size and, in some cases, also putting together pairs of very dissimilar objects.

A basic approach to define a distance is to consider *similar* those pairs of objects that follow approximatively the same spatiotemporal trajectory, i.e. at each time instant they are approximatively in the same place. Essentially, by clustering objects with such a distance, we can answer questions of the following kind:

$$\text{Which individuals of a population move together?} \quad (10.11)$$

Each cluster found will represent, depending on the context, a group of friends travelling together, an animal herd and so on. A small example is depicted in Fig. 10.1, where a set of trajectories is represented on space–time coordinates and forms two clusters plus two isolated trajectories. We note, in particular, that all the

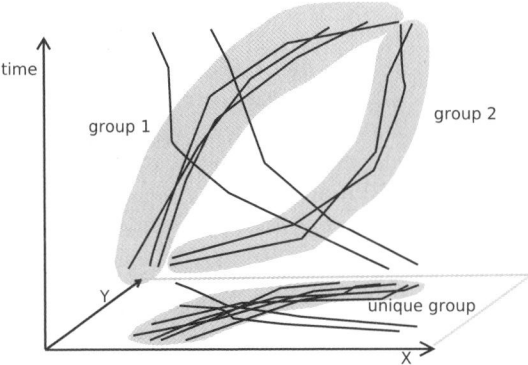

Fig. 10.1 Example for (10.11) and (10.12)

clustered trajectories follow similar paths, as can be seen by their spatial projection depicted at the bottom of the same figure, but with different speeds and, therefore, different timings: those in the first group start moving at a high speed and then slow down, while those in the second group start slowly and then accelerate the movement. A simple way to model this kind of comparison is to represent trajectories as fixed-length vectors of coordinates and then to compare such vectors by means of some standard distance measure used in the time-series literature, such as the Euclidean distance (the most common one) or any other in the family of p-norms. An alternative solution is given in [48], where the spatial distance between two objects is virtually computed for each time instant, and then the results are aggregated to obtain the overall distance, e.g. by computing the average value, the minimum or the maximum.

Moving in the same way at the same time is sometimes too restrictive to discover useful information, and thus the temporal constraint might be removed. In these cases, we could look for groups of objects that follow the same route (i.e. the temporally oriented spatial projection of a trajectory) but at any moment in time, thus formulating requests of the type

$$\text{Find groups of individuals moving along the same roads,} \quad (10.12)$$

for example, boats following some common itinerary to cross a sea, or cars following the same paths from home to workplace and reverse, etc. The bottom part of Fig. 10.1, where trajectories are spatially projected on the $X-Y$ plane, shows a simple example of that, and the result is a unique cluster of objects that follow the same path, even though at different times and speeds. Again, in the time-series literature, we can find some general methods that yield some similar result. One is the comparison of pairs of time series by allowing *(dynamic) time warping* (e.g. see [5, 65] for an efficient approximation), i.e. a non-linear transformation of time, so that the order of appearance of the locations in the series is kept, but possibly compressing/expanding the movement times. Another method, proposed in [2] and further studied in [7], consists in computing the distance as length of the *least common sub-sequence* (LCSS) of the two series, essentially formulated as an edit-distance problem.

A final step in loosening the constraints imposed to clusters consists in not requiring a strict co-location of trajectories/routes, but only asking to group objects that perform similar movements, like going in the same direction or performing the same turns (i.e. turns of the same amplitude, whatever the absolute direction). The first example can be simply modelled by defining as similar any couple of objects that follow approximatively the same path but allowing spatial translation, as proposed in [66] through a translation-invariant, non-metric extension of the above-mentioned LCSS. A step further is then accomplished in [64], where a distance that is also rotation-invariant is proposed, thus allowing us to answer requests of the type

$$\text{Find groups of objects that perform similar sequences of changes (or non-changes) in their direction.} \quad (10.13)$$

We conclude by mentioning the existence of other time series-based approaches that define distances between features extracted from series, rather than comparing the series themselves. For instance, we could extract all pairs of consecutive values in each series (in our context, consecutive locations within each trajectory) and then simply count the number of pairs shared by the two series compared, as proposed in [2]; or, as an alternative, we could extract a set of *landmarks* for each time series (i.e. local behaviours of the time series such as minima or maxima or, more specific to our context, changes of speed or direction) and compute the distance between the series by simply comparing their corresponding series of landmarks, as described in [51].

10.3.2 Trajectory-Specific Clustering

Purely distance-based methods impose some limitations both at the expressiveness level, i.e. some notions of clusters cannot be modelled in a handy way, and at the performance level, i.e. some opportunities for improving performances cannot be taken. That is mainly due to the strong separation between the dissimilarity criterion and the clustering schema that uses it without exactly knowing its semantics – usually only assuming the distance is a metric. For instance, any method based on some notion of *centre* or, more generally, *representative* of a cluster needs to compute it in a way that is coherent with – and thus dependent on – the distance function adopted. The most prominent example is the basic *k-means* algorithm, where the representative is usually computed as the object (possibly new, not yet present in the input data) that minimizes the average distance between itself and all the objects in the cluster. In other cases, the model of cluster requested is not based on any notion of distance at all – or, at least, not a distance between whole trajectories.

One example of model of cluster that does not require an explicit distance notion between trajectories is a generalization of (10.11), where the deviation between the trajectories in a cluster is expressed as noise in a probabilistic formulation. For example, the following problem

$$\text{Find groups of objects that follow a common trajectory, allowing a limited amount of random noise} \quad (10.14)$$

was essentially tackled by Gaffney and Smyth [19], who proposed a mixture model-based clustering method for continuous trajectories, which groups together objects that are likely to be generated from a common core trajectory by adding Gaussian noise. In a successive work [15], spatial and (discrete) temporal shifting of trajectories within clusters is also considered and integrated as parameters of the mixture model. Another model-based approach is presented in [3], where the representative of a cluster is not a trajectory but a Markov model that tries to explain the transitions between a position and the next one, positions being discretized a priori. More exactly, Hidden Markov models (HMMs) are used to model clusters, and a mixture model approach (and the EM algorithm, in particular) is adopted for the parameter estimation task.

An alternative approach is based on the search of sub-segments of trajectories that match sufficiently well. In [26], trajectories are represented as piece-wise linear, possibly with missing segments (e.g. due to disconnection of a phone from its cellular network). Then, a *close time interval* for a group of trajectories is defined as the maximal interval such that all individuals are pair-wise close to each other (w.r.t. a given threshold). Groups of trajectories are associated with a weight expressing the proportion of the time in which trajectories are close, and then the mining problem is to find all trajectory groups with a weight beyond a given threshold. We note, in particular, that in this approach a threshold is set at the beginning to define spatial closeness as a simple predicate. Then, the method tries to discover maximal-size, maximal-temporal extension clusters of close segments of trajectory. From this viewpoint, a similar but simplified objective is pursued in [41]. Here, an extension of *micro-clustering* (first defined for the BIRCH clustering algorithm) to moving objects is proposed, which groups together rectilinear segments of trajectories that lay within a rectangle of given size in some time interval. Even in this case, spatial closeness is decided through thresholds (the size of the rectangle), while group size and temporal extension are maximized, in this case restricting to consider only single time intervals. Finally, a different approach to a similar problem has been recently proposed in [40]. Trajectories are represented as sequences of points without explicit temporal information and a simplification heuristics is applied to partition each trajectory into a set of quasi-linear segments. Then, all such segments are grouped by means of a density-based clustering method, and at the end a representative trajectory is computed for each resulting cluster.

A trade-off between distance-based approaches and trajectory-specific ones occurs when the distance has to compare not the whole trajectories but only parts of them, focusing on a time interval that is given as a parameter of the distance function. That leads to tackle problems of the following type

$$\text{Find groups of objects that move together within some (unknown) time interval of minimum size } I \tag{10.15}$$

addressed, for instance, in [49]: here, trajectories are clustered by means of a generic density-based algorithm where the adopted distance is the average spatial distance between the trajectories within a given time interval, which is a parameter of the distance. Then, for each time interval T, the algorithm can be run focusing on the trajectory segments laying within T. The final objective is to discover which time interval T results in the clusters of best quality and then return these clusters together with T. A sample result of the process is given in Fig. 10.2, which depicts a set of trajectories forming three clusters (plus some noise) and shows the optimal time interval (that where the clusters are clearest) as dark trajectory segments. A similar objective is pursued in [28], but from a different perspective. Here, the authors consider moving objects as associated with a spatial position for a set of *time slices* and face the problem of discovering density-based spatial clusters that approximatively

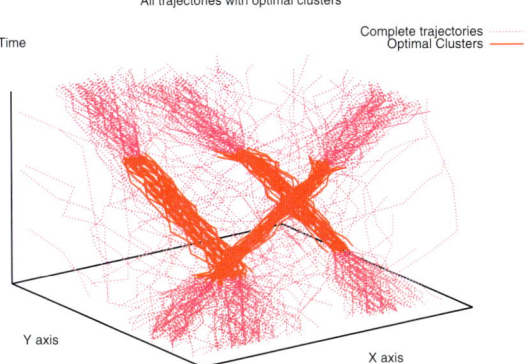

Fig. 10.2 Example of clusters over a time interval (10.15)

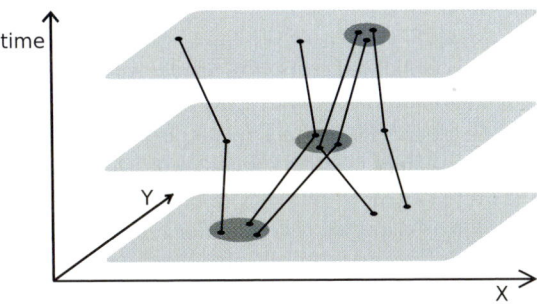

Fig. 10.3 Example of moving cluster

persist along several contiguous time slices, as exemplified in Fig. 10.3, where one cluster that persists for three units of time is found. Persistence of a cluster means that the individuals contained in a cluster in some time slice are approximatively the same that appear in a cluster in the next time slice. Note that the first slice of a moving cluster and its last one could share only a few objects (in our example, only one) or even no object at all, since gradual, step-wise migrations in clusters are allowed, and no global check is performed – i.e. here the focus is on the spatial regions covered by clusters and not on the trajectories they contain. For this reason, this approach can be considered a borderline case between clustering (aimed at finding a partitioning of objects) and frequent patterns (aimed at finding regularities that involve many individuals), the clusters being the (segments of) trajectories involved and the patterns being the spatiotemporal locations where the clusters lay.

All the approaches described above share a common property: they seek groups of objects that move essentially together, i.e. they tackle variants of (10.11). Relaxing the temporal requirements, we can search groups of objects only looking at the paths they follow. In one of the first works related to the topic, Ketterlin [32] proposes an approach of that kind: the author considers generic sequences (thus modelling trajectories as sequences of points) together with a conceptual hierarchy over the sequence elements, used to compute both the cluster representatives (i.e. the least common ancestor of the cluster elements on the hierarchy) and the distance between two sequences (essentially equal to the number of generalization steps needed to have the sequences coinciding). Therefore, trajectories are abstracted into sequences of regions as large as needed to obtain a match between different trajectories.

10.4 Spatiotemporal Local Patterns

In coherence with the notion of pattern that was introduced in Chap. 1, mining spatiotemporal patterns means searching for concise representations of interesting behaviours of single moving objects or groups of objects. In particular, in this section, we are interested in mining *local* patterns, that is to say, patterns that aim to characterize potentially small portions of the data space, such as sub-sets of individuals, small time intervals or limited regions of space. No effort is made to provide a complete characterization of the full data space (even though, in some situations, local patterns might also be exploited to do that), which is instead the aim of the so-called *global models*, such as clustering (treated in Sect. 10.3) and classification models (Sect. 10.5).

The kind of interesting behaviours that we want to discover strictly depends on the context and is usually specified by selecting a sub-set of all possible patterns and, possibly, enforcing some constraints on how patterns occur in the data, i.e. on the behaviours they summarize.

Apart from the specification of patterns, pattern mining depends on whether the specific focus of the task is on finding interesting patterns or on finding occurrences of the patterns (i.e. where and when they occur and who they involve). To some extent, the first case corresponds to *direct searches*, as introduced in Chap. 1, while the second case corresponds to *inverse searches*, though the distinction is not crisp. In a direct search, we may specify the hypothesis space H, the space of all patterns regarded in our search, which is usually very large, and aim at identifying all frequent or in another sense interesting patterns $h \in H$. Alternatively, we could specify a set of interesting patterns (or hypotheses) H in advance, H usually being relatively small, and ask for all occurrences that match such patterns in our data.

An additional characterization of the two kinds of problems is that patterns in direct searches are usually (but not necessarily always) quite simple and involve single individuals, and then include some constraints on the number of occurrences; on the contrary, patterns in inverse searches are usually rather complex and involve a set of individuals and the constraints are on the size or on the composition of such set.

The algorithms applied in pattern mining always depend on the input data being searched, which in our context can be (1) the movement trajectories themselves and/or (2) the information derived from them. Derived information can include sets/sequences of events (e.g. purely spatial events, such as the locations visited by the trajectory, or spatiotemporal events, such as the manoeuvres performed: U-turns, stops, extreme accelerations, etc.), static aggregates (e.g. the kilometres covered), dynamic aggregates (e.g. the velocity as function of time), etc. The examples seen above regard information extracted on a single individual basis and thus the patterns that can be extracted from them will describe individual behaviours. In other cases, we can derive information that describe a population, e.g. the traffic jams of last week in the town, thus focusing more on group behaviours. In some pattern discovery tasks, both kinds of features are involved, e.g. we may want to discover that a traffic jam here is often followed by a car accident nearby or that if somebody is stuck in a traffic jam, often he/she has an accident later. Note that working on derived information usually yields an easier problem, and such information can be processed by using generic pattern extraction approaches (i.e. not tailored around spatiotemporal data).

In the following, we will try to give a systematic, yet not exhaustive, overview of patterns and relate them to current literature, where available. We will begin with direct pattern search (namely, frequent pattern mining) and proceed with inverse search, distinguishing relations that are constrained by sets of events or sequences of events, and punctuating the direct use of trajectory data vs. approaches based on feature extraction.

10.4.1 Extracting Frequent Patterns

Frequent patterns are a basic element of data mining, and thus, quite naturally, a relatively great effort has been paid by researchers to study their spatiotemporal counterparts.

A simple, very common approach to mining frequent spatiotemporal patterns consists in a feature extraction solution: first, sets of features are derived from the data, yielding events, attribute values or, more generally, spatiotemporal predicates that describe each trajectory; then, generic mining algorithms are applied on the new feature-based representation of data, extracting frequent sets, association rules or frequent sequences of features. Following this approach, the semantics of spatiotemporal data are essentially taken into consideration once and for all during the pre-processing step, and then it is not involved in any way in the mining phase (the only limited exception being frequent sequences, since they are based on an order relation among events that is a very basic form of temporal information). However, the variety of frequent patterns we can mine with this simplification of the problem is still wide, ranging from simple rules of the form

$$\text{Length}(\text{trajectory}) > 50\,\text{km} \Rightarrow \text{average\_speed}(\text{trajectory}) > 60\,\text{km} \quad (10.16)$$

involving basic aggregation values, to patterns relating complex object behaviours, e.g. traffic jams as described in sample problem (10.3) at the beginning of this chapter. This approach essentially corresponds to the *spatiotemporal association rules* and *evolution rules* first suggested in [1]. More precisely, association rules express relations between simple spatial, non-spatial and temporal predicates, and evolution rules relate complex predicates that describe the spatiotemporal evolution of an object or a group of objects.

Obviously, the choice of the attributes to extract is a crucial aspect of the mining process, since it defines once and for all the pattern space to be searched. A basic family of features for trajectories of moving objects consists of individual-based features, i.e. those that describe the behaviour of each object separately from all the others. For example, we can have:

- Spatial and/or temporal aggregates (the length of the path covered, the amount of time spent in the city centre, the minimum/maximum/average velocity, the most frequent direction followed, etc.), as in example (10.16).
- Spatial events (visiting some pre-defined spatial regions or visiting twice the same place), as in (10.6). A similar approach was proposed in [46], where trajectories are expressed as simple sequences of spatial locations and standard sequential pattern mining tools are used.[1]
- Spatiotemporal events (temporally localized manoeuvres like performing U-turns, abrupt stops, sudden accelerations or longer-term behaviours like covering some road segment at some moment and then covering it again later in the opposite direction) as in sequences of the form

$$\text{visits}(x, \text{Market square}) \rightarrow \text{abrupt\_stop}(x) \rightarrow \text{perform\_U-turn}(x) \qquad (10.17)$$

that mixes spatial events (visiting a given region) with simple behaviours (stopping and U-turning). An example is given in Fig. 10.4b, where this same pattern is obtained by extracting spatial and spatiotemporal features from the trajectory in Fig. 10.4a. Possible events can also include predicates in the style of *spatiotemporal predicates*, as defined in [62], which allow to express some form of spatiotemporal topology between reference spatial regions and trajectories having a spatial

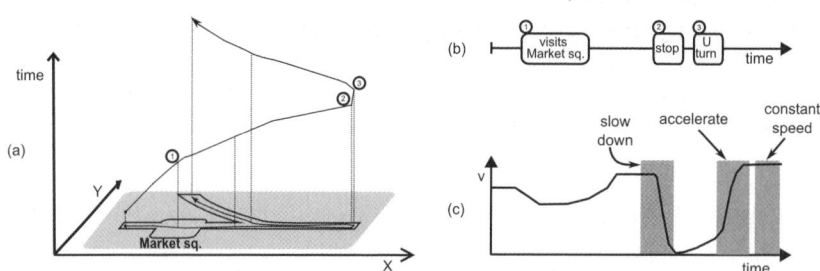

Fig. 10.4 A sample trajectory (**a**), derived spatial and spatiotemporal events (**b**) and speed variation with shape-based events (**c**)

---

[1] More exactly, here some of the spatial semantics is exploited in the mining process: only sequential patterns moving between contiguous regions are searched, and thus the topological relations between regions are used to prune the search space.

uncertainty (i.e. locations are not points but circular boundaries that contain the real position). A sample spatiotemporal predicate is Sometimes_Definitely_Inside (x,A), meaning that there is at least one time instant (Sometimes) such that object x is surely inside region A (Definitely_Inside), also taking into account uncertainty. Therefore we can mine rules of the form

$$\text{Sometimes\_Definitely\_Inside}(x, \text{hospital}) \Rightarrow \\ \text{Always\_Possibly\_Inside}(x, \text{city centre}), \quad (10.18)$$

meaning that people who certainly visit the hospital at least once usually appear to never leave the town.

In some contexts, more complex features are required, describing the behaviour of groups of moving objects. Examples of that are the already mentioned case of traffic jams, interactions between objects (e.g. at least n individuals meet somewhere) or, more generally, the occurrence of any pattern describing some pre-defined behaviour. These represent examples of *inverse pattern searches*, which will be discussed in detail in Sect. 10.4.2, here used to support the pre-processing phase. Similarly, a group-based feature can be defined as the co-occurrence of several simple events approximatively in the same location at the same time, thus relying on analysis tools that search for crowded regions in space–time, such as Kulldorff's spatial scan statistics [36], which searches spatiotemporal cylinders (i.e. circular regions considered within a time interval) where the density of events is higher than outside, and its extensions (e.g. [27], which considers square pyramid shapes in place of cylinders).

A limited degree of flexibility can be added to the feature-based approach by adding the full temporal information to the extracted features. That corresponds to time-stamping spatiotemporal events (whereas only their order was considered so far) and extracting *dynamic attributes*, i.e. attributes having time-dependent values. Time-stamps allow to extract more detailed patterns that describe also the temporal relations between events, e.g. sequences with characteristic transition times between consecutive events as proposed in [21] (see also example (10.7)), or more general sets of events with temporal constraints between them, such as chronicles [63]. We note that these methods are not specific for moving object data. Similarly, dynamic attributes essentially provide time series that can be mined by means of tools for extracting rules and sequences from them, for example mining associations between typical "shapes" occurring in the series, such as in the following example

$$\text{sharp\_slow down} \rightarrow \text{sharp\_acceleration} \rightarrow \text{constant\_speed}, \quad (10.19)$$

which is also depicted in Fig. 10.4c, in relation to the trajectory of Fig. 10.4a. For example, in [16] a similar approach is followed, where common shapes are extracted directly from data through a clustering step, instead of being pre-defined (as sharp_slow down, etc., in the previous example).

The opposite alternative to the feature-based approach to mining frequent patterns consists in directly analysing trajectories, for instance to discover paths frequently followed by cars in the city centre, frequent manoeuvres performed by animal predators or hunted preys, etc. That means, in particular, that no a priori discretization or other form of pre-processing of spatial and/or temporal information is performed, and therefore the spatiotemporal semantics of data can potentially play a role in the mining phase. A first consequence of this scenario is that the standard notion of frequent pattern borrowed from transactional data mining, i.e. a pattern that *exactly* occurs several times in the data, usually cannot be applied. Indeed, the continuity of space and time usually makes it almost impossible to see a configuration occurring more than once perfectly in the same way, and thus some kind of tolerance to small perturbations is needed.

The continuity problem mentioned above can be tackled in at least two complementary ways by (1) considering patterns that are in the form of trajectory segments and searching approximate instances in the data and (2) considering patterns that are in the form of moving regions within time intervals, such as spatiotemporal cylinders or tubes – that, in some sense, represent a segment of trajectory plus a bounded approximation/uncertainty – and counting as occurrences all trajectory segments fully contained in the moving regions. The work in [9] provides an example of the first approach: a trajectory is approximated by means of a sequence of spatial segments obtained through a simplification step and then patterns are extracted essentially in the form of sequences of contiguous spatial segments; in particular, each element of the sequence has to be similar to several segments of the input trajectory, similarity being defined w.r.t. three key parameters: spatial closeness, length and slope angle. Frequent sequences are then outputted as sequences of rectangles such that their width quantifies the average distance between each segment and the points in the trajectory it covers. Figure 10.5 depicts a simple pattern of this kind, formed of two segments and corresponding rectangles. A very similar simplify-and-aggregate approach was also followed recently in [40], yet limited to the aggregation of single segments and more focused on a clustering perspective (indeed, this work was also mentioned in Sect. 10.3.2 about clustering), without an explicit notion of frequency.

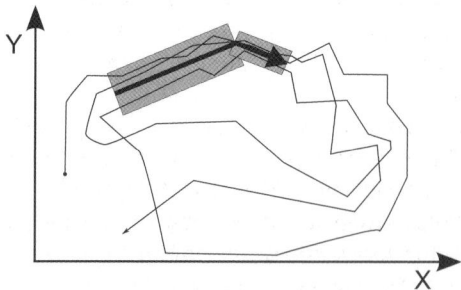

Fig. 10.5 A basic example of the spatiotemporal patterns of [9]

The second approach, based on moving regions, is followed by [28], which was already discussed in Sect. 10.3.2 as borderline example of a clustering task, and concerns the discovery of density-based spatial clusters that persist along several contiguous time slices. Finally, a similar goal, but focused on cyclic patterns, is pursued in [47]: the authors define the *spatiotemporal periodic pattern* mining problem (i.e. finding cyclic sequential patterns of given period) and propose an effective and fast mining algorithm for retrieving *maximal* periodic patterns. While time is simply assumed to be discrete, spatial locations are discretized dynamically through density-based clustering. Each time a periodic pattern is generated, in the form of a sequence of spatial *regions*, a check is performed to ensure that all regions in the pattern are dense – and then significant.

Some variants of the problems mentioned above have been proposed and studied in the recent years. One example is provided in [68], where patterns in the form of sequences of locations are mined, and also the uncertainty of object locations is considered from a probabilistic viewpoint (i.e. the position of each object follows a given probability distribution over space around the given location point). Here, candidate patterns are built over a pre-defined discretization of space (a grid) and time (fixed snapshots), and the support of a pattern is computed as its *expected support* w.r.t. the location distributions of the input objects. In [22], a different notion of trajectory pattern is introduced that integrates spatial and temporal information without a priori discretizations. A *T-pattern* is defined as a sequence of points in space with transition times, which express the time taken to move from each point to the next one in the sequence. An occurrence of a pattern of n elements in a data set of trajectories is any sub-sequence of n points of an input trajectory, such that each point in the sub-sequence approximately *matches* the corresponding point in the pattern (i.e. it falls within a spatial neighbourhood of such point), and the transition times are approximately the same as in the pattern (i.e. they correspond up to a given *time tolerance* threshold). Then, frequent T-patterns are extracted, by heuristically grouping close spatial points into (rectangular) regions, and representing sets of similar transition times through intervals. Another case is studied in [10], where each input object is associated to an object type (e.g. deers, pumas, vultures, etc.), and then patterns describing the proximity (i.e. collocation) between object types are mined. In particular, the pattern type mined is a sequence of collocation relations, each describing which pairs of object types were close to each other over a significant time window, such that these relations share a common object type, called *reference feature*. For example, we could obtain a pattern (deers, {pumas}) → (deers, {vultures}), meaning that frequently a deer stays close to a puma for sometime and later the same deer stays close to a vulture.

The examples discussed in this section show that the spatial information can play different roles in different types of patterns: in some cases, as in [9], the continuity of movement in space is taken into account and the patterns provide an approximate description of the whole movement within a time window; in other cases, as in [46] (previously mentioned in this section), the movement in space is discretized, but contiguity is preserved, thus obtaining sequences of form $A \rightarrow B \rightarrow C$, where A is adjacent to B and B is adjacent to C; finally, in cases as [28, 68] we obtain patterns describing (discrete) sequences of locations of form $A \rightarrow B \rightarrow C$ without any check on contiguity, thus not providing explicit information on the movements performed in the gap between two consecutive locations of the pattern, e.g. between A and B in $A \rightarrow B \rightarrow C$.

Finally, we note that, although frequency is the most common criteria used for filtering patterns, other measures of interest could be applied in conjunction with frequency, such as confidence, correlation and likelihood, or might even replace it, e.g. by focusing on infrequent or rare patterns. In the latter case, we obtain a problem similar to outlier detection, mentioned in Sect. 10.3.

10.4.2 Occurrence Retrieval

Contrary to the extraction of frequent patterns from the data, a user may already have some specific pattern in mind and ask for all of its occurrences. We therefore refer to this task as *occurrence retrieval* or *inverse query*. In Chap. 1 about basic concepts of movement data, we have seen that two types of queries can be distinguished: elementary and synoptic. While elementary queries represent movement behaviour of single entities, synoptic queries depict patterns of collective movement behaviour (for more detail, see Sect. 1.3). In this section, we study both retrieval tasks, but focus on synoptic queries. As we will see, elementary inverse queries are closely related to database literature, which is treated in Chaps. 5 and 6.

Inverse elementary queries involve patterns that can be answered from a single trajectory. For example, the query

$$\text{Find all trajectories that pass location } A \text{ between time } t_1 \text{ and } t_2 \quad (10.20)$$

may retrieve several trajectories. Yet, each trajectory by itself is sufficient to decide whether the pattern is fulfilled or not. In the above query, the location is specified explicitly while the temporal constraint corresponds to a range query. Note that the pattern does not involve sequential information. To include sequential information in our query, we could request that after location A, a second location B must be passed. In [24], this kind of query is termed *spatiotemporal pattern query* (STP) and is defined as a sequence of spatial predicates with either exact or relative temporal order. Alternately, we could include derived information about an object's velocity or direction of travel, or request some periodicity in an object's movement. In database literature, queries that concentrate on a single part of a trajectory are known as *coordinate-based queries*, while queries that rely on sequential information are called *trajectory-based queries* [53]. As both query types are discussed in Sect. 5.3.3, the reader is referred to that chapter for further details.

Inverse synoptic queries identify objects that conform to a specified collective behaviour. These patterns target simultaneous movements and the interaction between objects. They are also referred to as *group patterns* and may involve derived information concerning the whole group of objects (e.g. average speed). Intuitively,

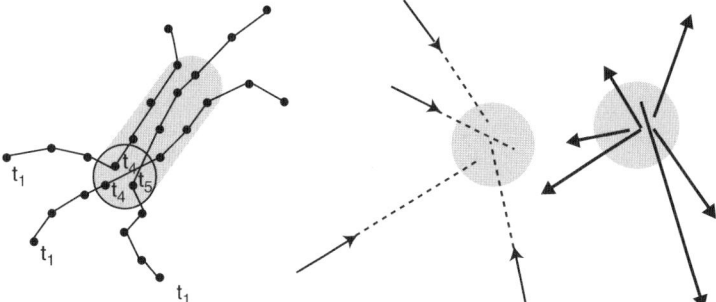

Fig. 10.6 Relative motion patterns showing leadership (*left*), convergence (*middle*) and divergence (*right*)

a group is formed by a number of objects that stay close in space for a meaningful period of time. In [67], physical proximity is delimited by a maximum distance threshold between each pair of objects. If k objects stay close for a given minimal threshold of time, they form a so-called k-group pattern. The algorithm of Wang et al. [67] discovers mobile group patterns on trajectory data where the location is recorded at fixed, regularly spaced points in time. A generalization to irregularly spaced trajectories, assuming linear movement, is provided in [26].

Apart from the general definition of spatiotemporal closeness, a group can be specified by some characteristics of its inner structure. For example, a group could be headed by some individual who anticipates the group motion. This pattern is called *leadership* (Fig. 10.6, left) and was first introduced by [38] under the general concept of *relative motion* (REMO). Other basic spatiotemporal group patterns of REMO are flock, convergence and divergence. A flock corresponds to a mobile group pattern as defined above, while convergence and divergence describe the simultaneous motion of objects to or from some point in space (see Fig. 10.6, middle and right). It is easy to think of an extension of these patterns to include further characteristics of motion. For instance, the speed of cars could be required to increase as they move away from Pisa as in example (10.8), thus covering to some extent the problem of trend detection. However, it is important to note that not all patterns, as specified by REMO and in related literature, are directly derived from trajectories. The retrieval of occurrences for convergence or encounter patterns is usually based on a single snapshot in time and assumes a constant speed and direction of objects [23, 39].

So far, all patterns rely on a stable group of objects. Yet, a pattern may continue over time, although its group members change. For example, a traffic jam can prevail for several hours while new cars continuously arrive at one end and escape at the other end. This phenomenon is called a *moving* cluster and refers to a cluster that retains its density (or other similar properties, like cluster size or diameter) although different objects participate in the cluster during its lifetime. Kalnis et al. [28] define this pattern in their already mentioned work on persistent clusters.

Obviously, group patterns are strongly connected to the clustering task and are in part generalized by query (10.11). However, clustering searches for a global model of the data and targets primarily the grouping of complete trajectories. Spatiotemporal patterns are local models that may cover only a small part of the data or even the trajectory itself.

10.5 Prediction

"Quo vadis?" is not only a common phrase among ancient Romans. In the age of just-in-time logistics, real-time traffic management, location-based services and GPS navigation, the reliable statement about an object's future position or destination plays a central role. Anticipating the motion of individuals or groups of objects enables these systems to take preparatory actions in case of delay, more favourable even to avoid conglomerations, or to deliver helpful information at the desired time. Thus, spatiotemporal data offer a wide perspective of predictive tasks, which include prediction of locations and trajectories, prediction of density, reach and events, as well as the classification of trajectories. This section studies these tasks in more detail.

10.5.1 Prediction of Locations and Trajectories

During the past years, the reliable prediction of future locations of moving objects has been of interest in mainly two research areas, namely database systems and wireless communication networks. Moving object databases employ future locations of objects for example in range or nearest-neighbour searches of *forecasting queries*. These queries require sophisticated structures for indexing future positions of moving objects. In wireless networks, the anticipation of future movement is important to enable an efficient allocation of network resources.

In database literature, forecasting queries rely on indexing structures for current positions and motion vectors. Given the current location l_c and velocity vector v_c of an object, the future position after time Δt can be computed as $l_f = l_c + v_c \Delta t$. The TPR-tree [55] and its optimized version TPR* [60] have been developed to handle predictive range queries [55], time-parameterized nearest-neighbour queries [59] or reverse nearest-neighbour queries [4] over the future positions of moving objects. The underlying assumption of all techniques is that the involved objects continue their motion with the given velocity vector until the ending time of the query interval. This assumption applies for linear movement in unobstructed spaces, as for example for ships, planes or weather phenomena. However, it is not reasonable for street networks where objects change their direction and speed within short-time intervals [61].

Such unstable conditions are met in wireless communication networks, where mobility management serves mainly two tasks. First, appropriate resources must be allocated to guarantee a smooth transfer of service if a user moves from one cell to the other. Second, when an incoming call arrives, the network should page as few cells as possible within a given location area. Both tasks require to anticipate the motion of users in the near future. Several algorithms have been investigated to accomplish this task. Biesterfeld et al. [6] and Liou and Huang [45] train neuronal networks based on the location area and x,y-coordinates, respectively. Liang and Haas [44] apply Gauss–Markov models based on the location and velocity of objects. A common approach for location prediction is to analyse historic trajectories, derive predominant patterns and apply the most similar pattern to the trajectory in question. Such an approach is followed by Katsaros et al. [30] and Yavas et al. [69], who apply clustering and sequential pattern mining, respectively, to extract patterns. A comprehensive study and comparison of methods for location prediction in wireless networks can be found in Chen et al. [12] and Song et al. [56].

In addition to location prediction in the near future, an important research task is to anticipate the most likely route and destination of a moving object. For example, location-based services can offer more sophisticated services when knowing which locations a user will pass and whether the user is on the way to work or to the supermarket. The general assumption behind the prediction of routes and destinations is that people follow daily or weekly routines. Usually, people visit only a few places frequently, as for example their home, workplace or favourite restaurant. In addition, people are creatures of habit and select their present route from a small set of candidate routes. Karimi and Liu [29] adapt a transition matrix to personal preferences and are thus able to predict the most likely route and destination of a single person within a given time frame. While Karimi and Liu [29] base their predictions solely on routing information, Laasonen [37] incorporates residence times into his model. The author first detects places where a user spends a comparatively large amount of time. These places form the set of all possible destinations and delimit individual routes. Similar to [30], Laasonen clusters historic routes and compares the obtained types with the present trajectory. The predicted destination belongs to the most similar trajectory type and can optionally be conditioned on the time of day and day of week.

10.5.2 Prediction of Density

The object density of some area is defined as the number of objects inside the area in proportion to the area's size at a given point in time. It is a global characteristic that emerges through the interaction of a number of objects and changes over time. The prediction of densities promises many benefits, especially in the traffic domain. For example, a traffic management system that is able to identify dense regions and impending bottlenecks may counteract those effects in time.

In [25], the notion of density is extended to an interval of time where the enumerator contains the minimum number of objects that are concurrently in the given area. To calculate densities, a spatiotemporal cube is created. Each cell contains the density for a given area (*x*- and *y*-axis) during some moment or interval of time (*z*-axis). For density prediction, Hadjieleftheriou et al. [25] assume a linear movement of objects and compute future densities by extrapolation. A different approach is taken by Sun et al. [57]. They regard the development of each spatial cell separately and calculate the very next density in time as weighted sum of previous densities. The weights follow an exponential smoothing process that emphasizes the influence of recent densities and reduces the weight of values further away in time.

10.5.3 Extrapolation and Prediction of Reach

Reach is a time-dependent measure about the publicity of a location within some population. Imagine that a new restaurant opens in the city centre. Within 1 day, 20% of the inhabitants will have noted the new location. After 1 week, the reach will be increased to 60%. If the restaurant had been opened in an outer quarter of the city, only 40% of the inhabitants would have passed by after 1 week. Reach is not restricted to a single location, but can span a network of sites. It is defined as the proportion of population that passes at least one of the locations within the network within a given period of time.

Given GPS trajectories of a group of people over several days, the number of contacts with a given network can easily be calculated. The challenge lies in the extrapolation of unbalanced and incomplete sample trajectories. If persons in the survey are not representative for the whole population, for example if they live mostly in one part of the city, the data sample needs to be stratified to provide an unbiased reach. Also, incomplete trajectories that originate in defect GPS devices, forgetfulness or drop outs of the survey, pose a serious problem as reach relies on a sequence of measurements on consecutive days. Within the Swiss Poster Research project (SPR) [18, 58], Fraunhofer IAIS applies methods from survival analysis to consider varying numbers of people in the data sample. Survival analysis [35] is often used in clinical studies, and censors test persons when no further tracks are available.

Fraunhofer IAIS tracks further research questions that concern the prediction of reach when only a limited number of measurements are available. A first example is the prediction of reach when the measurement period is shorter than the desired time interval. A second challenging question arises when no measurements at all are given and the reach of networks within one city needs to be inferred from the mobility of another (similar) city.

10.5.4 Prediction of Events

The work described in [8] studies the problem of predicting spatiotemporal events that are associated with other (a-spatial and a-temporal) features. For example, the authors predict the probability that some crime will be committed within a given region and time interval based on locations, times and socio-economic features of past incidents. The underlying rationale is the definition of a transition density model, which predicts the probability density in space–time given the historical data, coupled with density estimation processes for discovering significant (sets of) features and space–time locations.

10.5.5 Prediction in Geo-Referenced Time Series

In this section, we regard spatiotemporal data from a temporal point of view. The space–time cube then divides into a number of geo-referenced time series, one for each location or area. In contrast to ordinary time series, geo-referenced time series are not independent of each other but are spatially correlated. The general goal in time-series prediction is, given historic data at times $t_1,\ldots,t_n$, to derive the value of some variable at time t_{n+1}. Current approaches in literature, for example, merge results of individual time-series predictors to account for correlation between time series [42, 43]. Other approaches apply (non-)linear regression models and exploit spatiotemporal correlation of the regression residuals [54]. Spatial correlation can also be used to reduce computational costs [70].

The remaining section demonstrates the complexity of geo-referenced time-series analysis based on a project that is currently investigated by Fraunhofer IAIS. The Springer project [18] analyses the customer migration and interrelation between shops that sell newspapers. Consider the number of newspapers that are delivered to various shops for re-sale. The number must be calculated carefully, too few delivered newspapers result in a loss of profit, too many newspapers imply wasted resources. Given the number of newspapers sold in the past at each shop, how many newspapers should be delivered the next day? The sales figures of each location form time series that are spatially correlated to all other points of sale. Let us take a closer look into the variation and dependencies of newspaper figures. First, the time series are governed by a global trend. If an interesting headline raises the attention of the population, more newspapers will be sold throughout the country. Second, spatial and temporal variation on medium scale arises, for example due to the weather or vacations. Third, local trends exist within the time series that depend on the immediate environment of the point of sale. A location within a living area differs from the central station, and also the social background of the area plays an important role. Considering time, (periodic) differences take place during working days and the weekend.

Finally, the locations influence each other, which may pose the greatest challenge of all. If one location is sold out, customers are likely to obtain a newspaper at a nearby location. However, if a shop sells out frequently or goes into vacation, people will adapt their routes accordingly and may change their behaviour long term. They may also buy copies on their way to work or at a central point of sale and not within the immediate neighbourhood. These few examples already show the complex dynamic behaviour of moving objects that needs to be captured and predicted in spatiotemporal time series.

10.5.6 Classification of Trajectories

The classification of (parts of) trajectories offers a wide range of meta-data to be derived from and attached to trajectories. Imagine for example, how the sightseeing driven path of a tourist differs from the path of a working local. Such information can be used by location-based services to tailor their offers to the current need of a prospective customer. Another classification task is to infer the means of transportation from a trajectory. It allows to answer questions such as the following: What portion of a person's daily motion can be attributed to private vehicles? Which streets outside of the city centre are predominantly used by pedestrians?

To our best knowledge, no research has been conducted so far to classify trajectory data. However, similar problems exist in the area of time-series analysis. Keogh and Pazzani [31] use a piece-wise linear representation of time series and weight each segment according to its importance. This representation is used for classification, clustering and relevance feedback. In [20], time series are classified by the application of patterns as test criteria in decision trees. Each pattern thereby corresponds to a temporally confined constant model of the signal, which can, for example, represent the velocity of an object.

In general, trajectories can be classified using nearest-neighbour algorithms provided that an appropriate distance function is given. However, the definition of a distance function depends on the classification task and is not easily determined as various scaling, translation and noise effects have to be taken into account.

10.5.7 Open Issues

Today, the two predominant sources of trajectory data for moving objects are wireless networks and GPS. On one side, telecommunication companies accrue masses of cell-based movement data. On the other side, technologies like GPS provide a considerably more precise positioning. Yet, the trade-off for high-quality data lies in substantially reduced quantity as GPS data are not easily available. One challenging research task therefore concerns the combination of both data sources and the exploitation of synergetic effects to boost prediction accuracy.

Another challenge is the prediction within an unstable environment. Usually, predictions for traffic applications assume a fixed street network. Yet, roads may be closed due to reconstruction work or a traffic management system may be able to control the orientation of lanes during rush hour. Also, the mobile behaviour of individuals changes over time. A new place of work, opening and closure of shops or changed means of transportation naturally influence the mobile behaviour. It is therefore important that algorithms can easily incorporate structural changes and adapt to new patterns in mobile behaviour.

10.6 The Role of Uncertainty in Spatiotemporal Data Mining

Uncertainty is an inherent characteristic of spatiotemporal data. It arises due to physical and technical limitations during data collection and storage. While it can be broadly assumed that time is delivered with high accuracy, uncertainty of location varies with the applied technology between a few metres (GPS) and kilometres (GSM). In addition, the sampling rate possesses a great influence on accuracy. The faster an object moves, the more often must an object's location be reported to sustain a given level of spatial uncertainty.

Background knowledge as well as certain assumptions about movement behaviour help to reduce the uncertainty in data. For example, when tracking a vehicle, we can safely assume that all movements are restricted to the street network. Cars are unlikely to move through buildings. Another frequently made assumption is that of linear movement between two reported positions. In general, given two consecutive positions P_1 and P_2 at times t_1 and t_2 and a maximum speed, an object's position at each moment in time $t \in [t_1, t_2]$ is restricted to some area [52], see also Sect. 5.2 for details. If no further information is given, a uniform distribution of the objects within this area can be assumed.

While a significant amount of recent literature concerns the processing of uncertain queries [13, 62], research in the area of uncertain data mining is limited. Chau et al. [11] propose a clustering algorithm that optimizes intra-cluster distance based on the probability density function of an object's current position. They are thus able to express uncertainty of location based on an assumed maximum velocity. Another example is provided in [68], already mentioned in Sect. 10.4.1, where sequential patterns over uncertain locations are mined through a notion of expected support, defined w.r.t. the probabilistic distribution of locations of the input objects. A natural representation of uncertainty can be achieved using Bayesian models. Uncertainty in continuous data can be modelled, for example, with Bayesian Kriging [14, 50].

10.7 Conclusion

Data mining on spatiotemporal data, and in particular on trajectory data, is a largely unexplored area. In this chapter, we presented several classes of problems that so far have been studied very little or, in some cases, not at all. The problems considered have been organized along a classical taxonomy of data mining tasks inherited from standard contexts, which include clustering, classification and local pattern tasks.

Finally, because of their pervasive presence in the spatiotemporal context, also some problems raised by the uncertainty in the data have been briefly discussed.

The main problems and issues pointed out include the following:

- Which notions of similarity and distance are best suited for a specific (distance-based) clustering task? In particular, different settings may require different levels of *strictness* in comparing trajectories: from checking spatial and temporal coincidence (trajectories are similar if they visit the same places at the same times) to spatial-only coincidence (the order of visit might be important, but not the precise timings), similarity of relative motions (considering speed, direction, etc.) or simply similarity of general features (average speed, length, etc.).
- Which notions of cluster best model the concepts of group peculiar to trajectory data? The complex nature of trajectories can lead to concepts of groups and clusters related to the inner structure of the data, such as the movement information relative to specific sub-intervals of time or sub-regions of space or, in other cases, they can require to develop general models of the overall movement, such as probabilistic models.
- Which features best model the kind of events or characteristics for which we want to extract classical local patterns, such as frequent itemsets, sequential patterns and association rules? A wide range of alternatives is possible, in principle, ranging from simple aggregate information (e.g. length of the trajectory) to spatial or spatiotemporal descriptions of the movement (e.g. set of visited places, or manoeuvres like U-turns).
- Which notions of local patterns can best fit the trajectory data domain? So far, the available approaches mainly focus on the spatial component or try to adapt classical local patterns to this specific domain.
- Which are the best methods for predicting different kinds of phenomena, such as individual future position, future density of a region, events of various types, values of variables associated to spatial locations?
- Which features and methods are best suited for classifying objects from the trajectories that describe their movements? This is a largely unexplored field, and there are apparently no approaches at all specifically focusing on trajectory data.
- How to deal with uncertainty in the best way? Beside pre-processing methods, data mining algorithms might be guided by their knowledge about the approximation that affects the input data.
- Finally, how can background knowledge and ontologies – especially about the geographical space where trajectories exist – be used to extract more significant and more useful information? The integration of spatiotemporal pattern extraction and existing knowledge is an appealing, yet very little developed, line of research.

In addition to the issues listed above, we mention also the absence, so far, of a satisfactory theoretical framework able to give a well-structured, global picture of the problems and methods related to spatiotemporal data mining (and trajectory mining in particular). The classical taxonomy of data mining methods, also applied in this chapter, seems to be too limited for complex data domains, as proved by the existence of methods that cannot be neatly inserted into a single class – e.g. the moving clusters approach in [28], which is both a local pattern and a clustering method. We end this chapter by presenting a very preliminary proposal of an alternative taxonomy, based on the role played by the temporal information in the mining problem or method, the main aim of the discussion being to give a glimpse of the complexity of this kind of operation.

10.7.1 Towards a Taxonomy of Trajectory Mining Problems and Methods

In this chapter, we have seen many examples of spatiotemporal mining patterns. These tasks can be classified in several ways. A first, and traditional, way of classification is given by the sections in this chapter (clustering, spatiotemporal pattern discovery, prediction and classification). However, there are other possibilities of classification that are more closely related to the spatial, temporal and spatiotemporal nature of the patterns.

As an example, we here discuss a classification according to the use of time in the discovered pattern. Let us start by looking at examples (10.3)–(10.5). We observe that in traffic_jam(Pisa, 7.30AM) ⇒ traffic_jam(Lucca, 8.30AM) time is treated in a very precise way. If we assume that the data we mine on deal with one day, this pattern talks about two fixed moments in time. In the pattern traffic_jam(Pisa, t) ⇒ traffic_jam(Lucca, $t+1$ h), time is considered in a more relative way. In fact, for this pattern, only the time interval of 1 h is important. In the third pattern traffic_jam(Pisa) ⇒ traffic_jam (Lucca), time is treated even more freely, in the sense that time does not play a role at all.

We could formalize the role of time in these examples by the transformations of time that leave these patterns invariant. Definition 10.1 roughly captures this idea. Let us assume time to be a continuum (like **R**).

Definition 10.1. Let $\mathscr{G}$ be a group of transformations of time. A data mining task is called $\mathscr{G}$-*invariant* if for any collection D of trajectories it returns the same result on D and any $D' = \gamma(D)$, for any $\gamma \in \mathscr{G}$.

Using this definition, we observe that pattern (10.3) is $\mathscr{I}$-invariant, where $\mathscr{I} = \{\text{id}\}$, i.e. $\mathscr{I}$ only contains the identity transformation of time. Patterns (10.4) and (10.6) are invariant for the group of translations of time $\mathscr{T} = \{t \mapsto t+a \mid a \in \mathbb{R}\}$. Finally, pattern (10.5) is $\mathscr{P}$-invariant, where $\mathscr{P}$ could be the group of arbitrary permutations of time.

Sequential patterns, like $A \to B \to C$ (see (10.6)), or oriented path-based clustering (see (10.12)) are examples of patterns that are invariant under monotone increasing bijections of time, i.e. under the group $\mathscr{O} = \{t \mapsto f(t) \mid f \text{ a monotone increasing bijection}\}$ that reduces time to an ordering of events. However, example (10.7) is $\mathscr{T}$-invariant. Purely spatial patterns would be invariant under the group $\mathscr{C}_0 = \{t \mapsto 0\}$.

We could also imagine using a more flexible definition of invariance, where a data mining task is called $\mathscr{G}$-invariant if on any collection D of trajectories it returns the same result on D and any $D' = \{\gamma_d(d) \mid \gamma_d \in \mathscr{G} \text{ and } d \in D\}$. In this modified definition, each trajectory may be subject to a different transformation, whereas Definition 10.1 involves the same transformation γ for all trajectories. If we want to ignore small delays for each trajectory, for instance, this second definition – with the group $\mathscr{T}_\varepsilon = \{t \times d \mapsto t + a_d \mid a_d \in \mathbb{R} \wedge |a_d| < \varepsilon\}$, for some small ε – could model this by introducing temporal mis-alignments between trajectories. Note that the elements of this group are functions of time (t) and the input trajectory (d).

However, the notions of invariance introduced above are still limited, since they can easily capture global time shift and global timescaling, but they are apparently insufficient to capture local temporal transformations, like local timescaling (such as dynamic time warping or longest common sub-sequence), or local timescaling within a small time window. A deeper study in this direction is needed.

Apart from the traditional classification and a classification of patterns according to the use of time, other classifications can be imagined. Ultimately, we would like to classify patterns according to their invariance under transformations of both time and space.

References

1. T. Abraham. *Knowledge Discovery in Spatio-Temporal Databases*. Ph.D. Thesis, School of Computer and Information Science, Faculty of Information Technology, University of South Australia, 1999.
2. R. Agrawal, K.-I. Lin, H.S. Sawhney, and K. Shim. Fast similarity search in the presence of noise, scaling, and translation in time-series databases. In *Proceedings of 21st International Conference on Very Large Data Bases (VLDB'95)*, pp. 490–501. Morgan Kaufmann, Los Altos, CA, 1995.
3. J. Alon, S. Sclaroff, G. Kollios, and V. Pavlovic. Discovering clusters in motion time-series data. In *Proceedings of the 2003 Computer Society Conference on Computer Vision and Pattern Recognition (CVPR'03)*, pp. 375–381. IEEE, Los Alamitos, CA, 2003.
4. R. Benetis, C.S. Jensen, G. Karciauskas, and S. Saltenis. Nearest and reverse nearest neighbor queries for moving objects. *The Very Large Databases Journal*, 15(3):229–249, 2006.
5. D. Berndt and J. Clifford. Using dynamic time warping to find patterns in time series. In *Proceedings of Knowledge Discovery and Delivery Workshop*, pp. 359–370, 1994.
6. J. Biesterfeld, E. Ennigrou, and K. Jobmann. Neural networks for location prediction in mobile networks. In *Proceedings of the International Workshop on Applications of Neural Networks to Telecommunications (IWANNT'97)*, pp. 207–214, 1997.
7. T. Bozkaya, N. Yazdani, and Z.M. Özsoyoglu. Matching and indexing sequences of different lengths. In *Proceedings of the 6th International Conference on Information and Knowledge Management (CIKM'97)*, pp. 128–135, 1997.
8. D.E. Brown, H. Liu, and Y. Xue. Mining preferences from spatial-temporal data. In *Proceedings of the 1st International Conference on Data Mining (SDM'01)*, 2001.
9. H. Cao, N. Mamoulis, and D.W. Cheung. Mining frequent spatio-temporal sequential patterns. In *Proceedings of the 5th International Conference on Data Mining (ICDM'05)*, pp. 82–89. IEEE, New Orleans, LA, 2005.
10. H. Cao, N. Mamoulis, and D.W. Cheung. Discovery of collocation episodes in spatiotemporal data. In *Proceedings of the 6th International Conference on Data Mining (ICDM'06)*, pp. 823–827. IEEE, Hong Kong, China, 2006.

11. M. Chau, R. Cheng, B. Kao, and J. Ng. Uncertain data mining: An example in clustering location data. In *Proceedings of the 10th Pacific–Asia Conference on Knowledge Discovery and Data Mining (PAKDD'06)*, pp. 199–204. Springer, Berlin Heidelberg New York, 2006.
12. C. Cheng, R. Jain, and E. van den Berg. Location prediction algorithms for mobile wireless systems. In B. Furht and M. Ilyas, editors, *Wireless Internet Handbook: Technologies, Standards, and Applications*, pp. 245–263. CRC, Boca Raton, 2003.
13. R. Cheng, D.V. Kalashnikov, and S. Prabhakar. Querying imprecise data in moving object environments. *IEEE Transactions on Knowledge and Data Engineering*, 16(9):1112–1127, 2004.
14. J.-P. Chilès and P. Delfiner. *Geostatistics – Modeling Spatial Uncertainty*. Wiley, London, 1999.
15. D. Chudova, S. Gaffney, E. Mjolsness, and P. Smyth. Translation-invariant mixture models for curve clustering. In *Proceedings of the 9th International Conference on Knowledge Discovery and Data Mining (KDD'03)*, pp. 79–88. ACM, New York, 2003.
16. G. Das, K.-I. Lin, H. Mannila, G. Renganathan, and P. Smyth. Rule discovery from time series. In *Proceedings of the 4th International Conference on Knowledge Discovery and Data Mining (KDD'98)*, pp. 16–22. AAAI, New York, 1998.
17. M. Ester, H.-P. Kriegel, and J. Sanders. Algorithms and applications for spatial data mining. In H.J. Miller and J. Han, editors, *Geographic Data Mining and Knowledge Discovery*, pp. 160–187. Taylor & Francis, London, 2001.
18. Fraunhofer Institut Intelligente Analyse- und Informationssysteme (IAIS). http://www.iais.fraunhofer.de, 2007.
19. S. Gaffney and P. Smyth. Trajectory clustering with mixture of regression models. In *Proceedings of the 5th International Conference on Knowledge Discovery and Data Mining (KDD'99)*, pp. 63–72. ACM, New York, 1999.
20. P. Geurts. Pattern extraction for time series classification. In *Proceedings of the 5th European Conference on Principles of Data Mining and Knowledge Discovery (PKDD'01)*, pp. 115–127. Springer, Berlin Heidelberg New York, 2001.
21. F. Giannotti, M. Nanni, and D. Pedreschi. Efficient mining of temporally annotated sequences. In *Proceedings of the 6th International Conference on Data Mining (SDM'06)*, pp. 346–357. SIAM, Bethesda, MD, 2006.
22. F. Giannotti, M. Nanni, D. Pedreschi, and F. Pinelli. Trajectory pattern mining. In *Proceedings of the 13th International Conference on Knowledge Discovery and Data Mining (KDD'07)*. ACM, New York, 2007.
23. J. Gudmundsson, M.J. van Kreveld, and B. Speckmann. Efficient detection of motion patterns in spatio-temporal data sets. In *Proceedings of the 12th International Workshop on Geographic Information Systems (GIS'04)*, pp. 250–257. ACM, New York, 2004.
24. M. Hadjieleftheriou, G. Kollios, P. Bakalov, and V.J. Tsotras. Complex spatio-temporal pattern queries. In *Proceedings of the 31st International Conference on Very Large Data Bases (VLDB'05)*, pp. 877–888. ACM, New York, 2005.
25. M. Hadjieleftheriou, G. Kollios, D. Gunopulos, and V.J. Tsotras. On-line discovery of dense areas in spatio-temporal databases. In *Proceedings of the 8th International Symposium on Advances in Spatial and Temporal Databases (SSTD'03)*, pp. 306–324. Springer, Berlin Heidelberg New York, 2003.
26. S.-Y. Hwang, Y.-H. Liu, J.-K. Chiu, and E.-P. Lim. Mining mobile group patterns: A trajectory-based approach. In *Proceedings of the 9th Pacific–Asia Conference on Knowledge Discovery and Data Mining (PAKDD'05)*, pp. 713–718. Springer, Berlin Heidelberg New York, 2005.
27. V.S. Iyengar. On detecting space–time clusters. In *Proceedings of the 10th International Conference on Knowledge Discovery and Data Mining (KDD'04)*, pp. 587–592. ACM, New York, 2004.
28. P. Kalnis, N. Mamoulis, and S. Bakiras. On discovering moving clusters in spatio-temporal data. In *Proceedings of 9th International Symposium on Spatial and Temporal Databases (SSTD'05)*, pp. 364–381. Springer, Berlin Heidelberg New York, 2005.

29. H.A. Karimi and X. Liu. A predictive location model for location-based services. In *Proceedings of the 11th International Symposium on Geographic Information Systems (GIS'03)*, pp. 126–133. ACM, New York, 2003.
30. D. Katsaros, A. Nanopoulos, M. Karakaya, G. Yavas, O. Ulusoy, and Y. Manolopoulos. Clustering mobile trajectories for resource allocation in mobile environments. In *Proceedings of the 5th International Symposium on Intelligent Data Analysis (IDA'03)*, pp. 319–329. Springer, Berlin Heidelberg New York, 2003.
31. E. Keogh and M. Pazzani. An enhanced representation of time series which allows fast and accurate classification, clustering and relevance feedback. In *Proceedings of the 4th International Conference on Knowledge Discovery and Data Mining (KDD'98)*, pp. 239–241. ACM, New York, 1998.
32. A. Ketterlin. Clustering sequences of complex objects. In *Proceedings of the 3rd International Conference on Knowledge Discovery and Data Mining (KDD'97)*, pp. 215–218. AAAI, New York, 1997.
33. K. Koperski and J. Han. Discovery of spatial association rules in geographic information databases. In *Proceedings of the 4th International Symposium on Advances in Spatial Databases (SSD'95)*, pp. 47–66. Springer, Berlin Heidelberg New York, 1995.
34. K. Koperski, J. Han, and N. Stefanovic. An efficient two-step method for classification of spatial data. In *Proceedings of the 8th International Symposium on Spatial Data Handling (SDH'98)*, pp. 45–55, 1998.
35. A.P. Kragh, B. Ornulf, G.D. Richard, and K. Niels. *Statistical Models Based on Counting Processes*. Springer Series in Statistics. Springer, Berlin Heidelberg New York, 1993.
36. M. Kulldorff. A spatial scan statistic. *Communications in Statistics: Theory and Methods*, 26(6):1481–1496, 1997.
37. K. Laasonen. Clustering and prediction of mobile user routes from cellular data. In *Proceeding of 9th European Conference on Principles and Practice of Knowledge Discovery in Databases (PKDD'05)*, pp. 569–576. Springer, Berlin Heidelberg New York, 2005.
38. P. Laube and S. Imfeld. Analyzing relative motion within groups of trackable moving point objects. In *Proceedings of 2nd International Conference on Geographic Information Science (GIS'02)*, pp. 132–144. Springer, Berlin Heidelberg New York, 2002.
39. P. Laube, M. van Kreveld, and S. Imfeld. Finding REMO – Detecting relative motion patterns in geospatial lifelines. In *Proceedings of 11th International Symposium on Spatial Data Handling (SDH'04)*, pp. 201–214. Springer, Berlin Heidelberg New York, 2004.
40. J.-G. Lee, J. Han, and K.-Y. Whang. Trajectory clustering: A partition-and-group framework. In *Proceedings of the 2007 ACM SIGMOD International Conference on Management of Data (SIGMOD'07)*, pp. 593–604. ACM, New York, 2007.
41. Y. Li, J. Han, and J. Yang. Clustering moving objects. In *Proceedings of the 10th International Conference on Knowledge Discovery and Data Mining (KDD'04)*, pp. 617–622. ACM, New York, 2004.
42. Z. Li, M.H. Dunham, and Y. Xiao. STIFF: A forecasting framework for spatio-temporal data. In *Mining Multimedia and Complex Data*, pp. 183–198. Springer, Berlin Heidelberg New York, 2002.
43. Z. Li, L. Liu, and M.H. Dunham. Considering correlation between variables to improve spatiotemporal forecasting. In *Proceedings of the 7th Pacific–Asia Conference on Advances in Knowledge Discovery and Data Mining (PAKDD'03)*, pp. 519–531. Springer, Berlin Heidelberg New York, 2003.
44. B. Liang and Z.J. Haas. Predictive distance-based mobility management for multidimensional PCS networks. *IEEE/ACM Transactions on Networking*, 11(5):718–732, 2003.
45. S.C. Liou and Y.M. Huang. Trajectory predictions in mobile networks. *International Journal of Information Technology*, 11(11):109–122, 2005.
46. S. Ma, S. Tang, D. Yang, T. Wang, and J. Han. Combining clustering with moving sequential pattern mining: A novel and efficient technique. In *Proceedings of the 8th Pacific–Asia Conference on Knowledge Discovery and Data Mining (PAKDD'04)*, pp. 419–423. Springer, Berlin Heidelberg New York, 2004.

47. N. Mamoulis, H. Cao, G. Kollios, M. Hadjieleftheriou, Y. Tao, and D. Cheung. Mining, indexing, and querying historical spatiotemporal data. In *Proceedings of the 10th International Conference on Knowledge Discovery and Data Mining (KDD'04)*, pp. 236–245. ACM, New York, 2004.
48. M. Nanni. *Clustering Methods for Spatio-Temporal Data*. Ph.D. Thesis, Computer Science Department, University of Pisa, 2002.
49. M. Nanni and D. Pedreschi. Time-focused density-based clustering of trajectories of moving objects. *Journal of Intelligent Information Systems*, 27(3):267–289, 2006.
50. G. Paaß and J. Kindermann. Current approaches to spatial statistics and Bayesian extensions. Technical Report, GMD – Forschungszentrum Informationstechnik, 2000.
51. C. Perng, H. Wang, S. Zhang, and S. Parker. Landmarks: A new model for similarity-based pattern querying in time series databases. In *Proceedings of the 16th International Conference on Data Engineering (ICDE'00)*, pp. 33–42. IEEE, San Diego, CA, 2000.
52. D. Pfoser and C.S. Jensen. Capturing the uncertainty of moving-object representations. In *Proceedings of the 6th International Symposium on Advances in Spatial Databases (SSD'99)*, pp. 111–132. Springer, Berlin Heidelberg New York, 1999.
53. D. Pfoser, C.S. Jensen, and J. Theodoridis. Novel approaches in query processing for moving object trajectories. In *Proceedings of the 26th International Conference Very Large Databases (VLDB'00)*, pp. 395–406. Morgan Kaufmann, Los Altos, CA, 2000.
54. D. Pokrajac and Z. Obradovic. Improved spatial-temporal forecasting through modelling of spatial residuals in recent history. In *Proceedings of the 1st International Conference on Data Mining (SDM'01)*, 2001.
55. S. Saltenis, C.S. Jensen, S.T. Leutenegger, and M.A. Lopez. Indexing the positions of continuously moving objects. In *Proceedings of the International Conference on Management of Data (SIGMOD'00)*, pp. 331–342. ACM, New York, 2000.
56. L. Song and X. He. Evaluating next-cell predictors with extensive Wi-Fi mobility data. *IEEE Transactions on Mobile Computing*, 5(12):1633–1649, 2006.
57. J. Sun, D. Papadias, Y. Tao, and B. Liu. Querying about the past, the present, and the future in spatio-temporal databases. In *Proceedings of the 20th International Conference on Data Engineering (ICDE'04)*, pp. 202–213. IEEE, Los Alamitos, CA, 2004.
58. Swiss Poster Research Plus. http://www.spr-plus.ch, 2007.
59. Y. Tao and D. Papadias. Time-parameterized queries in spatio-temporal databases. In *Proceedings of the International Conference on Management of Data (SIGMOD'02)*, pp. 334–345. ACM, New York, 2002.
60. Y. Tao, D. Papadias, and J. Sun. The TPR*-tree: An optimized spatio-temporal access method for predictive queries. In *Proceedings of 29th International Conference on Very Large Data Bases (VLDB'03)*, pp. 790–801. Morgan Kaufmann, Los Altos, CA, 2003.
61. Y. Tao, J. Sun, and D. Papadias. Analysis of predictive spatio-temporal queries. *ACM Transactions on Database Systems*, 28(4):295–336, 2003.
62. G. Trajcevski, O. Wolfson, K. Hinrichs, and S. Chamberlain. Managing uncertainty in moving objects databases. *ACM Transactions on Database Systems*, 29(3):463–507, 2004.
63. A. Vautier, M.-O. Cordier, and R. Quiniou. An inductive database for mining temporal patterns in event sequences. In *ECML/PKDD Workshop on Mining Spatial and Temporal Data*, 2005.
64. M. Vlachos, D. Gunopulos, and G. Das. Rotation invariant distance measures for trajectories. In *Proceedings of the 10th ACM International Conference on Knowledge Discovery and Data Mining (KDD'04)*, pp. 707–712. ACM, New York, 2004.
65. M. Vlachos, M. Hadjieleftheriou, D. Gunopulos, and E.J. Keogh. Indexing multi-dimensional time-series with support for multiple distance measures. In *Proceedings of the 9th ACM International Conference on Knowledge Discovery and Data Mining (KDD'03)*, pp. 216–225. ACM, New York, 2003.
66. M. Vlachos, G. Kollios, and D. Gunopulos. Discovering similar multidimensional trajectories. In *Proceedings of the 18th International Conference on Data Engineering (ICDE'02)*, pp. 673–684. IEEE, San Jose, CA, 2002.

67. Y. Wang, E.-P. Lim, and S.-Y. Hwang. On mining group patterns of mobile users. In *Proceedings of the 14th International Conference on Database and Expert Systems Applications (DEXA'03)*, pp. 287–296. Springer, Berlin Heidelberg New York, 2003.
68. J. Yang and M. Hu. TrajPattern: Mining sequential patterns from imprecise trajectories of mobile objects. In *Proceedings of 10th International Conference on Extending Database Technology (EDBT'06)*, pp. 664–681. Springer, Berlin Heidelberg New York, 2006.
69. G. Yavas, D. Katsaros, Ö. Ulusoy, and Y. Manolopoulos. A data mining approach for location prediction in mobile environments. *Data and Knowledge Engineering*, 54(2):121–146, 2005.
70. P. Zhang, Y. Huang, S. Shekhar, and V. Kumar. Correlation analysis of spatial time series datasets: A filter-and-refine approach. In *Proceedings of the 7th Pacific–Asia Conference on Advances in Knowledge Discovery and Data Mining (PAKDD'03)*, pp. 532–544. Springer, Berlin Heidelberg New York, 2003.

Chapter 11
Privacy in Spatiotemporal Data Mining

F. Bonchi, Y. Saygin, V.S. Verykios, M. Atzori, A. Gkoulalas-Divanis, S.V. Kaya, and E. Savaş

11.1 Introduction

Privacy is an essential requirement for the provision of electronic and knowledge-based services in modern e-business, e-commerce, e-government, and e-health environments. Nowadays, service providers can easily track individuals' actions, behaviors, and habits. Given large data collections of person-specific information, providers can mine data to learn patterns, models, and trends that can be used to provide personalized services. The potential benefits of data mining are substantial, but it is evident that the collection and analysis of sensitive personal data arouses concerns about citizens' privacy, confidentiality, and freedom.

When addressed at a technical level, privacy awareness fosters the dissemination and adoption of emerging knowledge-based applications. Obtaining the potential benefits of data mining with a privacy-aware technology can enable a wider social acceptance of a multitude of new services and applications based on the knowledge discovery process. Source data of particular importance include, for instance, biomedical patient data, Web usage log data, mobility data from wireless and sensor networks; in each case there exist substantial privacy threats, as well as a potential usefulness of knowledge discovered from these data.

The awareness that privacy protection in data mining is a crucial issue has captured the attention of many researchers and administrators across a large number of application domains. Consequently, *privacy-preserving data mining* (PPDM) [6, 21, 60, 82], i.e., the study of data mining side effects on privacy, has rapidly become a hot and lively research area. This is evident from the fact that major companies, including IBM, Microsoft, and Yahoo, allocate significant resources to study this problem. However, despite such efforts, we agree with [19] that a common understanding of what is meant by "privacy" is still missing. As a consequence, there is a proliferation of many completely different approaches of privacy-preserving

F. Bonchi
KDD Laboratory, ISTI-CNR, Pisa, Italy, e-mail: francesco.bonchi@isti.cnr.it

F. Giannotti and D. Pedreschi (eds.) *Mobility, Data Mining and Privacy.*
© Springer-Verlag Berlin Heidelberg 2008

data mining[1]: some aim at *individual privacy*, i.e., the protection of sensitive individual data, while others aim at *corporate privacy*, i.e., the protection of strategic information at organization level; the latter is more a secrecy, rather than a privacy, issue. To make the scene more complex, often, the need to guarantee individual privacy for an organization leads to adoption of corporate privacy policies. The scene is even more complex when dealing with *spatiotemporal data mining*.

Spatiotemporal, geo-referenced data sets are growing rapidly and will be more so in the near future. This phenomenon is mostly due to the daily collection of telecommunication data from mobile phones and other location-aware devices. The increasing availability of these forms of geo-referenced information is expected to enable novel classes of applications, where the discovery of consumable, concise, and applicable knowledge is the key step (as those described in Chap. 2 of this volume). As a distinguishing example, the presence of a large number of location-aware wireless-connected mobile devices presents a growing possibility to access space–time trajectories of these personal devices and their human companions: trajectories are indeed the traces of moving objects and individuals. These trajectories contain detailed information about personal and vehicular mobile behavior, and therefore offer interesting practical opportunities to find behavioral patterns, to be used, for instance, in traffic and sustainable mobility management, e.g., to study accessibility to services. Clearly, in these applications privacy is a concern.

One approach to avoid the privacy threats could be to suppress the identities of individuals before the data are released. Unfortunately, this is not sufficient as spatiotemporal trajectories can easily be linked to individuals using publicly available information such as home and work addresses. Therefore, new techniques for de-identifying or anonymizing spatiotemporal data are needed if the data are handed over to a third party. The issue of spatiotemporal data privacy and anonymization was addressed in Chap. 8. In addition to the research work on privacy of spatiotemporal data, we need to develop privacy-preserving data mining techniques. Time-stamped location observations of an object cannot be regarded as normal, nonsequential, tabular data as spatiotemporal observations of an object are not independent. Therefore, employing the existing privacy-preserving data mining techniques as they are would not be sufficient to solve our problem.

How can trajectories of mobile individuals be analyzed without infringing personal privacy rights? How can, out of privacy-sensitive trajectory data, patterns that are demonstrably anonymity preserving be extracted? Unfortunately, only little work is available on these issues so far. To our knowledge, only one work that addresses directly the privacy preservation issues in a spatiotemporal data mining context has been published [39]. Therefore, the objective of this chapter is twofold: on the one hand, we aim at providing a classification of the actual state of the art of the research in privacy-preserving data mining; on the other hand, we aim at collecting open research issues to define a road map toward privacy-preserving spatiotemporal data mining.

---

[1] See http://www.cs.umbc.edu/~kunliu1/research/privacy_review.html for an updated bibliography on privacy-preserving data mining.

11 Privacy in Spatiotemporal Data Mining

The chapter is organized in four sections corresponding to the four main approaches that we have found in the literature on privacy-preserving data mining, and a final section containing the research road map. Our classification of the various different approaches is based on the following questions: *how are the data organized* (centralized or distributed), *what is disclosed* (the data or the knowledge extracted from the data), *what is hidden* (the original data, such as the identity of the individuals recorded in the data, or some strategic knowledge that could be extracted from the data). In Fig. 11.1 a decision tree representation of our classification is provided. As previously stated, the research in privacy-preserving data mining is made of many different approaches, with different objectives and application scenarios, and thus the classification we propose is not exhaustive. However, we believe that this is a rather simple classification, capturing most of the research work developed so far in this field.

Another taxonomy tree is provided in Fig. 11.2: This tree takes a completely different viewpoint, focusing more on ethical aspects. In particular, it distinguishes

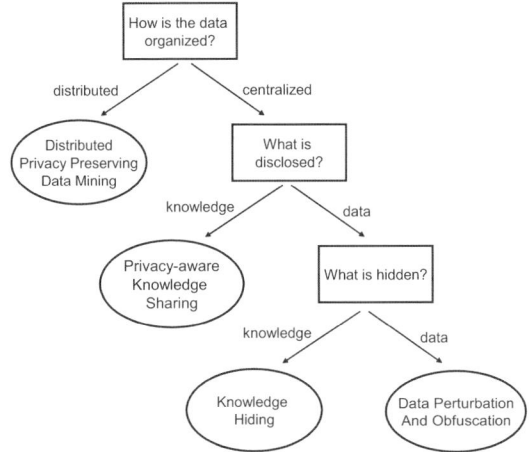

Fig. 11.1 Classification of different approaches to privacy-preserving data mining

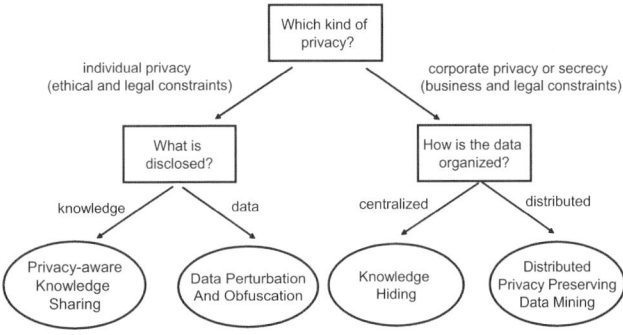

Fig. 11.2 Another classification of different approaches to privacy-preserving data mining

between the kinds of privacy pursued by the various approaches: the case where the focus is on the privacy of the individuals whose data are collected in the database in analysis, and the case where the privacy is intended more as corporate *secrecy*, and thus, the objective is to avoid disclosure of business sensitive data or knowledge.

While this chapter provides sufficient background to be self-contained, there are propaedeutic chapters in this volume. In particular, Chap. 4 discusses the privacy regulations, which can be considered as specifications and constraints for the development of privacy-preserving data mining techniques; Chap. 8 presents the state of the art of anonymization techniques on spatiotemporal data; finally, Chap. 10 describes the state of the art of spatiotemporal data mining.

11.2 Data Perturbation and Obfuscation

One of the techniques for privacy-preserving data mining is data perturbation, which is based on either adding noise to the original database or randomizing it. Data perturbation techniques were initially used for statistical disclosure control [1] and later on for privacy-preserving data mining [6]. The seminal research on perturbation-based privacy-preserving data mining is conducted by Agrawal and Srikant [6]. The following data perturbation techniques are investigated in the context of privacy-preserving data mining:

1. Class membership
2. Data distortion
3. Data swapping

In case of class membership, the original values of an attribute are partitioned into a set of disjoint classes (i.e., groups of values) and the class label instead of the actual value is disclosed. A given taxonomy or some discretization techniques can be used for that purpose. In case of data distortion, a random number from a certain distribution is added to the actual value and the result is disclosed. Data swapping is used to replace a value in a tuple by a value from another tuple.

In the following subsections, we provide the state of the art in perturbation techniques for privacy-preserving data mining in general and perturbation techniques for spatiotemporal trajectories in particular.

11.2.1 Data Perturbation by Adding Noise

The initial work done by Agrawal and Srikant [6], which defined the research on privacy-preserving data mining, is actually based on data perturbation by adding noise. In this work, the motivating scenario is that the data owners or data collectors do not have the necessary know-how for data mining; therefore, the data mining task needs to be outsourced to a third party. However, the data themselves contain confidential attributes that need to be kept secret such as the salaries of employees, or diagnosis of patients. In this scenario, data are assumed to be stored in a centralized environment and transferred to a third party for data mining. Authors identify class membership and data distortion as two relevant techniques for privacy-preserving data mining. For class membership they assume a simple discretization technique where the domain is partitioned into fixed size intervals, and the actual values are replaced by the interval representatives. For data distortion, they add a random value to the original data values from a predefined distribution. They consider uniform and Gaussian distribution in their work. To show the effectiveness of the perturbation methods, a privacy metric is defined. According to the privacy metric, if they can estimate that a data value x lies in the interval $[x_1, x_2]$ with $c\%$ confidence then they have a privacy at $c\%$ confidence level. For example, in case of discretization, if we replace value 12 by an interval $[10, 20]$, then an adversary knows with 100% confidence that the actual value is in the range $[10, 20]$ and s(he) can estimate that the actual value is in the interval $[10, 15]$ with 50% confidence. So, in order for discretization to have the same level of privacy as data distortion using random values with uniform distribution from the interval $[-a, +a]$, the interval length chosen for discretization should be $2a$. For large values of a, the intervals for discretization will also be large, and may cover the whole domain, i.e, all values should be replaced by a single representative interval, which is the whole domain. Therefore, Agrawal and Srikant argue that data distortion techniques using uniform and Gaussian distribution are better candidates for data perturbation to achieve privacy-preserving data mining.

Before the data are transferred to a third party, confidential attributes are altered by adding a random number following certain distribution (uniform and Gaussian distribution are proposed in the paper). The tricky part is that, upon receiving the perturbed data, the third party should be able to reconstruct the probability distribution of original data but not the actual data values. They use Bayes rule to estimate the probability distribution function of original values, given the cumulative distribution function for the random values used for perturbation. They show that this can be done with a limited error range for a large sample set.

Probability distribution of the data is sufficient to construct some data mining models such as decision trees. Agrawal and Srikant show through experiments that the probability distribution of the original data can be reconstructed by an iterative algorithm and it is sufficient to build an accurate decision tree. Decision trees can be constructed efficiently with a top–down divide and conquer algorithm. In case of numerical attributes, the split point needs to be determined, and the data values need to be partitioned with respect to the split point. Gini index is used to determine the split point as knowing the probability distribution of the class values is sufficient to find the Gini index.

There is a tradeoff between accuracy and privacy in the sense that, if you perturb more, you will have more privacy, but the resulting data mining models will be less accurate. Another important issue raised in a following paper by Agrawal and Aggarwal [3] is that the accuracy of the estimates of the original probability distributions is sensitive to the choice of the reconstruction algorithm. A reconstruction algorithm may converge but may not provide an accurate enough estimate, or in the worst case it may not converge at all. Agrawal and Aggarwal propose an algorithm that converges to the maximum likelihood estimate of the data distribution [3].

The initial privacy metric for quantification of privacy proposed in [6] is based on confidence interval, which was proven in [3] to be inadequate when additional knowledge is available. To overcome the limitation they propose an information theoretic privacy metric based on mutual information. However, the privacy provided by these techniques are still questionable. In fact, the work by Kargupta et al. [44] showed that the original values from the perturbed data set can be recovered by a random matrix-based spectral filtering technique. They also demonstrated this fact with experiments on various data sets.

11.2.2 Data Perturbation by Randomization

Another interesting data perturbation technique is randomization. Du and Zhan [26] propose using randomized response techniques for privacy-preserving data mining. They concentrate on a specific data mining model, decision trees. We quote from [26] the privacy problem of building a decision tree on private data mining in the following paragraphs:

Party A wants to collect data from users and form a central database, then wishes to conduct data mining on this database. Party A sends out a survey containing N questions; each customer needs to answer those questions and sends back the answers. However, the survey contains some sensitive questions, and not every user feels comfortable to disclose his/her answers to those questions. How could A collect data without learning too much information about the users, while still being able to build reasonably accurate decision tree classifiers?

The technique proposed by the authors to solve the above problem is based on scrambling the data so that the data collector cannot tell whether the answers to the survey are actually true above a certain confidence threshold. They claim that for large number of users, the aggregate information for the users can be estimated with an accuracy that is sufficient to build a decision tree model over the data. The related-question model described in [86] is used to tackle the problem. In related-question model, let us say we ask the user to answer the question "Do you have sensitive attribute A"? Instead of asking a single question, we ask two related true–false questions, "I have the sensitive attribute A," and "I do not have the sensitive attribute A". Answers to these questions should be opposite to each other. A randomization device is used by the respondent to decide which question to answer, and the data collector does not know which question is answered by the respondent to preserve his/her privacy. Let us say the probability of choosing the first question is P, then the probability of choosing the second question is $1 - P$. The data collector gets the answer "yes" or "no" but he/she only has the probability for asking each question. Given that probability, the data collector, constructing the decision tree, can estimate the actual probability of the "yes" and "no" answers up to a certain precision.

The work by Du and Zhan [26] assumes the related question model where the questions are about the same subject, but the answers are the opposite. The unrelated question model for surveys was assumed in more recent work by Zhan and Matwin in [90], where the authors propose techniques for building a Naive Bayesian classifier from randomized data.

Randomization techniques are also used for privacy-preserving association rule mining problem in two concurrent papers [31] and [72]. Association rule mining deals with finding frequent associations in the form of $A \to B$, where A and B are sets of items where the support (frequency) and confidence (strength expressed in the form of a conditional support) of the item sets are above a certain predefined threshold. Finding the frequent item sets is the main problem in association rule mining. A randomization-based technique proposed in [72] is based on distorting/randomizing the transactional data to preserve the privacy, and featuring a reconstruction algorithm to find the association rules from randomized data. The transactional data are visualized as a data matrix where rows represent transactions and columns represent items. A value of 1 at row i and column j of the matrix tells us that the transaction i contains item j. A value of 0 tells us that the item does not exist in the transaction. As in the case of randomized response technique, the matrix is randomized by replacing 1s with 0s and vice versa with a certain probability, say p which may vary between 0.1% and 90%. Reconstruction is done by a combination of the probability of flipping and conditional probabilities. Distortion probabilities may be tuned to reach the desired level of privacy and accuracy.

Although perturbation-based privacy-preserving techniques were studied and privacy metrics were developed for measuring their effectiveness, the work by Kargupta et al. [44] challenged the effectiveness of randomization-based privacy-preserving data mining, which won the best paper award at the conference. Authors in this paper show that the original data matrix can be obtained from the randomized data matrix using random matrix-based spectral filtering technique. This shows that a lot of work is still needed for achieving privacy-preserving data mining.

11.2.3 Spatiotemporal Data Perturbation

Although there is a lot of research going on in the area of privacy-preserving data mining, most of the work is conducted on general, tabular data sets. Spatiotemporal data mining itself is still in its infancy; therefore, privacy issues in the context of spatiotemporal data were not researched yet. Space and time attributes in databases have always been treated separately, leading to temporal databases and spatial databases. Data mining community also needs to concentrate on the different aspects of spatial and temporal components.

Simple data transformation or data perturbation may not be sufficient for the case of spatiotemporal data. We need to map the data to a different domain, and furthermore we need to perturb the spatiotemporal components.

In [37], the authors propose the "path confusion" algorithm for perturbing object trajectories so that if the proximity of two nonintersecting paths falls below the threshold called perturbation radius, these paths are crossed and their ids are interchanged after the intersection point. The key idea is that an adversary cannot identify whether these two paths were intersecting in the original data set or not, since path confusion is only applied to nonintersecting paths. Kido proposes two obfuscation methods for hiding the current location and the complete trajectory of a user [45]. In these methods users send fake location messages together with the exact location to location-based service (LBS) provider and choose the appropriate message among the responses of the provider without disclosure of sensitive location information. A similar approach in [27] builds a graph of locations connected to the user's location and chooses fake messages from this graph. Perturbation of spatiotemporal trajectories is explained in this chapter in more detail.

In previous subsections, we explained the data perturbation techniques specifically tailored for privacy-preserving data mining. Current techniques for spatiotemporal data perturbation do not consider privacy from a data mining perspective. Therefore, the existing perturbation techniques need to be evaluated in terms of their affect to the data mining results. In the context of spatiotemporal data, we think that k-anonymity [73, 77] could be used as a standard for defining the level of privacy. Several approaches are adopted for providing location privacy by using anonymization. Using pseudonyms instead of real user identities is one of the methods to provide anonymity. However, we can easily show that using pseudonyms is not enough to protect the privacy of individuals. Let us assume that the trajectories are anonymous, i.e., personally identifying information is removed from the database. With data mining tools one can identify frequent trajectories by the same person. A simple reasoning could be as follows: the starting point can be inferred as the home address and the end point would probably be the work address. Home addresses and work addresses of people are public information that can be used to link the trajectories to individuals therefore violating the privacy. Data generalization (e.g., replacing actual date of birth with a more general value such as year of birth, cities with states, etc.) could be used to make sure that such inference channels are blocked. Data generalization should also consider data mining results as the main quality metric. Generalization should be done in a way to facilitate data-mining algorithms to produce meaningful results.

11.3 Knowledge Hiding

The process of uncovering hidden patterns from large databases was first indicated as a threat to database security by O'Leary in [60]. Since then, many different approaches for knowledge hiding have emerged over the years. In what follows, we concentrate on methodologies that rely on *data sanitization* to accomplish knowledge hiding. Data sanitization involves altering the original data set, concealing existing data, or introducing additional spurious data that distorts the mining process in a way that sensitive knowledge is protected.

Knowledge hiding differs from *data perturbation* and *obfuscation* techniques, since its target is to hide sensitive patterns depicting knowledge withdrawn from the data set, rather than individual records existing in the data set. However, since these patterns are constructed due to the statistical significance under which some attributes' values appear in the individual records, hiding of knowledge always results in the modification of a portion of transactions' attributes in the data set in order to lower the observed patterns' significance in the sanitized outcome.

It is important to mention at this point that the knowledge hiding process should always take into consideration the way the data mining algorithms seek for knowledge patterns, when applied on the data, so that the sanitized outcome it produces still preserves all these statistical properties that will allow it to remain useful (for the nonsensitive part of knowledge), when mined by this set of algorithms.

In what follows, we focus our presentation on methodologies regarding two primary dimensions, namely *association rule hiding* and *classification rule hiding*. Because of the lack of space, we deliberately leave investigation of other methodologies such as privacy-preserving mining of time-series to be addressed in future work.

11.3.1 Association Rule Hiding

Association mining is the process of discovering sets of items that frequently co-occur in a transactional database and produce *association rules* that hold for these item sets at a rate higher than two prespecified thresholds: *support* and *confidence*. The reader is encouraged to refer to [4, 5] for a detailed overview of association rule mining. Knowledge hiding, in the context of association rule mining, aims at sanitizing the original data set in a way that a set of rules, denoted as *sensitive*, cannot be discovered in the released data set at prespecified thresholds of confidence and support.

On the basis of their nature, existing approaches can be partitioned into three categories: *heuristic-based*, *miscellaneous*, and *evaluation techniques and frameworks*. Heuristic-based approaches are commonly incorporated in data mining tasks since they allow for efficient solutions, with a potentially small deviation in the quality of the final outcome. Most knowledge hiding approaches are based, in one way or another, on heuristics. On the other hand, *miscellaneous approaches* group together a set of methodologies that experience some innovative characteristics, thus require separate examination. Finally, the demand for thorough knowledge hiding solutions for use in business, and the need for a standardization in the field of privacy-preserving data mining, led researchers to the incorporation of numerous *sanitization techniques* and *evaluation methodologies* into unified frameworks. These frameworks are examined in the final category of our partitioning.

11.3.1.1 Heuristic-Based Approaches

Heuristic-based approaches rely on efficient, fast algorithms for selecting transactions to sanitize and the sequence in which the sanitization process will occur. Existing heuristic approaches are either targeted towards sensitive *frequent item set* or *association rule* hiding. Since both directions are similar in nature, in what follows, we present them together.

A simple and effective way to hide sensitive rules is by decreasing the support of the frequent item sets that lead to their production. This approach was introduced by Atallah et al. in [8]. In this work, the authors examine the problem of limiting disclosure of sensitive rules by reducing their significance, while leaving unaltered or minimally affecting the significance of other, nonsensitive rules. One of the most important contributions of this paper is the proof that finding an optimal sanitization of a data set is *NP-hard*. A heuristic is thus proposed aiming at sanitizing the database, based on the notion of the frequent item sets' *lattice*. Given a set of sensitive item sets, the algorithm sorts them based on support and proceeds to hide them in a one-by-one fashion. A greedy search is performed for each sensitive item set through its ancestors, selecting at each level the parent with the maximum support and setting it as the new item set to be hidden. At the end of the process, a large 1-item set is selected. The algorithm searches through the common list of transactions that support both the selected item and the initial sensitive item set to identify the transaction that affects the minimum number of 2-item sets, and removes the selected item from this transaction. In the sequel, it propagates the results of this modification to the graph of large item sets.

In the work of Dasseni et al. in [22], the target is to hide individual sensitive rules instead of all rules produced by some sensitive item sets. The authors propose three strategies that aim at either hiding the frequent sets that participate in these rules or reducing the rules' importance by setting their confidence below the minimum confidence threshold. The decrement of the confidence of a rule is achieved by either increasing the support of the rule's antecedent through transactions that partially support it or decreasing the support of the rule's consequent in transactions supporting both the antecedent and the consequent. For all three approaches, the authors make the assumption that only rules supported by disjoint item sets are to be hidden.

In [74, 75], Saygin et al. introduce the notion of "*unknowns*" to prevent discovery of association rules. The goal of the three presented algorithms is to obscure a given set of sensitive rules from being identified, by replacing known values in transactions with unknowns, and then appropriately adjusting the values of these unknowns to minimally affect the nonsensitive rules. Because of the incorporation of unknowns, the definitions of support and confidence are modified and become intervals rather than single values. Hiding is achieved either by reducing the minimum support of the item sets that generated these rules or reducing the minimum confidence of the rules. All approaches incorporate the use of a *safety margin*, which captures how much below the minimum thresholds should the new support and confidence of a sensitive rule be, in order for this rule to be safely hidden.

In [65], Oliveira et al. propose an efficient, scalable, one-scan, heuristic algorithm called *sliding window algorithm* (SWA) that achieves to hide sensitive rules while balancing privacy and knowledge discovery in association rule mining. SWA presents a significant improvement over previous approaches, such as [22, 63, 64, 74, 75], since it requires only *one* pass over the database, regardless of its size or the number of sensitive rules that need to be protected. An important contribution of this work is the extension of the notion of a *disclosure threshold*. Instead of using a common threshold ψ for the entire sanitization process, the authors propose the incorporation of a distinct threshold ψ_i for each sensitive association rule i.

An extension to the work of Dasseni et al. in [22] is presented by Verykios et al. in [83]. Two heuristic algorithms are introduced to protect sensitive rules from disclosure by reducing the support of the rules' generating item sets through transactions that fully support them. The first scheme hides sensitive rules by selecting the item with maximum support and removing it from the minimum-length transaction. The generating item sets of the sensitive rules are hidden one by one (starting from the largest) by lowering their support. Similarly, the second algorithm sorts the items of the sensitive item set in descending order of support and the item set's supporting transactions in ascending order of size. Each sensitive item set is then hidden by removal of its items from supporting transactions, in a *round robin* fashion. The intuition behind the idea of hiding in round robin fashion is fairness. However, this algorithm is not as sophisticated as other techniques in [22, 83] and is more targeted toward providing the baseline for conducting experiments.

Pontikakis et al. in [70] propose two *distortion-based* heuristic techniques for selectively hiding sensitive rules. The hiding process may introduce a number of side effects, either by generating rules that were previously unknown or by eliminating existing nonsensitive rules. Both proposed schemes use effective data structures for the representation of the association rules and strongly rely on the prioritization in the selection of transactions for sanitization. The first algorithm, called *priority-based distortion algorithm*, reduces the confidence of a rule by reversing ones to zeros in items belonging in its consequent. The second algorithm, called *weight-based sorting distortion algorithm* [70, 71], concentrates on the optimization of the hiding process in an attempt to achieve the least side effects and the minimum complexity. This is achieved through the use of *priority values*, assigned to transactions based on weights.

11.3.1.2 Miscellaneous Approaches

In this section, we address techniques that experience some special, innovative characteristics in the way they formulate the hiding process; thus they need to be examined separately. In what follows, we discuss two *algebraic approaches* [48, 84], one *border-based approach* [76] and an *integer programming technique* [55], all aiming at hiding sensitive patterns extracted via association rule mining.

A technique for hiding *maximal* sensitive patterns using a correlation matrix was introduced by Lee et al. in [48]. Instead of selecting individual transactions and sanitizing them, the authors propose a methodology for directly constructing a sanitization matrix M by observing the relationship that holds between sensitive patterns and nonsensitive ones. After this matrix is constructed, it is being multiplied (adhering to a new definition of matrix multiplication) by the original database D, yielding a new *sanitized* database D', which achieves to address the privacy concerns. Three algorithms are presented as part of this work, for the construction of the correlation matrix.

The use of *border* in frequent item set hiding was first introduced by Sun et al. in [76]. In this paper, the authors use the notion of the *border* [53] of the nonsensitive frequent item sets to track the impact of altering transactions. To do so, they compute both the *positive* and the *negative* borders of the item sets' lattice. During the hiding process, instead of considering every nonsensitive frequent item set, the proposed methodology focuses on preserving the quality of the border, which directly reflects the quality of the sanitized database that is produced. The algorithm is heuristic in nature. The key step for its success lies on the efficient identification of the candidate whose hiding will lead to the minimum impact on the border.

A novel methodology for frequent item set hiding based on borders' quality preservation is also analyzed by Menon et al. in [55]. In this paper, the authors propose an integer programing optimization algorithm for discovering the minimum number of transactions that need to be sanitized to hide the sensitive item sets. To avoid the *NP-hardness* issue encountered in solving the entire problem, the authors decided to reduce the problem size to capture only the sensitive item sets, requesting that their support remains below the minimum support threshold. To drive the optimization process, a criterion function, inspired by the measure of *accuracy* [48], is being used. The constraints imposed in the integer programming formulation depict the number of supporting transactions that need to be sanitized for each sensitive item set in order to be hidden. After that, a heuristic sanitization approach is used to identify the actual transactions within the database that will be sanitized and perform the sanitization.

Wang et al. in [84] extend the work of [48] by addressing the *Forward-Inference (F-I) Attack* problem, also discussed in [67]. Previous algorithms, such as [63–65], fail to address this type of attack. The *F-I* attack states that if a pattern is hidden in a set of modified patterns that have to be published, but all subpatterns of the pattern are still frequent in the set, then the attackers can easily infer that the pattern is hidden. To avoid *F-I* attack problems, at least one subpattern of this pattern having length 2 should be removed, or the hidden pattern will be inferred recursively. By observing the relations that hold between sensitive and nonsensitive patterns, the authors construct a sanitization matrix M and multiply the original database by it. Since M is properly constructed, the produced sanitized database can successfully resist the *F-I* attack.

11.3.1.3 Evaluation Techniques and Frameworks

A number of *Privacy-Preserving Data Mining* (PPDM) frameworks were recently proposed in the literature. Among them, three are attributed to the work of Oliveira et al. [62–64,66] and one to the work of Bertino et al. in [13]. The former works aim toward providing a set of *sanitization techniques* for knowledge hiding along with a set of *performance evaluation metrics* for assessment of the quality of the hiding process. On the other hand, the work of Bertino et al. in [13] aims at capturing the several evaluation dimensions applicable to PPDM algorithms and targets toward a standardization of the evaluation criteria, under which the quality of a PPDM technique should be assessed.

Oliveira and Zaiane in [62,63] examine the trade-off between *privacy* and *accuracy* in the context of frequent item set hiding. They propose a framework consisting of a *transactional database*, an *inverted file index*, a *transaction retrieval engine*, and a set of *sanitization algorithms*. The retrieval mechanism identifies transactions supporting sensitive item sets and the sanitization algorithms modify these transactions to produce the privacy-aware resulting database that can be safely released. The developed sanitization approaches aim at identifying a balance between hiding sensitive patterns and disclosing nonsensitive ones. They are classified as either *pattern restriction-based*, which remove complete sensitive patterns from the supporting transactions, or *item restriction-based approaches*, which selectively remove some items from sensitive transactions. Finally, the authors propose a pattern restriction-based algorithm and three item restriction-based schemes, all relying on removal of items from transactions.

An extension to the framework of Oliveira et al. in [62,63] is presented in [64]. The new framework incorporates a set of *performance measures* and two novel sanitization schemes. The performance measures quantify the notion of *hiding failure* (i.e., the percentage of restrictive patterns that are discovered from the sanitized database), *misses cost* (i.e., the percentage of nonrestrictive patterns that are hidden after the sanitization process), and *artifactual patterns* (i.e., the percentage of discovered patterns that are artifacts). Moreover, the new algorithms require only two scans regardless of the database size and the number of restrictive association rules that must be protected. This represents a significant improvement over previous algorithms, such as [22,74,75], which were CPU-intensive and required multiple scans depending on the number of association rules to be hidden.

A framework for evaluating PPDM algorithms is proposed by Bertino et al. in [13]. The authors aim at providing a standardization of the PPDM approaches. The contribution of this paper is twofold: first the authors recognize the various parameters characterizing a PPDM algorithm; then, they propose a set of metrics to assess the quality and compare PPDM techniques, according to a fixed set of evaluation criteria. Their framework captures the following evaluation dimensions: *efficiency*, *scalability*, *data quality*, *hiding failure*, and *privacy level*. Since there are several different types of techniques for privacy preservation, some dimensions may be more important than others when evaluating a PPDM approach. An interesting discussion on this observation can be sought in [28].

Oliveira et al. in [66] provide an extension to the work presented in [64]. As a first step, SWA [65] is integrated as part of the new framework. Moreover, the authors propose three new heuristics (two for *data sharing* and one for *pattern sharing*) for protecting sensitive rules, a new sanitization algorithm, called *Downright Sanitizing Algorithm*, and two novel pattern sharing-based measures. These measures are used to quantify not only how much sensitive knowledge has been disclosed, but also measure the effectiveness of the proposed algorithms in terms of *information loss* and nonsensitive rules removed as a side effect of the transformation process. Finally, as part of this work, two novel pattern sharing-based measures are introduced: the *side effect factor*, measuring the number of nonsensitive association rules removed as a side effect of the sanitization process, and the *recovery factor*, which expresses the possibility of an adversary recovering a sensitive rule based on nonsensitive ones.

11.3.2 Classification Rule Hiding

The goal of classification mining is to discover patterns that classify objects (or tuples, in the context of the relational data model), into a set of predefined classes [69]. A classification model is first trained and then used to classify previously unseen data. The effectiveness of the classification task is captured by the metrics of *accuracy* and *error rate*. Similar to the setting of association rule hiding, in classification rule hiding, we consider some rules as *sensitive*, since they may reveal sensitive information concerning a target group or *class*. To protect such knowledge, a sanitization procedure needs to be enforced. In what follows, we partition existing approaches into two classes: *suppression-based* and *reconstruction-based* schemes; we briefly review the most interesting among them.

11.3.2.1 Suppression-Based Approaches

Suppression-based approaches aim at reducing the *confidence* of a sensitive classification rule by distorting some of the attributes in the data set, corresponding to transactions related to its existence. By reducing the confidence of a rule, one can achieve to minimize the data miner's belief related to this rule's actual existence in the data set; thus, the rule is properly protected. In what follows, we review the works of [16, 18, 41, 85] in this field of study.

Chang et al. in [16] are the first to address the inference problem caused by downgrading data in the context of decision rules. The inference problem is properly defined in the work of Morgenstern in [57]. The authors' paradigm uses decision trees to solve the inference problem, and the proposed methodology, called *parsimonious downgrading*, aims at securing a data set in a way that inference channels are blocked. To achieve this goal, each produced rule is considered as having a *confidence* value related to its existence. The notion of confidence is not materialized in the paper; however, the reader can think of it as the belief related to the rule's holding, given the data. The application of parsimonious downgrading occurs using a novel theory, called the *thermodynamic approach*, which aims at reducing the confidence in rules that are considered as sensitive. The *thermodynamic approach* is directly based on the notion of *entropy*, a well-known measure in the field of classification and information theory. Raising the entropy has the immediate effect of lowering the confidence associated to a rule. The proposed perturbation technique does not introduce erroneous data; rather, it selectively deletes data so that missing values appear in the released data set.

Johnsten et al. in [41] investigate issues pertaining to the assessment of the impact of classification mining on database security. In particular, the authors examine the security threat presented by a category of classification mining algorithms, known as *decision-region based*. Classification mining algorithms may use sensitive data to rank objects; each group of objects has a description given by a combination of nonsensitive attributes. The sets of descriptors obtained for a certain value of the sensitive attribute are referred to as *description space*. For *decision-region based* algorithms, the description space generated by each value of the sensitive attribute can be determined a-priori. Moreover, the predicted accuracy of assigning an object, satisfying a description, to a certain class, is dependent on the distribution of the particular class label relative to all the class labels associated with the objects that satisfy the same description.

Inspired by the theory of pattern recognition, Clifton in [18] proposes an open methodology, which achieves to confuse an adversary regarding the quality of the learned knowledge after the application of a rule mining algorithm. The proposed methodology carefully limits the sample size that can be safely released and mined in a way that allows the owner of the data to state clear limits on what can be learned from the sample. These limits are in the form of expected error on what is learned. One of the most interesting characteristics of this approach is that its success is irrespective of the quality of the applied data mining algorithm, since it guarrantees that no matter how good the data mining is, any result that is mined will be wrong $\varepsilon\%$ of the time with probability δ.

Wang et al. in [85] specify sensitive inferences among data patterns, using the notion of "*privacy templates*." Each template specifies the sensitive information to be protected, a set of identifying attributes, and the maximum association between the two. Templates are specified by the user, who also provides the maximum *confidence* associated to each sensitive inference. The authors present a template-based, privacy preservation approach, called *Progressive Disclosure Algorithm*, to protect sensitive knowledge induced in classification rules. The proposed approach is heuristic and scalable. It is based on the technique of *suppression of domain values*, applied in an effective way, and targets along two directions: *preservation of the information for a wanted classification analysis* and *limitation of the usefulness of unwanted sensitive inferences derived from the data*. It achieves to eliminate *all* sensitive inferences, including those with a low support, and efficiently handles overlapping inference rules.

11.3.2.2 Reconstruction-Based Approaches

A new direction to classification rules hiding, inspired by the work of [17, 72], is introduced in the paper of Natwichai et al. [58]. Instead of selecting transactions and suppressing values of some of their attributes to reduce the confidence of the produced rules, the authors propose a reconstruction approach based on the classification rules that have been checked and agreed by the data owner for releasing. This methodology first performs a rule-based classification of the original data set, and then presents a set of classification rules to the data owner to obtain the sensitive rules that need to be hidden. After that, a new decision tree is built from scratch, constituting only of nonsensitive rules approved by the data owner. Finally, a new data set adhering to the rules of this decision tree is reconstructed. The data set is similar to the original one in terms of knowledge, except the sensitive part.

Natwichai et al. in [59] present a new approach for hiding classification rules, based on the same premises as [58]. The new algorithm aims at improving the quality of the reconstructed data set by minimizing the side effects of the hiding process. Thus, it uses more extracted characteristics information from the original data set with regard to the classification issue, and improves the decision tree building procedure. To improve the usability, in the proposed scheme, not only the extracted rules are considered as the characteristics of data sets, but also the *gain ratio* of each attribute. Attributes are selected using the gain ratio, instead of least common attribute (used in [58]). Therefore, attributes with higher gain are put in higher levels of the constructed tree. With the aid of *information gain*, the usability of the released data set is highly improved.

11.4 Distributed Privacy-Preserving Data Mining

Distribution of the data introduces new problems and makes the privacy more challenging. The scenario is the following: There are data holder sites and they do not wish to share their data with the others due to regulations or due to the commercial value of the data. Another reason is that merging all these data sources may itself be a privacy threat. Therefore, individual sites are not willing to disclose their data. But at the same time, mining the data together will produce more meaningful results and all the parties are interested in obtaining the global data mining models. An often made assumption, which is, nevertheless, subject to further research, is that the global data mining models are assumed to be public, and they do not reveal private information.

Tools that can be used for privacy-preserving distributed mining of spatiotemporal data include *secure multi-party computation (SMC)*, which is a protocol in which participants of the protocol can compute some common function of local inputs, without revealing the local inputs to each other. Each participant participates in the protocol by supplying one or more inputs. At the end of the protocol, the only additional information each participant learns is the result of the computation and nothing else. A special case of SMC is secure two-party computation, epitomized by Yao's millionaires problem, whose goal is to determine which participant is richer, without revealing to each other how much money each party has.

In the general case, the task is to evaluate a function $y := F(x_1, x_2, \ldots, x_n)$ securely with inputs x_i's coming from the participants. The SMC protocol must be *fair* in the sense that each participant will be able to extract the correct result from the function evaluation. A generic SMC protocol that can be applied to any function exists, but usually renders an inefficient solution to the problem. Tailored solutions for specific applications are, therefore, preferred.

Different models have been proposed for efficient SMC protocols: (1) one approach is to employ a third party(TP), in which the actual computation is performed at an external party, on which all the participants rely to a certain extent. Note that, in third party model, participants do not reveal their local inputs to TP. The role of the TP in this model is to oversee the fair execution of the protocol steps. (2) In the second model, there is no TP and its functionality is simulated by participants of the protocol, resulting in a more distributed computation. In addition, other models as defined in [15] are needed to formulate the behaviors of the participants, which must be considered as an assumption in the protocol design. A *malicious party* may get involved in extra-protocol activities to deceive other parties and to get an upper hand during the protocol. Its goals may range from learning inputs of other parties to prevent other parties from learning the true result of protocol. A malicious party does not have to follow the protocol steps and can mount different attacks to achieve its goals. The *semihonest model*, on the other hand, assumes that the participants are honest but curious. A semi-honest party follows the protocol steps correctly. Nevertheless, it may want to extract more information from the protocol steps than it is entitled to. For instance, a semihonest participant is very curious, indeed, about the local inputs of the other parties. But, its activities are passive and the leakage of extra information to a semihonest party can be prevented with a carefully designed protocol.

The selection of a specific model depends on the application and the specific function to be computed to a great extent. In data mining applications, where parties are called data holders, the semihonest model is usually preferred. The semihonest model is particularly suitable for data mining applications, where the data holders refrain from active attacks and extra-protocol activities since following the protocol steps is usually in the best interest of all participants. Their behavior, therefore, cannot be disruptive. A semihonest TP is more in line with reality since internal operation of TP is strictly regulated.

SMC protocols for data mining applications can be implemented by using different tools and approaches. There are basically two main approaches that SMC protocols use for privacy-preserving data mining. We can classify them as cryptographic and noncryptographic approaches. In the following subsections, these approaches will be explained in detail.

11.4.1 Cryptographic Approaches

Cryptography offers many tools and primitives, which play an extremely essential role in many SMC applications in data mining. The conventional cryptographic primitives such as encryption functions (in particular *commutative* and *homomorphic* encryption), cryptographic hash functions, and pseudo-random number generators are especially important in data mining algorithms such as privacy-preserving association rule mining, privacy-preserving clustering, etc. Encryption is a mathematical transformation of a message to a random looking bit stream called ciphertext under a secret key. The ciphertext can be back-transformed into the original message only if the secret key is known. Pseudo-random number generators (PRNG) can be thought as deterministic finite state machine, which generates random-looking bit sequences. Its output is unpredictable so far as the initial state (seed) is kept secret. PRNGs are especially useful when more than one parties have to generate the same random sequence. Homomorphic encryption basically uses the property that operations can be performed over encrypted values in such a way that the resulting value is the ciphertext of the result we actually aim to compute. Homomorphic encryption can be formulated as $E(X) \cdot E(Y) = E(X \cdot Y)$, where $E(X)$ and $E(Y)$ are encrypted values of X and Y, respectively. Commutative encryption has the property that $E_2(E_1(X)) = E_1(E_2(X))$ such that E_1 and E_2 are commutative encryption schemes with different keys. Decryption operation also has the same property as does commutative encryption. This way multiple encryption and decryption operations can be performed over a value without any restriction on the order of these operations.

The seminal paper by Kantarcioglu and Clifton [42] on association rule mining over horizontally partitioned data heavily relies on the aforementioned cryptographic primitives. In [42], three different protocols over distributed data are proposed for securely calculating *sum* and *set union*. In the computation of secure sum, the initiator party adds random values (generated by PRNG) to its local support values, resulting in a randomized running sum. All other parties add their support values to the running sum, which finally reaches to the initiator. The initiator subtracts the random number from the randomized running sum and extracts the sum of the support values. In secure computation of set union, each party encrypts its local value using its personal secret key and pass it to the next party, which in turn does the same using its own secret key. In the end of the protocol, each local value is encrypted by each party using its corresponding local keys. When all local values are encrypted by every party, the ciphertexts are accumulated in one place. Since the encryption function used is commutative, the same local values result in identical ciphertexts after encrypted by all parties. Therefore, union of the local values can easily be calculated by decrypting the ciphertexts by all the parties.

Secure sum [42] is an information-theoretically secure protocol, where the security does not rely on an assumption where adversaries are computationally bounded as in the case of homomorphic encryption-based techniques.

Another class of data mining algorithms that utilizes pseudo-random numbers are privacy-preserving expectation-maximization clustering (PPEMC) [20, 50]. The goal in PPEMC is to cluster data points distributed at multiple parties to find global clustering result. For that purpose, E-Step and M-Step of EM algorithm can be formulated in such a way that each step can be performed by secure summation protocol, based on pseudo-random number generator.

Homomorphic encryption also has applications in privacy-preserving data mining protocols. Additive homomorphic schemes are due to ElGamal [29] and Paillier [68], where multiplication of ciphertexts results in the addition of plaintexts. In [88], the authors take advantage of ElGamal homomorphic encryption scheme for privacy-preserving naïve Bayes learning protocol. Using additive homomorphic property of ElGamal scheme such that $E(X) * E(Y) = E(X+Y)$, Yang et al. [88] propose privacy-preserving frequency mining protocol. The protocol uses ElGamal cryptosystem, which allows a central party to aggregate encrypted results received from the participants into the desired sum of frequencies. Aggregation of these secrets at a central party using homomorphic property reveals the result. The primitive for frequency mining protocol is used to design a privacy-preserving naïve Bayes learning protocol. Yang et al. [88] also shows that privacy-preserving association rule mining and decision tree learning protocols can be designed in a similar fashion. Another application of homomorphic encryption in SMC-based privacy-preserving data mining is employed in [91], where privacy-preserving association rule mining over vertically partitioned data is performed using additive property of homomorphic encryption.

Secure scalar product protocol is a building block for several privacy-preserving data mining protocols, which also employ cryptographic primitives. Du and Atallah [24] propose two methods to achieve secure scalar product. The first method heavily depends on 1-out-of-N oblivious transfer protocol and works as the following: Alice has vector X and Bob has vector Y. What Alice and Bob are trying to compute securely is the scalar product $X \cdot Y$. Alice starts the protocol and sends N vectors to Bob, merely one of which is Alice's original vector X; however, Bob does not know which one is the real X. Then Bob computes scalar product of his own vector Y with each vector sent by Alice. Bob sends the resulting N values to Alice who can choose the correct result within those values. The second method [24] proposes for secure scalar product uses permutation and homomorphic encryption. The idea is that scalar product of two vectors X and Y is equal to scalar product of permuted values of X and Y with respect to the same permutation scheme Π, namely $X \cdot Y = \Pi(X) \cdot \Pi(Y)$. However, if Alice has vector X and Bob has vector Y and Π is only known to Bob, then the question is how Alice obtains $\Pi(X)$ and $\Pi(Y)$ without learning permutation scheme Π. Homomorphic encryption is used for that purpose and the values sent to Alice are basically $\Pi(X+R)$ and $\Pi(Y)$. The random value R is deleted at the end of the protocol and the remaining value is the result.

A secure scalar product protocol based on cryptographic primitives is applied in various privacy-preserving data mining protocols such as privacy-preserving naïve Bayes classifier protocol for vertically partitioned data [81], privacy preserving decision-tree classification [25], and privacy-preserving k-means clustering over vertically partitioned data [80].

Another cryptographic primitive that finds usage in privacy preserving distributed data mining protocols is Shamir's secret sharing scheme. Emekci et al. [30] apply secret sharing to privacy preserving decision tree learning protocol by developing a secure summation protocol based on the following setting: each party creates a random polynomial of degree k, which all parties agreed on, with the condition that constant coefficient of each party's polynomial is equal to its (secret) input value. Then, for each party an evaluation point is chosen and broadcast to the group. In the next step, each party evaluates its own polynomial with the evaluation point of other parties and sends evaluations to the corresponding parties. Receiving evaluations from all other parties, each party can learn sum of all values but the individual values. Secret sharing based secure summation protocol is used to achieve privacy-preserving decision tree learning [30].

Yao's millionaires' problem (a.k.a. greater than function) is a famous problem proposed by Yao [89], in which two parties try to compare their wealth and conclude which one is richer without revealing any information on their wealth to each other. Many solutions to this problem have been proposed, based on different methods such as oblivious transfer [36], quadratic-residuosity [32], and encoding schemes [49]. Yao's millionaires' problem is applicable to several privacy-preserving data mining protocols. Lindell and Pinkas [51] conclude that private distributed decision tree learning can be reduced to find the attribute with the maximum information gain privately, which can be accomplished by employing Yao's millionaires' problem using oblivious transfer.

11.4.2 Noncryptographic Approaches

Besides cryptographic protocols, various noncryptographic approaches are applied in privacy-preserving data mining protocols. One of the mostly used representative of this approach is perturbation method [6]. Perturbation method relies on the fact that data mining results are about aggregation of the individual values for certain functions. Hence, what really matters is the behavior of the aggregated data not the individual values. From that perspective, if we can perturb individual values while preserving the distribution of the aggregated data, data mining protocols can still reach the same result as before the perturbation. On the other hand, individual privacy is preserved as the original values are not revealed. Agarwal and Srikant [6] state a privacy-preserving decision-tree classifier applying two different methods for perturbation. The first method is value-class membership method, which partitions values for an attribute into disjoint intervals, and each individual is assigned the interval it belongs to as the new value. The second perturbation method is value distortion. This method distorts individual values by adding a random value drawn from a distribution such as Gaussian or uniform. Liu et al. [52] propose a different perturbation method called random projection-based method to transform individual data while preserving certain statistical characteristics of the data. It shows application of this method to several protocols such as scalar product protocol, Euclidean distance estimation protocol, and k-means clustering protocol from distributed data.

Privacy-preserving association rule mining protocol for vertically partitioned data proposed in [79] uses a different approach, which relies on the idea that to be able to get a unique solution of a linear system of n unknowns at least n equations are needed. Hence, a secure scalar product protocol is developed by alteration of input vectors of each party into a linear system.

Model sharing and aggregation is another noncryptographic approach welcomed in privacy-preserving data mining. Merugu and Ghosh [56] and Klusch et al. [47] propose model-based solutions for the privacy-preserving clustering problem. Data holder parties build local models of their data, which is subject to privacy constraints. Then a third party builds a global model from these local models and clusters the data generated by this global model.

To sum up, the techniques applied in privacy-preserving distributed data mining can be of great help to find out new protocols that meet the needs of spatiotemporal privacy-preserving data mining. One example is privacy-preserving spatiotemporal clustering protocol proposed by İnan and Saygin [39], which takes advantage of pseudo-random numbers to develop a secure difference protocol to form the dissimilarity matrix used in many clustering algorithms. Details of this protocol are explained in Sect. 11.4.3. In a similar fashion, the existing methods for secure multiparty computation can be applied to privacy-preserving spatiotemporal data mining.

11.4.3 Application of SMC on Privacy-Preserving Mining of Spatiotemporal Data

To our knowledge, privacy issues in spatiotemporal data from the data mining perspective have not been addressed except for one paper published in 2006 [39]. They introduce a secure multiparty solution to privacy problems in spatiotemporal data without any loss of accuracy. In the following paragraphs, we describe that work in detail and show how SMC tools can be used to develop privacy-preserving mining techniques for spatiotemporal data.

Before, however, giving the details of [39], it is useful to introduce trajectory clustering, an application of trajectory tracking, and finally privacy problem in the application. Trajectory T of a moving object X is a set of location observations in the form $O = \{t, d\}$, where t represents time and d location of X at time t, respectively. Data holders want to cluster trajectories of moving objects, which requires grouping of objects that are similar with respect to location observations. An example application of trajectory clustering is in the traffic modeling and congestion analysis. Traffic control offices want to cluster trajectories of drivers to solve traffic congestions. However, the required location observations for drivers are not readily available, but can be collected from GSM operators. GSM operators are not always willing to share their data because of privacy concerns and regulations, since context of location observations can reveal sensitive information about the drivers such as where they live and where they work. Therefore, privacy-preserving clustering algorithms should be utilized to get the desired result without revealing individual location observations.

In [39], clustering is chosen as the target data mining model, which is the process of grouping similar objects together. To group similar objects, a similarity metric is needed for comparing two trajectories. Similarity of records with numerical data is simple and there are robust methods to do that. Unfortunately, for trajectory comparison, this is not the case. This is due to the fact that observation intervals are nonoverlapping, sampling rates are different and there are time shifts. Trajectory comparison functions are still being studied to overcome the previously mentioned difficulties.

If it is assumed that trajectories are of the same length and time-stamps of the corresponding elements are equal, then things are simpler. In such a case, the distance between two trajectories can easily be computed using Euclidean distance via summing up the distances over all elements with equal time-stamps. Most trajectory comparison functions stem from four basic algorithms: (1) Euclidean distance, (2) longest common subsequence (LCSS), (3) dynamic time warping (DTW), and (4) edit distance. Significance of the proposed privacy-preserving trajectory comparison protocol in [39] is due to the fact that it can be applied to all comparison functions listed above. In addition to that, the protocol does not sacrifice accuracy against privacy.

Assume that there are data holders that track locations of moving objects (people, vehicles, etc.) with unique object ID's. Data holders would like to cluster the trajectories of moving objects that are distributed at multiple sites. While doing so, sensitive location information should be kept confidential, and clustering results are published to each data holder at the end of the protocol.

Storage space and computation power is provided by a distinct third party denoted by TP. Semihonest model for the parties involved is the key assumption in this protocol. In semihonest model, all the parties including the TP follow the protocol, but they are also curious in the sense that they may store information they receive, which can be used to infer private data. Another key assumption is that parties do not share private information with each other, i.e., they are noncolluding.

TP plays the following roles in the protocol:

- Managing the communication between data holders
- Constructing the global dissimilarity matrix in a privacy-preserving manner
- Clustering the trajectories using the dissimilarity matrix
- Publishing the results to the data holders

There are two phases of the protocol: communication phase and the computation phase.

In the communication phase, data holders exchange data among themselves and the third party (TP), who will carry out the computation phase and publish the clustering results. Prior to the communication phase, we assume that every involved party, including the third party, has already generated pair-wise keys. These keys are used as seeds to pseudo-random number generators, which disguise the exchanged messages. Diffie–Hellman key exchange protocol is perfectly suitable for key generation [23]. Dissimilarity matrix is an object-by-object structure. In case of spatiotemporal data, an entry $D[i][j]$ of the dissimilarity matrix D is the distance between trajectories of objects i and j calculated using any comparison function. The proposed privacy-preserving comparison protocols are suitable for all comparison functions. If trajectories of both i and j are held by the same site, this site can calculate their distance locally and send it to the third party. However, if trajectories of i and j are at separate sites, these sites should run the protocol explained later. Assuming K data holders, $C(K,2)$ runs are required, one for each pair of data holders. Suppose that two data holders, DHA and DHB, with size(A) and size(B) trajectories, respectively, want to compare their data. Assume that the protocol starts with DHA. For each trajectory T in DHA's database, two pseudo-random number generators are initialized, rngAB and rngAT. The seed for rngAB is the key shared with DHB and the seed for rngAT is the key shared with TP. Then, for each spatial dimension of T's elements (i.e., x and y), DHA disguises its input as follows: if the pseudo-random number generated by rngAB is odd, DHA negates its input and increments it by the pseudo-random number generated by rngAT. Finally, DHA sends the disguised values to DHB.

Upon receiving data from DHA, DHB initializes a matrix M of dimension of size(A)Xsize(B), which will be DHB's output. For each trajectory T in its database, DHB initializes a pseudo-random number generator rngAB with the key shared with DHA and negates its inputs in a similar fashion. This time negation is done when the generated number is even. DHB then starts filling values into M. An entry $M[i][j][m][n]$ of M is DHA's jth trajectory's nth observation compared with DHB's ith trajectories mth observation. DHB simply adds its input to the input received from DHA. At the end, M is sent to TP by DHB.

TP subtracts the random numbers added by DHA using the pseudo-random number generator, rngAT, initialized with the key shared with DHA. Now, absolute value of any entry $M[i][j][m][n]$ is $|DHA[j][n] - DHB[i][m]|$. These values are all that are needed by any comparison function to compute the distance between trajectories i and j.

The third party can compute pair-wise trajectory distances for data holder sites A and B, once the comparison matrix M is built through the protocol. If the comparison function measures distances using real penalty, then $M[i][j][m][n]$ is the cost for nth observation of A's jth trajectory with respect to the mth observation of B's ith trajectory. Otherwise, if a quantized penalty comparison function is to be employed, TP simply checks whether $M[i][j][m][n] < \varepsilon$ to match these two observations. What remains is performing comparisons of the form $M[i][j]$, where both i and j are trajectories of the same data holder site. In such cases, another privacy-preserving protocol is not required to compute these values, since conveying local dissimilarity matrices to TP does not leak any private information, as proven in [61]. To build the dissimilarity matrix, TP must ensure that every data holder site has sent its local dissimilarity matrix and run the pair-wise comparison protocol with every other data holder.

After gathering comparison results for all pairs of trajectories, TP normalizes the values in the dissimilarity matrix. These normalized distances are the only required input for most clustering algorithms, such as k-medoids, hierarchical, and density-based clustering algorithms. Another key observation here is that using this protocol, TP may use any clustering algorithm depending on the requirements of the data holders. At the end of the clustering process, the third party sends the clustering results to the data holders. The results are in the form of lists of object identifiers, since publishing the dissimilarity matrix would cause private information leakage. The third party can also publish clustering quality parameters, if requested by the data holders.

A basic assumption in the secure comparison protocol discussed earlier is that, any information that can be computed from the dissimilarity matrix can be revealed without any privacy breach. Such an argument was also made by Oliveira and Zaiane [61] for the case when the attacker does not know the domain of the data. This assumption may be too strong in some applications and this is left for further investigation in [39].

For the privacy analysis of the protocol in [39], we assume that the players do not collude and all communication channels are encrypted. Since the initiator of the protocol never receives any information, clearly he cannot breach the privacy of any entity in the database of the follower. The follower only receives data of the form $(-1)^{mg_{AB}} DH_A[i] + rng_{AT}(i)$, for $i = 1, \ldots, \text{size}(DH_A)$, which is a random number, when $rng_{AT}(i)$ is a random number chosen uniformly over the entire domain. When random numbers rng_{AT} are chosen independently, the numbers $DH_A[i] + rng_{AT}(i)$ are also independent, and so the follower cannot breach the privacy of any entity in the database of the initiator. This, of course, only holds under the assumption that real independent random numbers are chosen, but the assumption is that the random number generator is cryptographically secure, which means that no one can distinguish it from a true random source. Finally, the third party receives data of the form $(-1)^{mg_{AB}} (DH_A[i] - DH_B[j]) + rng_{AT}(i)$, for $i = 1, \ldots, \text{size}(DH_A)$, and $j = 1, \ldots, \text{size}(DH_B)$. The random number $rng_{AT}(i)$ is chosen independently of the data, so it does not give any information to the third party. The number $(-1)^{mg_{AB}} (DH_A[i] - DH_B[j])$ is exactly the (i, j)th entry of the dissimilarity matrix (up to the sign $(-1)^{mg_{AB}}$), which is the legal output of the third party. Since the sign $(-1)^{mg_{AB}}$ is chosen independently of the data, it does not provide any information about the data. Thus, none of the three players can cause a privacy breach.

11.5 Privacy-Aware Knowledge Sharing

Previous approaches (as described in the previous sections of this chapter) were focused on producing a valid mining model without accessing the original data, or on building a common mining model in a distributed environment without disclosing each party's private data, or producing databases such that some given patterns are hidden. But they all still leave a privacy question open [43]: *do the original data mining results themselves violate privacy?*

Starting from this question, a novel approach to privacy-preserving data mining has emerged in the last years. In several real-world scenarios, we are willing to disclose interesting patterns mined from possibly sensitive data. What we are interested in is, therefore, producing such patterns in the form of general descriptions of the underlying data distributions without releasing information that can be maliciously used to breach privacy.

The generic problem definition is thus how to produce and publish valid mining models without disclosing "private" information. As usual this issue has been investigated in completely different ways.

The work in [43] complements the line of research in distributed privacy-preserving data mining described in Sect. 11.4, but focusing on the possible privacy threat caused by the data mining results. In particular, the authors study the case of a classifier trained over a mixture of different kind of data: *public* (known to every one including the adversary), *private/sensitive* (should remain unknown to the adversary), and *unknown* (neither sensitive nor known by the adversary). The authors propose a model for privacy implication of the learned classifier, based on a black box classifier, i.e., the user willing to classify new instances can get the class label without directly accessing the internal structure of the classifier. Classifiers are, therefore, functions from tuples to class labels. Private/public/unknown data may appear in the input of the classifier function, in the output (class labels), or both, leading to different privacy concerns. Within this model, authors study possible ways in which the classifier can be used by an adversary (that is, the user of the classifier) to compromise privacy.

The work in [67] has some common aspects with the line of research in knowledge hiding of Sect. 11.3. But this time, instead of the problem of sanitizing the data, the problem of *association rule sanitization* is addressed. The data owner, rather than sharing the data, prefers to mine it and share the discovered association rules. As usual for all works in intensional knowledge hiding, the data owner knows a set of restricted association rules that he does not want to disclose. The authors propose a framework to sanitize a set of association rules protecting the restricted ones by blocking some inference channels.

In [34], a framework for evaluating classification rules in terms of their perceived privacy and ethical sensitivity is described. The proposed framework empowers the data miner with alerts for sensitive rules, which can be accepted or dismissed by the user as appropriate. Such alerts are based on an aggregate *Sensitivity Combination Function*, which assigns to each rule a value of sensitivity by aggregating the sensitivity value (an integer in the range $0\ldots 10$) of each attribute involved in the rule. The process of labeling each attribute with its sensitivity value must be accomplished by the domain expert, which knows what is sensitive and what is not.

11.5.1 k-Anonymity of Data Mining Results

A common aspect of the three approaches described earlier is that they all require some a priori knowledge of what is sensitive and what is not. In the last two years, a novel approach has emerged, which concentrates on individual privacy, in the strict sense of *nonidentifiability*, as prescribed by the European Union regulations on privacy, as well as US rules on protected health information (HIPAA rules). Privacy is regulated at the European level by Directive 95/46/EC (October 24, 1995) and Regulation (EC) No 45/2001 (December 18, 2000). In such documents (see also Chap. 4 for further details), general statements about identifiability of an individual are given, such as

> To determine whether a person is identifiable, account should be taken of all the means likely to be reasonably used either by the controller or by any person to identify the said person. The principles of protection should not apply to data rendered anonymous in such a way that the data subject is no longer identifiable.

According to this perspective, we should ask ourselves whether the disclosure of patterns extracted by data mining techniques may open up the risk of privacy breaches that may reveal individual identity. In [11], it has been shown that the answer to this question is, unfortunately, positive: data mining results can indeed violate anonymity. This observation originates the following research problem:

> is it possible to devise well-founded definitions and measures of privacy that provably prevent the violation of anonymity in the mining results, hence guaranteeing non-identifiability?

A solution to this problem would be an enabling factor of many emerging applications, based on knowledge discovery, as the typical GeoPKDD applications. The trajectories of moving individuals contain detailed information about personal and vehicular mobile behavior, and therefore offer interesting practical opportunities to find behavioral patterns, to be used for instance in traffic and sustainable mobility management, e.g., to study the accessibility to services. Clearly, in these applications privacy is a concern. As an example, an overspecific pattern may reveal the behavior of an individual (or a small group of individuals) – even if the identity of the individual is concealed, it can be reconstructed by knowing that there is only one person living in a specific place that moves on a certain route at the specific time. Thus, anonymity is not only related to identity hiding, but also to the size of population identified by a pattern, either directly or indirectly. How can trajectories of mobile individuals be analyzed without infringing personal privacy rights? How can, out of privacy-sensitive trajectory data, patterns be extracted that are demonstrably anonymity-preserving?

Atzori et al. in [10–12] concentrate on the notion of *individual privacy* protection in frequent item set mining and provide a formulation that achieves to shift the concept of *k-anonymity* from source data to the extracted patterns. The notion of *k*-anonymity is of crucial importance in spatiotemporal environments, since it proves to be an effective way for protecting individuals' *location privacy* [35]. It was first introduced in the context of databases in the work of Samarati and Sweeney [73,78], where the authors demonstrated that protection of individual sources is insufficient to guarantee privacy when these sources are cross-examined. Thus, the objective of k-anonymity is to eliminate such risks of inferring private information through *cross linkage* by ensuring that each shared information regards at least k individuals. Data mining models and patterns (such as frequent item sets), in order to ensure the minimum required statistical significance for the production of rules, aim at representing a large number of individuals; thus, they are expected to conceal individuals' identities. However, as shown in [10, 11], this belief is ill-founded. Depending on the special characteristics of the data set, minimum support, and privacy thresholds used, it may be possible for an adversary to identify individuals "hidden" within certain frequent patterns.

Example 11.1. Consider the following association rule:

$$a_1 \wedge a_2 \wedge a_3 \Rightarrow a_4 \quad [\text{sup} = 80, \text{conf} = 98.7\%],$$

where sup and conf are the usual interestingness measures of *support* and *confidence* as defined above. Since the given rule holds for a number of individuals (80), which seems large enough to protect individual privacy, one could conclude that the given rule can be safely disclosed. But, is this all the information contained in such a rule? Indeed, one can easily derive the support of the premise of the rule:

$$\text{sup}(\{a_1, a_2, a_3\}) = \frac{\text{sup}(\{a_1, a_2, a_3, a_4\})}{\text{conf}} \approx \frac{80}{0.987} \approx 81.05.$$

Given that the pattern $a_1 \wedge a_2 \wedge a_3 \wedge a_4$ holds for 80 individuals and that the pattern $a_1 \wedge a_2 \wedge a_3$ holds for 81 individuals, we can infer that in our database there is just one individual for which the pattern $a_1 \wedge a_2 \wedge a_3 \wedge \neg a_4$ holds. The knowledge inferred is a clear threat to the anonymity of that individual: on the one hand, the pattern identifying the individual could itself contain sensitive information; on the other hand, it could be used to reidentify the same individual in other databases. It is worth noting that this problem is very general: the given rule could be, instead of an association, a classification rule or the path from the root to the leaf in a decision tree, and the same reasoning would still hold. Moreover, it is straightforward to note that, unluckily, the more accurate is a rule, the more unsafe it may be w.r.t. anonymity.

The authors formalize the threats to anonymity by means of inference channels through frequent item sets, and provide practical algorithms to efficiently and effectively identify these threats and eliminate them by means of controlled *pattern distortion*. The goal in this procedure is that produced k-anonymous patterns need to be as close as possible to the real patterns holding in the data. The overall framework provides comprehensive means to reason about the desired tradeoff between anonymity preservation and quality of the collection of patterns, as well as the distortion level required to block the threatening inference channels. An interesting observation in this work is that the trivial solution of first k-anonymizing the data and then mining the patterns is effective in eliminating inference threats, but can lead to the production of patterns that are impoverished by information loss, intrinsic in the generalization and suppression techniques. Since the objective is to extract valid and interesting patterns, k-anonymization needs to be postponed to occur after the actual mining step. The authors propose two algorithms for the detection of inference channels, which may lead to anonymity threats in the output frequent item set patterns. Finally, they propose two techniques, namely *additive* and *suppressive* sanitization, for blocking all potential inference channels.

In [33], the problem of k-anonymity is shifted to decision trees. The authors propose an ID3-like algorithm to induce k-anonymous decision trees, i.e., classifiers that guarantee not releasing personal information about people whose data are mined. Basically, at each iteration, it finds the best split according to some gain metric. Then, it verifies that the split respect the k-anonymity property. If k-anonymity is violated, the splitting is rolled back and the algorithm proceeds to consider the attribute with the next largest gain.

The work [40] moves the framework of [10–12] to a distributed scenario; thus this work collocates on the border between distributed privacy-preserving data mining and privacy-aware knowledge sharing. Two algorithm for two-party computations are presented: one is based on secure distributed k-anonymization of data, followed by secure mining; the other one is instead based on the secure computation of inference channels. Both algorithms end up with a set of k-anonymous frequent item sets and their supports.

11.5.2 Work Related to Spatiotemporal Knowledge Sharing

Privacy-aware knowledge sharing is the youngest issue in privacy-preserving data mining and, as mentioned earlier, very few work is currently available on this topic. To the best of our knowledge, no work addressing spatiotemporal knowledge sharing on has been published yet. In the following, we discuss some work related to spatiotemporal data privacy, which are likely to be extended to obtain first techniques for spatiotemporal knowledge sharing.

In [14, 54], k-anonymity techniques are adapted to handle different time granularities, motivated by location-based service applications. Limitation of straightforward application of existing techniques are shown, and a time-generalization algorithm for anonymizing database tables with time-dependent tuples is provided.

Another recent work on anonymization of spatiotemporal data is described in [35]. The authors describe algorithms and ad-hoc data structures to generalize effectively spatiotemporal data coming from mobile users of location-based services. Both space and time issues are addressed and performance are measured through simulations.

We believe that privacy-aware mining techniques in [12, 40] can be extended to spatiotemporal mining by exploiting the existing work developed in the context of databases. Another way we want to investigate to achieve privacy-preserving spatiotemporal mining is turning recent spatiotemporal data mining algorithms (such as the trajectory-based approach described in [38]) into privacy-preserving ones, again exploiting literature on data privacy and privacy-aware knowledge sharing.

11.6 Roadmap Toward Privacy-Aware Mining of Spatiotemporal Data

Spatiotemporal, geo-referenced data sets are growing rapidly, and will be more so in the near future. This phenomenon is mostly due to the daily collection of telecommunication data from mobile phones and other location-aware devices. The increasing availability of these forms of geo-referenced information is expected to enable novel classes of applications, based on the extraction of behavioral patterns from mobility data. Such patterns could be used for instance in traffic and sustainable mobility management, e.g., to study the accessibility to services. Clearly, in these applications privacy is a concern, since for instance an overspecific pattern may reveal the behavior of groups of few individual.

Spatiotemporal data sets present a new challenge for the privacy-preserving data mining community because of their spatial and temporal characteristics. So far privacy issues in spatiotemporal data mining have not been addressed except for one paper published in 2006 [39], which we have discussed in Sect. 11.4.3. Therefore, there is plenty of room for research in this interesting and challenging area. In the following, we will draw a road map for the research toward privacy-aware mining of spatiotemporal data.

11.6.0.1 Data Perturbation and Obfuscation

In a GeoPKDD-like application scenario, a telecommunication company is willing to mine the mobility data it owns, but since the company lacks the needed expertise, it is forced to give the data to a third party for mining. In this scenario, both privacy constraints imposed by the legal regulations and secrecy constraints imposed by business strategies may hold. In this case, a solution could be to apply data perturbation and/or obfuscation techniques to the data in such a way that the identification of the original trajectories is not possible, but it is still possible to extract valid and useful mining models and patterns. Obviously, the standard data perturbation and obfuscation techniques must be adjusted to cope with spatiotemporal mining tasks.

Data perturbation techniques for privacy-preserving spatiotemporal data mining are yet to be studied. Perturbation techniques of spatiotemporal trajectories need to be devised so that the privacy of the individuals to whom those trajectories belong is protected, while global models such as clustering, classification, and frequent trajectories can still be constructed on perturbed trajectories. A naive method would be to consider trajectories as spatiotemporal data points. But we believe that this method would not be sufficient since trajectories need to be considered as whole objects. Randomization of trajectories can be done as in the case of mixed zones discussed previously in this book. But this issue needs to be investigated from a data mining perspective.

11.6.0.2 Knowledge Hiding

Spatiotemporal privacy-preserving rule hiding methodologies will allow for sanitizing selected transactions from the original data set and prohibit leakage of sensitive rules related to "sensitive" spatial and/or temporal information. Hiding spatiotemporal patterns imposes greater challenges than the traditional knowledge hiding approaches. To address these extra requirements, new algorithms need to be devised that will accomplish to make use of both the *spatial* and *temporal* dimensions of the data, and achieve to identify all important correlations existing within the data sets; these correlations constitute rules that potentially depict behavioral aspects of the underlying subjects. It is, therefore, reasonably expected that a portion of the produced rules will contain sensitive knowledge that needs to remain confidential when sharing the data with other, untrusted third-parties.

Imagine the scenario under which data are stored in a transactional database, reflecting people's daily movements from one place to another. These data are properly time-stamped to reflect the temporal and contextual information related to each subject. Because of its sequential nature, special handling is required for its proper mining. Regular association rule mining emphasizes solely on items' co-occurrence in a data set to identify and create rules. Thus, it fails to properly address the *sequential nature* of spatiotemporal data. To overcome this defect, prior to mining trajectory data, one needs to reconstruct the actual *user-trajectories* (or parts of them in specific regions of interest) by linking distinct records referring to the same individual. This linking process is essential to ensure proper mining and needs to be performed in a carefully crafted manner. As a second step, *generalization* techniques need to be encompassed to allow the broadening of the rule's capacity by including more subjects. Such techniques need to be incorporated in the data mining process and have to be taken into account by the knowledge hiding approach that will be used to mine the data set.

Hiding sensitive knowledge depicted in the form of *association* or *classification* rules in a spatiotemporal environment may allow for the incorporation of different types of techniques apart from *support/confidence reduction* (in the case of association rule mining) or *suppression/reconstruction* (in the case of classification rule mining). Open issues to be addressed in the future include, but not limited to, are the discovery of the sequential patterns and the protection of the anonymity of these patterns, usually by means of *generalization*.

11.6.0.3 Distributed Privacy-Preserving Data Mining

In a possible GeoPKDD-like application scenario, we can image two or more telecommunication companies willing to build a mining model from the union of the mobility data that each company holds, for a common benefit, e.g., for building a predictive model of accessibility to services with the aim of bandwidth optimization. This is the typical SMC application scenario, where each party does not communicate to the other parties its own data, but it is still possible to build a valid model from the union of all the data. However, the peculiarities of spatiotemporal data call for ad-hoc techniques. Any possible spatiotemporal data mining task, as those one individuated in Chap. 10, can be addressed also in a distributed scenario. For instance, we could develop ad-hoc techniques for the mining of frequent trajectories on horizontally partitioned data.

As mentioned earlier, [39] reports, to our knowledge, on the first *distributed* privacy-preserving data mining method applied to the spatiotemporal data, which is distributed over a set of users. The basic tools extensively used in [39] are *blinding* with pseudo-random numbers and secure multiparty computation over the blinded data. The method computes secure difference of numerical attributes common in spatiotemporal databases, in which parties only learn the difference with the aid of a third party. One concern with the technique is its slightly high communication and computation costs. Furthermore, the assumption that the parties share secret keys with each other in addition to secret keys shared with the third party leads to one of the fundamental problems of modern day cryptography: *secure key distribution and management*. In addition, distributed data mining of spatiotemporal data may necessitates computation of other (arithmetic) functions than secure difference, which is not feasible in the given framework in [39].

Consequently, given the problems related to distributed privacy preserving mining of spatiotemporal data and the shortcomings of the existing solutions, possible research directions can be itemized as in the following:

- Efficient secure difference protocols with lower computation and communication costs
- Architectures, in which the parties share secret keys only with the third parties not with each other, thereby alleviating the key distribution and management problems
- New computation models where there are actually two third parties, which are not necessarily trusted but with conflicting interests, therefore never colluding
- The application of secure secret sharing protocols, where parties apply data mining operations on partial data, which does not carry any information (whatsoever) on the actual spatiotemporal data. The actual results can be obtained when the partial results are combined
- Extending secret sharing techniques to threshold schemes
- Algorithms for simulating the behavior of one or both third parties, thereby replacing them with regular participants of the data mining computation
- Techniques and algorithms for secure computation of other functions besides difference, which may be of high importance in spatiotemporal data mining

11.6.0.4 Privacy-Aware Knowledge Sharing

When the problem is to publish not the data but the extracted patterns and models, the problems described in Sect. 11.5 still hold in a spatiotemporal context, and they are even more challenging.

We believe that the anonymity-preserving data mining framework recently proposed in [10, 11] based on algorithms could be applied for handling pattern anonymization in spatiotemporal environments. Now, we can briefly investigate some issues emerging from this approach that are expected to be part of future research conducted in this field. One promising path of research regards the mapping of the above-mentioned theoretical framework to the more concrete case of categorical data, originating from relational tables. In this context, one could exploit the semantics of the attributes in order to apply generalization techniques, similar to what is done by classical k-anonymization. This would be very useful in the context of ST data, since we would be able to preserve privacy by specifying a grained level of detail for locations and times instead of completely removing such information.

Another open problem regards the release of multiple collections of k-anonymous patterns, extracted from the same source but at different support thresholds. A malicious adversary receiving more than one of these collections can easily violate the k-anonymity defense. Using database k-anonymization would solve this problem at the cost of a higher information loss [2]. However, new techniques for database anonymity have been developing recently [7, 9, 46, 87] in order to minimize the information loss. A study of their feasibility in ST privacy-preserving data mining is currently missing.

Finally, a worthwhile research direction regards the extension of the framework in [11] to capture other forms of patterns and models, such as *classification* or *clustering* models in the spatiotemporal scenario. Possible directions to enhance existing algorithms for item sets mining are the following:

- Extending the concept of k-anonymity, from frequent item sets to frequent trajectories
- Privacy-aware trajectories classifiers and next-step predictors
- Privacy-aware trajectories clustering

We believe that through the adaptation of the support and confidence concepts defined in frequent item set mining it will be possible to easily extend k-anonymous patterns to ST data mining scenarios.

11.7 Conclusions

In this chapter, we have presented a thorough examination and an elaborate classification of the work that have been performed in the privacy-preserving data mining area. Although similar studies have been published in the past, our work is unique in paving the way from today's conventional forms of data to tomorrow's multimodal forms of data. The analysis that is presented is starting out with a comprehensive description of the state of the art in privacy-preserving data mining and sets out a plan for moving toward in addressing the needs for privacy preservation originating from the interrogation of data produced by global scale and ubiquitous applications. Even though a lot of effort has been placed in this work, we are certain that it is far from being complete, since our main focus was on presenting a condensed classification covering only the majority of the works that appeared from the inception of this field to today. To provide a balance to this incompleteness, we have provided a detailed plan to lead the way from today to the short term future. Whether this plan will make it through the high waves of competitiveness is a matter of time.

References

1. N.R. Adam and J.C. Wortmann. Security-control methods for statistical databases: A comparative study. *ACM Computing Surveys*, 21(4):515–556, 1989.
2. C.C. Aggarwal. On k-anonymity and the curse of dimensionality. In *Proceedings of the 31th International Conference on Very Large Databases (VLDB'05)*, pp. 901–909, 2005.
3. D. Agrawal and C.C. Aggarwal. On the design and quantification of privacy preserving data mining algorithms. In *Proceedings of the 20th Symposium on Principles of Database Systems (PODS'01)*, pp. 247–255, 2001.
4. R. Agrawal, T. Imielinski, and A. Swami. Mining association rules between sets of items in large databases. In *Proceedings of International Conference on Management of Data (SIGMOD'93)*, pp. 207–216, 1993.
5. R. Agrawal and R. Srikant. Fast algorithms for mining association rules in large databases. In *Proceedings of the 20th International Conference on Very Large Databases (VLDB'94)*, pp. 487–499, 1994.
6. R. Agrawal and R. Srikant. Privacy-preserving data mining. In *Proceedings of International Conference on Management of Data (SIGMOD'00)*, pp. 439–450, 2000.
7. S. Agrawal and J.R. Haritsa. A framework for high-accuracy privacy-preserving mining. In *Proceedings of the 21st International Conference on Data Engineering (ICDE'05)*, pp. 193–204, 2005.
8. M. Atallah, E. Bertino, A. Elmagarmid, M. Ibrahim, and V.S. Verykios. Disclosure limitation of sensitive rules. In *Proceedings of the Knowledge and Data Engineering Exchange Workshop (KDEX'99)*, pp. 45–52, 1999.
9. M. Atzori. Weak k-anonymity: A low-distortion model for protecting privacy. In *Proceedings of the 8th International Information Security Conference (ISC06)*, pp. 60–71, 2006.
10. M. Atzori, F. Bonchi, F. Giannotti, and D. Pedreschi. Blocking anonymity threats raised by frequent itemset mining. In *Proceedings of the 5th International Conference on Data Mining (ICDM'05)*, pp. 561–564, 2005.
11. M. Atzori, F. Bonchi, F. Giannotti, and D. Pedreschi. k-Anonymous patterns. In *Proceedings of the 9th European Conference on Principles and Practice of Knowledge Discovery in Databases (PKDD'05)*, pp. 10–21, 2005.
12. M. Atzori, F. Bonchi, F. Giannotti, and D. Pedreschi. Anonymity preserving pattern discovery. *Very Large Data Bases Journal*. To Appear.
13. E. Bertino, I.N. Fovino, and L.P. Povenza. A framework for evaluating privacy preserving data mining algorithms. *Data Mining and Knowledge Discovery*, 11(2):121–154, 2005.
14. C. Bettini and S. Mascetti. Preserving k-anonymity in spatiotemporal datasets and location-based services. In *First Italian Workshop on PRIvacy and SEcurity (PRISE)*, 2006.

15. R. Canetti, U. Feige, O. Goldreich, and M. Naor. Adaptively secure multi-party computation. In *Proceedings of the 28th Annual Symposium on Theory of Computing (STOC'96)*, pp. 639–648. ACM Press, 1996.
16. L. Chang and I.S. Moskowitz. Parsimonious downgrading and decision trees applied to the inference problem. In *Proceedings of the Workshop on New Security Paradigms (NSPW'98)*, pp. 82–89, 1998.
17. X. Chen, M. Orlowska, and X. Li. A new framework of privacy preserving data sharing. In *Proceedings of the 4th IEEE International Workshop on Privacy and Security Aspects of Data Mining*, pp. 47–56, 2004.
18. C. Clifton. Using sample size to limit exposure to data mining. *Journal of Computer Security*, 8(4):281–307, 2000.
19. C. Clifton, M. Kantarcioglu, and J. Vaidya. Defining privacy for data mining. In *Natural Science Foundation Workshop on Next Generation Data Mining*, pp. 126–133, 2002.
20. C. Clifton, M. Kantarcioglu, J. Vaidya, X. Lin, and M.Y. Zhu. Tools for privacy preserving distributed data mining. *ACM SIGKDD Exploration Newsletter*, 4(2):28–34, 2002.
21. C. Clifton and D. Marks. Security and privacy implications of data mining. In *Proceedings of International Conference on Management of Data (SIGMOD'96)*, pp. 15–19, 1996.
22. E. Dasseni, V.S. Verykios, A.K. Elmagarmid, and E. Bertino. Hiding association rules by using confidence and support. In *Proceedings of the 4th International Workshop on Information Hiding (HI'01)*, pp. 369–383, 2001.
23. W. Diffie and M.E. Hellman. New directions in cryptography. *IEEE Transactions on Information Theory*, IT-22(6):644–654, 1976.
24. W. Du and M.J. Atallah. Privacy-preserving statistical analysis. In *Proceedings of the 17th Annual Computer Security Applications Conference (ACSAC'01)*, pp. 102–110, 2001.
25. W. Du and Z. Zhan. Building decision tree classifier on private data. In *Proceedings of the International Conference on Privacy, Security and Data Mining (CRPITS'02)*, pp. 1–8, 2002.
26. W. Du and Z. Zhan. Using randomized response techniques for privacy-preserving data mining. In *Proceedings of the 9th International Conference on Knowledge Discovery and Data Mining (KDD'03)*, pp. 505–510, 2003.
27. M. Duckham and L. Kulik. A formal model of obfuscation and negotiation for location privacy. In *Proceedings of the Third International Conference on Pervasive Computing (Pervasive'05)*, pp. 152–170, 2005.
28. C. Dwork and K. Nissim. Privacy preserving data mining in vertically partitioned databases. In *Proceedings of the 24th International Conference on Cryptology (CRYPTO'04)*, pp. 528–544, 2004.
29. T. ElGamal. A public key cryptosystem and a signature scheme based on discrete logarithms. *IEEE Transactions Information Theory*, 31:469–472, 1985.
30. F. Emekci, O.D. Sahin, D. Agrawal and A. El Abbadi. Privacy preserving decision tree learning over multiple parties. Data & Knowledge Engineering. 63(2):348–361, 2007.
31. A. Evfimievski, R. Srikant, R. Agrawal, and J. Gehrke. Privacy preserving mining of association rules. In *Proceedings of the 8th International Conference on Knowledge Discovery and Data Mining (KDD'02)*, pp. 343–364, 2002.
32. M. Fischlin. A cost-effective pay-per-multiplication comparison method for millionaires. *Lecture Notes in Computer Science*, 2020:457, 2001.
33. A. Friedman, A. Schuster, and R. Wolff. k-Anonymous decision tree induction. In *Proceedings of the 10th European Conference on Principles and Practice of Knowledge Discovery in Databases (PKDD'06)*, pp. 151–162. Springer-Verlag, 2006.
34. P. Fule and J.F. Roddick. Detecting privacy and ethical sensitivity in data mining results. In *Proceedings of the 22nd Workshop on Australasian Information Security, Data Mining and Web Intelligence, and Software Internationalisation*, pp. 159–166, 2004.
35. B. Gedik and L. Liu. Location privacy in mobile systems: A personalized anonymization model. In *Proceedings of the 25th International Conference on Distributed Computing Systems (ICDCS'05)*, pp. 620–629, 2005.

36. O. Goldreich, S. Micali, and A. Wigderson. How to play any mental game or a completeness theorem for protocols with honest majority. In *Proceedings of 19th Annual Symposium on Theory of Computing (STOC'87)*, pp. 218–229, 1987.
37. B. Hoh and M. Gruteser. Location privacy through path confusion. In *Proceedings of IEEE/CreateNet International Conference on Security and Privacy for Emerging Areas in Communication Networks (SecureComm'05)*, 2005.
38. S.-Y. Hwang, Y.-H. Liu, J.-K. Chiu, and E.-P. Lim. Mining mobile group patterns: A trajectory-based approach. In *Proceedings of the 9th Pacific-Asia Conference on Knowledge Discovery and Data Mining (PAKDD'05)*, pp. 713–718, 2005.
39. A. Inan and Y. Saygin. Privacy-preserving spatio-temporal clustering on horizontally partitioned data. In *Proceedings of 8th International Conference on Data Warehousing and Knowledge Discovery (DaWaK'06)*, Vol. 4081. *Lecture Notes in Computer Science*, pp. 459–468. Springer, 2006.
40. W. Jiang and M. Atzori. Secure distributed k-anonymous pattern mining. In *Proceedings of the 6th International Conference on Data Mining (ICDM'06)*. pp. 319–329.
41. T. Johnsten and V.V. Raghavan. Impact of decision-region based classification mining algorithms on database security. In *Proceedings of the IFIP TC13 WG11.3 13th International Conference on Database Security*, pp. 177–191, 2000.
42. M. Kantarcioglu and C. Clifton. Privacy-preserving distributed mining of association rules on horizontally partitioned data. In *In The ACM SIGMOD Workshop on Research Issues on Data Mining and Knowledge Discovery (DMKD'02)*, 2002.
43. M. Kantarcioglu, J. Jin, and C. Clifton. When do data mining results violate privacy? In *Proceedings of the 10th International Conference on Knowledge Discovery and Data Mining (KDD'04)*, pp. 599–604, 2004.
44. H. Kargupta, S. Datta, Q. Wang, and K. Sivakumar. On the privacy preserving properties of random data perturbation techniques. In *Proceedings of the 3rd International Conference on Data Mining (ICDM'03)*, pp. 99, 2003.
45. H. Kido. *Location Anonymization for Protecting User Privacy in Location-Based Services*. MS Thesis. 2006.
46. D. Kifer and J. Gehrke. Injecting utility into anonymized datasets. In *Proceedings of International Conference on Management of Data (SIGMOD'06)*, pp. 217–228, 2006.
47. M. Klusch, S. Lodi, and G. Moro. Distributed clustering based on sampling local density estimates. In *Proceedings of Internatational Joint Conference on Artificial Intelligence*, pp. 485–490, 2003.
48. G. Lee, C.-Y. Chang, and A.L.P. Chen. Hiding sensitive patterns in association rules mining. In *Proceedings of 28th Annual International Computer Software and Applications Conference (COMPSAC'04)*, pp. 424–429, 2004.
49. H.Y. Lin and W.G. Tzeng. An efficient solution to the millionaires' problem based on homomorphic encryption. In *Proceedings of Third International Conference on Applied Cryptography and Network Security (ACNS'05)*, Vol. 3531. *Lecture Notes in Computer Science*, pp. 456–466, 2005.
50. X. Lin, C. Clifton, and M. Zhu. Privacy preserving clustering with distributed EM mixture modeling. *Knowledge and Information Systems*, 8:68–81, 2005.
51. Y. Lindell and B. Pinkas. Privacy preserving data mining. *Lecture Notes in Computer Science*, 1880:36–52, 2000.
52. K. Liu, H. Kargupta, and J. Ryan. Random projection-based multiplicative perturbation for privacy preserving distributed data mining. *IEEE Transactions on Knowledge and Data Engineering*, 18(1):92–106, 2006.
53. H. Mannila and H. Toivonen. Levelwise search and borders of theories in knowledge discovery. *Data Mining and Knowledge Discovery*, 1(3):241–258, 1997.
54. S. Mascetti, C. Bettini, X.S. Wang, and S. Jajodia. k-Anonymity in databases with times-tamped data. In *Proceedings of the Thirteenth International Symposium on Temporal Representation and Reasoning (TIME'06)*, pp. 177–186. IEEE Computer Society, 2006.
55. S. Menon, S. Sarkar, and S. Mukherjee. Maximizing accuracy of shared databases when concealing sensitive patterns. *Information Systems Research*, 16(3):256–270, 2005.

56. S. Merugu and J. Ghosh. Privacy-preserving distributed clustering using generative models. In *Proceedings of the 3rd International Conference on Data Mining (ICDM'03)*, p. 211. IEEE Computer Society, 2003.
57. M. Morgenstern. Controlling logical inference in multilevel database and knowledge-base systems. In *Proceedings of the Symposium on Security and Privacy*, pp. 245–255. IEEE, 1988.
58. J. Natwichai, X. Li, and M. Orlowska. Hiding classification rules for data sharing with privacy preservation. In *Proceedings of the 7th International Conference on Data Warehousing and Knowledge Discovery (DaWaK'05)*, pp. 468–477, 2005.
59. J. Natwichai, X. Li, and M. Orlowska. A reconstruction-based algorithm for classiciation rules hiding. In *Proceedings of the 17th Australasian Database Conference (ADC'06)*, pp. 49–58, 2006.
60. D.E. O'Leary. Knowledge discovery as a threat to database security. In G. Piatetsky-Shapiro and W.J. Frawley (eds.), *Knowledge Discovery in Databases*, pp. 507–516. AAAI/MIT Press, 1991.
61. S. Oliveira and O. Zaiane. Privacy preserving clustering by object similarity-based representation. In *Proceedings of the Workshop on Privacy and Security Aspects of Data Mining*, pp. 40–46, 2004.
62. S.R.M. Oliveira and O.R. Zaiane. *A Framework for Enforcing Privacy in Mining Frequent Patterns*. Technical report, Computer Science Department, University of Alberta, 2002.
63. S.R.M. Oliveira and O.R. Zaïane. Privacy preserving frequent itemset mining. In *Proceedings of the International Conference on Privacy, Security and Data Mining (CRPITS'02)*, pp. 43–54, 2002.
64. S.R.M. Oliveira and O.R. Zaiane. Algorithms for balancing privacy and knowledge discovery in association rule mining. In *Proceedings of the International Database Engineering and Applications Symposium (IDEAS'03)*, pp. 54–63, 2003.
65. S.R.M. Oliveira and O.R. Zaïane. Protecting sensitive knowledge by data sanitization. In *Proceedings of the 3rd International Conference on Data Mining (ICDM'03)*, pp. 211–218, 2003.
66. S.R.M. Oliveira and O.R. Zaiane. A unified framework for protecting sensitive association rules in business collaboration. *International Journal of Business Intelligence and Data Mining*, 1(3):247–287, 2006.
67. S.R.M. Oliveira, O.R. Zaïane, and Y. Saygin. Secure association rule sharing. In *Proceedings of the 8th Pacific-Asia Conference on Knowledge Discovery and Data Mining (PAKDD'04)*, pp. 74–85, 2004.
68. P. Paillier. Public-key cryptosystems based on composite degree residuosity classes. *Lecture Notes in Computer Science*, 1592:223–238, 1999.
69. G. Piatetsky-Shapiro, U.M. Fayyad and P. Smyth. From data mining to knowledge discovery: An overview. In *Advances in Knowledge Discovery and Data Mining*, pp. 1–34. AAAI Press, 1996.
70. E.D. Pontikakis, A.A. Tsitsonis, and V.S. Verykios. An experimental study of distortion-based techniques for association rule hiding. In *Proceedings of the 18th Conference on Database Security (DBSEC'04)*, pp. 325–339, 2004.
71. E.D. Pontikakis, V.S. Verykios, and Y. Theodoridis. On the comparison of association rule hiding heuristics. In *Hellenic Database Management Symposium*, 2004.
72. S. Rizvi and J.R. Haritsa. Maintaining data privacy in association rule mining. In *Proceedings of the 28th International Conference on Very Large Databases (VLDB'02)*, 2002.
73. P. Samarati and L. Sweeney. *Generalizing Data to Provide Anonymity When Disclosing Information*. Technical report, 1998. Available at http://www.sld.sri.com/papers/344/.
74. Y. Saygin, V.S. Verykios, and C. Clifton. Using unknowns to prevent discovery of association rules. *ACM SIGMOD Record*, 30(4):45–54, 2001.
75. Y. Saygin, V.S. Verykios, and A.K. Elmagarmid. Privacy preserving association rule mining. In *Proceedings of the International Workshop on Research Issues in Data Engineering: Engineering E-Commerce/E-Business Systems (RIDE'02)*, 2002.

76. X. Sun and P.S. Yu. A border-based approach for hiding sensitive frequent itemsets. In *Proceedings of the 5th International Conference on Data Mining (ICDM'05)*, pp. 426–433, 2005.
77. L. Sweeney. Datafly: A system for providing anonymity in medical data. In *Proceedings of the IFIP TC11 WG11.3 11th International Conference on Database Security*, pp. 356–381, 1998.
78. L. Sweeney. k-Anonymity: A model for protecting privacy. *International Journal on Uncertainty Fuzziness and Knowledge-based Systems*, 10(5), 2002.
79. J. Vaidya and C. Clifton. Privacy preserving association rule mining in vertically partitioned data. In *Proceedings of the 8th International Conference on Knowledge Discovery and Data Mining (KDD'02)*, pp. 639–644, 2002.
80. J. Vaidya and C. Clifton. Privacy-preserving k-means clustering over vertically partitioned data. In *Proceedings of the 9th International Conference on Knowledge Discovery and Data Mining (KDD'03)*, pp. 206–215, 2003.
81. J. Vaidya and C. Clifton. Privacy preserving naïve bayes classifier for vertically partitioned data. In *Proceedings of the International Conference on Data Mining (SDM'04)*, 2004.
82. V.S. Verykios, E. Bertino, I.N. Fovino, L.P. Provenza, Y. Saygin, and Y. Theodoridis. State-of-the-art in privacy preserving data mining. *ACM SIGMOD Record*, 33(1):50–57, 2004.
83. V.S. Verykios, A.K. Emagarmid, E. Bertino, Y. Saygin, and E. Dasseni. Association rule hiding. *IEEE Transactions on Knowledge and Data Engineering*, 16(4):434–447, 2004.
84. E.T. Wang, G. Lee, and Y.T. Lin. A novel method for protecting sensitive knowledge in association rules mining. In *Proceedings of 29th Annual International Computer Software and Applications Conference (COMPSAC'05)*, pp. 511–516, 2005.
85. K. Wang, B.C.M. Fung, and P.S. Yu. Template-based privacy preservation in classification problems. In *Proceedings of the 5th International Conference on Data Mining (ICDM'05)*, pp. 466–473, 2005.
86. S. Warner. Randomized response: A survey technique for eliminating evasive answer bias. *Journal of The American Statistical Association*, 60(309), 1965.
87. X. Xiao and Y. Tao. Anatomy: Simple and effective privacy preservation. In *Proceedings of the 32th International Conference on Very Large Databases (VLDB'06)*, 2006.
88. Z. Yang, S. Zhong, and R.N. Wright. Privacy-preserving classification of customer data without loss of accuracy. In *The 2005 SIAM International Conference on Data Mining (SDM'05)*, 2005.
89. A.C. Yao. Protocols for secure computations. In *Proceedings of 23th Annual Symposium on Foundations of Computer Science (FOCS'82)*, pp. 160–164. IEEE Computer Society, 1982.
90. J. Zhan and S. Matwin. Privacy-preserving data mining in electronic surveys. *International Journal of Network Security*, 4(3):318–327, 2007.
91. J. Zhan, S. Matwin, and L. Chang. Privacy-preserving collaborative association rule mining. In *Proceedings of the 19th Annual IFIP Conference on Data and Applications Security (DBSEC'05)*, Vol. 3654. *Lecture Notes in Computer Science*, pp. 153–165, 2005.

Chapter 12
Querying and Reasoning for Spatiotemporal Data Mining

G. Manco, M. Baglioni, F. Giannotti, B. Kuijpers, A. Raffaetà, and C. Renso

12.1 Introduction

In the previous chapters, we studied movement data from several perspectives: the application opportunities, the type of analytical questions, the modeling requirements, and the challenges for mining. Moreover, the complexity of the overall analysis process was pointed out several times. The analytical questions posed by the end user need to be translated into several tasks such as choose analysis methods, prepare the data for application of these methods, apply the methods to the data, and interpret and evaluate the results obtained. To clarify these issues, let us consider an example involving the following analytical questions:

- *Describe the collective movement behavior of the population (or a given subset) of entities during the whole time period (or a given interval)*
- *Find the entity subsets and time periods with the collective movement behavior corresponding to a given pattern*
- *Compare the collective movement behaviors of the entities on given time intervals*

It is evident that there is a huge distance between these analytical questions and the complex computations needed to answer them. In fact, answering the above questions requires combining several forms of knowledge and the cooperation among solvers of different nature: we need spatiotemporal reasoning supporting deductive inferences along with inductive mechanisms, in conjunction with statistical methods.

Mining tasks and techniques are implemented by means of ad hoc algorithms, whose results in most cases are not directly useful for analysis purposes: they often need a tuning phase, in which they are interpreted and refined. Even when results are clear and easy to understand, the interpretation and usefulness of such results may not be immediate. In addition, most data mining tools and methods require

---

G. Manco
ICAR-CNR, Cosenza, Italy, e-mail: manco@icar.cnr.it

deep technical and statistical knowledge by the data analyst, in addition to a clear comprehension of the data.

Thus, there is an apparent dichotomy. Analyzing data requires a (special purpose) language support to ease the burden of turning analytical questions into calls to specific algorithms and mining tasks and to turn mining results into actionable knowledge. However, the aforementioned issues should convince that the design of an effective language support for the KDD process – a data mining equivalent of SQL – is hard to achieve.

Nevertheless, this issue has received a deep and recurring attention by data mining researchers, as well as data mining software producers since the seminal paper by Imielinski and Mannila [41]. We shall discuss in this chapter many of the proposals that were put forward since then. Today, the language issue for KDD process is still an open problem with several different proposals, but without a predominant one: this might suggest that the problem is too complex or too general to admit a solution. Notwithstanding, the motivations for such an enterprise are so strong that many researchers continue to search for language constructs and/or problem restrictions that bring the state of the art one step further – in the very same spirit of database research in the 1970s before the advent of the relational model and SQL.

As seen in the previous chapters, movement data are complex: spatiotemporal objects may be characterized by a *geometrical representation*, a *position* at a given time and a nonspatiotemporal description. Moreover, some attributes (such as orientation, area) and some spatiotemporal relations (e.g., topological relations) are *implicit*. Hence, handling moving objects requires a higher level of reasoning with respect to the traditional data stored in databases. For example, many spatial data are interrelated, i.e., the presence of a highly polluting factory can reduce the price of the houses close to it.

This chapter investigates the research issues arising by the quest toward a language framework, capable of supporting the user in specifying and refining mining objectives, combining multiple strategies, and defining the quality of the extracted knowledge in the specific context of movement data.

As a step in this direction, the chapter concentrates on requirements and open issues of a language framework that serves as interface between the analyst and the underlying computational support – trying to bridge the gap from analytical questions and the complexity of the KDD process.

The fundamental aspects that the advocated language framework has to consider are the various forms of knowledge to represent, the repertoire of analytical questions to support, and the type of analytical users to address.

- We will see that several forms of knowledge may be represented. The *source data or primary knowledge*, i.e., the data to be mined; the *background or domain knowledge*, i.e., the rules that enrich data semantics in a specific context; the *mined or extracted knowledge*, i.e., the patterns or models mined from the source data.
- The analytical questions, as illustrated in Chap. 1, suggest the characteristics of the moving entities, the type of reasoning needed to express relations among the moving entities, and which patterns and models are expected.

- There are two extreme types of analytical users: the domain expert and the data mining expert. The domain expert has to be supported in specifying and refining the analysis goals by means of highly expressive declarative queries. On the contrary, the data mining expert masters the KDD process and aims at constructing complex vertical analytical solutions, since he/she has to be supported in specifying and refining the analysis goals by means of procedural abstractions to control the KDD process.

A combination of design choices, according to the options offered by the above three aspects, defines the requirements of the language support for knowledge discovery. The multiplicity of such options highlights the complexity of the scenarios as well as explains the high number of existing proposals, both in research and in industry.

The rest of this chapter is organized as follows. Section 12.2 concentrates on the relevant dimensions and the design issues of a DMQL. Section 12.3 surveys the most significant proposals in the literature for relational and spatial data. Section 12.4 illustrates some spatiotemporal query languages with a focus on the qualitative predicates for complex reasoning on movement data. Section 12.5 is dedicated to present a road map of research raised by the issues of the chapter, and Sect. 12.6 draws some conclusions.

12.2 Elements of a Data Mining Query Language

Roughly, the process of analyzing (spatiotemporal) data can be represented as an interaction between the data mining engine and the user, where the user formulates a query describing the patterns of his/her interest, and the mining engine (the algorithm) returns the patterns by exploiting either domain-specific or physical optimizations. Research on data mining has mainly focused on the definition and implementation of specific data mining engines. Each of these engines describes a specific interaction in a given language. However, independent from the way queries can be executed, there are some main components (*dimensions* in a knowledge discovery process) that characterize a data mining query: the source data to explore, the patterns to discover, the criterion for determining the usefulness of patterns, and the background knowledge. We now discuss each dimension in more detail. Consider the following analysis problem as a running example:

> *Among all the movement patterns characterizing (i.e., occurring with high frequency within) polluted areas during rush hours, which ones involve the intersection of a river within/after X time units from their start?*

We can detect the following aspects that characterize it.

Primary knowledge sources. The first step in a knowledge discovery task is to identify the data to analyze. In the above example, we are interested in movement patterns, i.e., trajectories (or portions of) describing the movement behavior of some entities of interest. Thus, the source data is the set of all the trajectories describing the movements of entities during rush hours in polluted areas. These data represent the main input that the mining algorithm has to take into account. Let Σ denote the primary knowledge sources. Σ is a set of entities, equipped with a set of characterizing properties that are of interest for the purpose of our analysis. Identifying relevant entities and properties is a crucial task, since it requires a combination of basic operations, which include data integration and selection (i.e., the identification of the data relevant to the analysis task, and its retrieval from a set of possibly heterogeneous data sources), and data cleaning and transformation (i.e., the processing of noisy, missing, or irrelevant data and its transformation into appropriate forms). Although the quality of the extracted patterns strongly relies on the effectiveness of such operations, poor attention has been devoted to such a delicate task in the current literature: as a result, the management of knowledge sources is still informal and ad-hoc, and the current data mining tools provide little support to the related operations. Thus, typically a knowledge expert has the burden to manually study and preprocess the data with several loosely coupled tools, and to select the most appropriate reorganization of such data that is suitable for the extraction of patterns of interest.

Pattern search space. Patterns are sentences about primary knowledge sources, expressing rules, regularities, or models that hold on the entities in Σ. Patterns are characterized by a language (denoted by $\mathscr{L}$ in the following), which is fixed in advance, describing the set of all possible properties that hold in the primary knowledge sources. In this respect, a data mining algorithm describes a way to explore this set of all possible patterns and to detect the patterns that are of interest according to given criteria.

Patterns characterize a data mining task or method and they can be categorized as follows:

- *Descriptive* patterns, when they describe or summarize the entities within the source data. Examples of descriptive patterns are clusters of trajectories, or frequent trajectories, like in the above example.
- *Predictive* patterns, when they characterize a property or a set of properties of an entity according to the values exhibited by other properties, e.g., "good"/"bad" trajectories, such as trajectories that are likely to end-up in a traffic jam.

Search criteria. A search criterion is defined on the involved entities and determines whether a given pattern in the pattern language $\mathscr{L}$ is *potentially useful*. In [52], this construct is captured by a Boolean constraint predicate q, which can be described as relative to some pattern $l \in \mathscr{L}$ and possibly to a set Σ of entities. If $q(l, \Sigma)$ is true, then l is potentially useful. The KDD task then is to find the set $\{l \in \mathscr{L} \mid q(l, \Sigma) \text{ is true}\}$. An example definition of q is checking whether patterns involving trajectories cross a river at a given time. It can be computed as a result of a post-processing operation (among all the patterns discovered thus far, which are the ones satisfying such a condition?), or they can be pushed into a specialized mining engine in order to solve them.

Background Knowledge. This term denotes the domain knowledge that is already known in advance and that can be incorporated in each of the previous dimensions.

It can be used to enrich the primary knowledge or to derive good initial hypotheses to start the pattern search. In the example, we also specify geographic entities, such as river and areas, and they are useful to better characterize input data. Background knowledge can be exploited in the search strategy for pruning the pattern space or it can serve as a reference for interpreting discovered knowledge. Furthermore, since it is difficult to define adequate statistical measures for subjective concepts like novelty, usefulness, and understandability, background knowledge can be helpful in capturing such concepts more accurately.

The design of a data mining query language aims at formally characterizing all these aspects and at providing both the theoretical and methodological grounds for the investigation of the upcoming issues. Within a simple language perspective, a data mining query language should ease the process of describing the entities of interest and their properties and the pattern language upon which data mining algorithms should rely on. However, there are several further problems that affect the data mining process. Such issues require a structured, formal approach, as opposed to the informal and ad-hoc approach, which still nowadays describes the knowledge discovery discipline.

From a methodological viewpoint, data mining can be seen as advanced querying: *given a set of objects of interest (Σ and a set of properties $\mathscr{L}$), which are the properties within $\mathscr{L}$ of interest according to the search criteria q?* Under this perspective, there are some preliminary expressiveness issues to be tackled. The first one concerns the choice of the most appropriate structuring of the knowledge sources that is suitable for the extraction of patterns of interest.

Most classical data mining techniques assume that such a structuring has taken place as a separate process. As a consequence, they work on a materialized single table. Such techniques can only discover patterns involving entities whose properties are of the form "attribute = constant," and they are commonly called *propositional* because the solved problems can be expressed in propositional logic. In propositional learning, every example (case or instance) corresponds to a fact or, equivalently, to a tuple.

Figure 12.1 describes a "high-level" representation of trajectories, together with their properties. A trajectory is characterized by an identifier, an actor (further characterized as being the driver of a car, a pedestrian, etc.), the starting and ending place, the duration of the trajectory, and the number of traversed places (called *Cells*). The list of these places is modeled by the relation *Traverses*. Finally, places can exhibit properties and relationships.

A propositional representation requires that examples are described by a view over the data, which is fixed in advance, as expressed in a Datalog-like jargon in the following:

$$Q_1(Start, SType, STime, End, EType, ETime) \leftarrow Trajectory(I, \_, \_, \_, S, E, \_),$$
$$Place(S, Start, SType), Traverses(I, S, STime),$$
$$Place(E, End, EType), Traverses(I, E, ETime)$$

Trajectory	ID	Actor	Type	Cells	Start	End	Time
	#1	Jef	Vehicle	7	#P1	#P5	58m
	#2	Bart	Pedestrian	5	#P2	#P3	1h 32m
	#3	Ned	Vehicle	8	#P3	#P6	18m
	#4	Jef	Vehicle	6	#P6	#P1	27m
	...						

Traverses	TrajectoryID	PlaceID	Time
	#1	#P1	1:00pm
	#1	#P5	1:58pm
	#1	#P4	1:28pm
	#2	#P2	12:00am
	#2	#P3	1.32pm
	#3	#P6	1.32pm
	...		

Place	ID	Name	Type
	#P1	King's Road	Road
	#P2	Carnaby Street	Road
	#P3	Salisbury Hill	Square
	#P4	Janet's Place	Building
	#P5	Piccadilly Circus	Square
	#P6	Dawson's Road	Road
	#P7	Thames	River
	...		

Intersects	PlaceA	PlaceB
	#P6	#P3
	#P1	#P7
	#P2	#P3
	...	

Contains	PlaceA	PlaceB
	#P3	#P4
	...	

Pollution	PlaceID	Surveying Date
	#P3	12/09/2006
	#P5	12/09/2006
	#P5	13/09/2006
	#P5	14/09/2006
	#P7	14/09/2006
	...	

Actor	Name	Age	Sex	Address
	Jef	30	Male	London
	Bart	28	Male	Liverpool
	Ned	45	Male	London
	...			

Fig. 12.1 Multirelational representation of trajectories

Here, each example is a trajectory, which is described by a starting point and an endpoint, each of them with an associated type and traversal time. Clearly, different characterizations of trajectories can be used. For instance, the following view describes trajectories that traverse exactly three places:

$$Q_2(Start, Middle, End) \leftarrow Trajectory(I, \_, \_, \_, S, E, \_),$$
$$Place(S, Start, \_), Traverses(I, S, \_),$$
$$Place(M, Middle, \_), Traverses(I, M, \_),$$
$$Place(E, End, \_), Traverses(I, E, \_),$$
$$S \neq M \neq E, \neg Other(I, S, M, E)$$

$$Other(I, S, M, E) \leftarrow Traverses(I, P, \_),$$
$$P \neq S, P \neq M, P \neq E.$$

Propositional representations may be cumbersome in domains where examples are (or are better modeled as) structured objects. In the above example, although we can extend view Q_2 to represent trajectories traversing a larger number of places, we cannot generalize the view to include, within a single tuple, a description of trajectories with an arbitrary number of places. In other words, each possible restructuring of the view would only allow the inclusion of a fixed number of traversed places, whereas trajectories may indeed traverse an arbitrary number of objects.

In addition, in the spatiotemporal domain, moving objects may be characterized by their correlation with other objects: for example, in Fig. 12.1, relation *Intersects* specifies the intersection between a street and a square, or between a street and a river. It is impractical or even infeasible to capture in a single tuple the relative positioning of objects to each other.

In this context, another research direction named *multirelational* (or simply *relational*) considers the possibility of upgrading knowledge discovery to first-order logic, i.e., the same mining task can tackle the problem of finding which is the best reorganization of source data that guarantees the extraction of patterns of interest [19].

In relational (or first-order) learning, every example is a set of facts or, equivalently, a (small) relational database. An identifier is used to distinguish examples. The essence is that the analysis must define an entity to be analyzed. While in classical approaches an entity is represented by means of a single tuple (which raises the need to summarize all the entity's properties by means of tuple attributes), in multirelational approaches an entity can be described as an object exhibiting more complex properties, modeling, e.g., relationships with other entities. Thus, referring to the relation of Fig. 12.1, while in propositional representation the knowledge source is represented by a single view (such as Q_1 or Q_2), in relational learning the knowledge source is represented by the whole relational database, and it is the task of the mining algorithm to find the best reorganization of the data that describes a pattern of interest.

A further expressiveness issue is related to the problem of choosing an appropriate language for modeling patterns. The expressiveness of a pattern language can influence the resulting patterns. In traditional data mining tasks, patterns can be expressed as propositional formulae. For example, in view Q_1, a possible pattern could be the following

$$SEnd = \texttt{square} \wedge ETime = \texttt{10:00 a.m.},$$

describing a set of trajectories that end up in a square at 10:00 a.m. Clearly, this pattern involves only constants among the various attributes. By assuming, however, that the pattern language does not restrict itself to propositional formulae, more expressive patterns can be obtained, such as, e.g.,

$$SEnd = \texttt{square} \wedge ETime = STime + 2\,\text{h}.$$

The latter describes all those trajectories that end up in a square after 2 h from their start.

Patterns may involve either single or multiple relations. The latter are usually stated in a more expressive language than patterns defined on a single data table. An example relational pattern is

$$Trajectory(X, \_, \_, \_, \_, \_, \_), Traverses(X, Y, \_),$$
$$Intersects(Y, Z), Place(Z, \_, river), Pollution(Z, \_)$$

describing all the trajectories intersecting a polluted river.

A natural question is whether data mining can be put in the same methodological grounds as databases. Relational databases, in this respect, represent the paradigmatic example, where a simple formalism merges rich expressiveness and optimization opportunities. The set of mathematical primitives and their closure property allows to express a wide set of queries as composition of such primitives. The same formalism enables query execution optimization such as query decomposition, constraint pushing, advanced data structures, indexing methods. Thus, putting data mining in the same methodological grounds essentially means being capable of decoupling the specification from the execution of a data mining query.

From an optimization perspective, the challenge is how to merge the efficiency of DBMS technologies with data mining algorithms, more specifically, how to integrate data mining more closely with traditional database systems, above all with respect to querying.

A further aspect to be investigated is the process-oriented nature of knowledge discovery. Typically, when analyzing data one needs to derive good initial hypotheses to start the pattern search, to incrementally refine the search strategy, and to interpret discovered knowledge. Each of the above issues is concerned with a specific dimension and can be described in a different language. However, using different languages results in an "impedance mismatch" between dimensions, which can be a serious obstacle to efficiency and efficacy. A major challenge in building data mining systems concerns the smooth cooperation between different dimensions.

Thus, a coherent formalism, capable of dealing uniformly with all dimensions, would represent a breakthrough in the design and development of decision support systems in diverse application domains. The advantages of such an integrated formalism include the ability to formalize the overall KDD process, and the possibility to tailor a methodology to a specific application domain.

12.3 DMQL Approaches in the Literature

There has been a proliferation of approaches with different focuses. Among them, we detected some common criteria that characterize the proposals according to the way they support the user in specifying and executing data mining tasks. In a first research direction, the focus is to provide an interface between data sources and data mining tasks. Under this perspective, a DMQL is seen as a standard mean for specifying data sources, patterns of interest, and properties characterizing them. Interestingly, there are several different research objectives that can be pursued here. For example, one could be interested in providing minimal extensions to SQL, which allow to specify mining queries, or alternatively looking for more expressive languages such as first-order logic. Orthogonally, the interest could be focused on a minimal set of mining primitives – constraints over the patterns – upon which more complex mining tasks can be defined, by means of composition operators.

In a second direction, a DMQL is meant to support the design of specific procedural workflows, which integrate reasoning on the mining results and possibly define ad-hoc evaluation strategies and activations of the data mining tasks. Therefore, the idea here is to embody data mining query languages in a more general framework, where effective support to the whole knowledge discovery process is provided.

It is worth noticing that these lines of research are, in a sense, orthogonal: indeed, they raise from different and not necessarily contrasting needs, and in general propose solutions which, in principle, could be integrated. In the next two sections, we analyze the various proposals in the literature according to the presented directions, whereas in Sect. 12.3.3 we illustrate an interesting example of DMQL for spatial data.

12.3.1 DMQL as Interface to Algorithms

The problem of providing an effective interface between data sources and data mining tasks has been a primary concern in data mining. There are several perspectives upon which this interface is desirable, mainly (1) to provide a standard formalization of the desired patterns and the constraints they should obey to and (2) to achieve a tighter integration between the data sources and the relational databases (which likely accommodate them). Notice that the coupling with relational databases also raises the important question on whether the results of mining (the extracted patterns) should be amalgamated within the relational DBMS. In particular, the question is whether the closure principle should be pursued, thus allowing that the results of mining are queried and investigated using a standard database jargon.

The approaches outlined in this section mostly concentrate on the support given to the user in specifying the mining step of the KDD process, without the need of covering execution details. However, the supported pattern languages and constraints over them are limited. Most of the proposals deal only with association rules and no support to different mining tasks is provided. The point is that the relational representation of association rules is quite natural, but other models such as decision trees or clustering results are more difficult to represent. In addition, these data mining query languages only provide an interface to mining algorithms, with little support to the preprocessing and evaluation phase. As a matter of fact, the KDD process is quite complex, and clearly a richer environment capable of effectively supporting all its aspects could improve the productivity of a data miner.

12.3.1.1 SQL-Based Approaches

The common ground of most approaches is in adopting an SQL-based style for creating and manipulating data mining models, thus abstracting away from the algorithmic particulars. On the other side, approaches differ on which patterns are of interest and which constraints the patterns should satisfy.

One of the first proposals in the literature is *MINE RULE* [53]. The proposal is specifically tailored for extracting association rules by extending SQL with a specific operator. Within *MINE RULE*, primary knowledge sources are represented by propositional views specified via SQL queries. As a consequence, source data are represented as a single relational table. Consider, for example, the database of Fig. 12.1, and suppose you are interested in association rules about roads frequently traversed together. The corresponding *MINE RULE* is

```
MINE RULE SimpleAssociations AS
SELECT DISTINCT 1..n Place.Name AS BODY, 1..1 Place.Name AS HEAD,
                SUPPORT, CONFIDENCE
FROM Traverses, Place
WHERE Traverses.PlaceID = Place.ID
AND   Place.Type = "Road"
GROUP BY TrajectoryID
EXTRACTING RULES WITH SUPPORT: 0.1, CONFIDENCE: 0.2.
```

There are three components plugged within the query. First, the `FROM` and `WHERE` clauses specify the selection of the primary knowledge, using standard SQL syntax.

Second, within the `SELECT` clause, the user specifies the pattern search space. The example describes rules exhibiting many elements in the body and a single element in the head. As a result, the `SELECT` clause produces a new table named `SimpleAssociations`, and composed of four attributes, namely `BODY`, `HEAD`, `SUPPORT`, and `CONFIDENCE`, which represent extracted rules.

The `EXTRACTING RULES WITH` clause allows to specify support and confidence constraints. Other constraints can be specified in other parts of the query: for example, the following query looks for patterns relating roads to squares.

```
MINE RULE FilteredOrderSets AS
SELECT DISTINCT 1..n Place.Name AS BODY,
1..n Place.Name AS HEAD,
SUPPORT, CONFIDENCE
WHERE BODY.type = "Road" AND HEAD.type = "Square"
FROM Traverses, Place
WHERE Traverses.PlaceID = Place.ID
GROUP BY TrajectoryID
EXTRACTING RULES WITH SUPPORT: 0.1, CONFIDENCE: 0.2.
```

Within *MINE RULE*, the specification of background knowledge relies on SQL. The following example describes association rules involving areas eventually surveyed as polluted:

```
MINE RULE GeneralizedRules AS
SELECT DISTINCT 1..n Name As BODY,
```

```
                  1..n Name As HEAD,
                 SUPPORT, CONFIDENCE
FROM (SELECT Traverses.TrajectoryID AS ID, Place.Name AS Name
      FROM Traverses, Place, Pollution
      WHERE Traverses.PlaceID = Place.ID
      AND Place.ID = Pollution.PlaceID)
GROUP BY ID
EXTRACTING RULES WITH SUPPORT:0.3,
                    CONFIDENCE:0.5.
```

Another interesting proposal is the *DMQL* language [40]. Different from *MINE RULE*, the DMQL language focuses on several mining tasks: characteristic, discriminant, classification, and association rules. In addition, DMQL allows the direct specification of concept hierarchies to be exploited by specific mining algorithms. A first kind of specification acts on a given relational schema. A group of attributes can be generalized by removing some attributes from the group:

```
DEFINE HIERARCHY FOR Address:
   {City, Province, Country} < {Province, Country}.
```

Hierarchies can be alternatively specified in values by explicitly describing which values are generalized by another value:

```
DEFINE HIERARCHY FOR Address:
   {King's Road, Carnaby Street, Salisbury Hill} < {London}
DEFINE HIERARCHY FOR Address:
   {London, Liverpool} < {England}.
```

In this case, the *Address* attribute enumerates several locations, some of which are related as described: e.g., London includes King's Road, Carnaby Street, and Salisbury Hill.

Imielinkski and others [42] propose a data mining query language (*MSQL*) capable of generating association rules and further query them. Rules in MSQL are of the form $A_1, A_2, \ldots, A_n \rightarrow A_{n+1}$, where each A_i has typically the form *Attribute=value*. Rules are generated by means of a GetRules statement which, apart from syntax issues, has similar features as MINE RULE and DMQL. In addition, MSQL allows for nested queries. Suppose you are interested in associations involving the *Type* attribute in the body, which are also maximal: a rule R is maximal if no other rule exists whose antecedent contains the antecedent of R. The following query formalizes this, where the maximality condition is expressed within the subquery:

```
GetRules(Trajectory) R1
where Body has {Type = *} and Support > .05
      and Confidence > .7
      and not exists ( GetRules(Trajectory) R2
                       where Support > .05 and Confidence > .7
                       and R2.Body HAS R1.Body).
```

The extracted rules are stored in a *RuleBase* and then they can be further queried by means of the SelectRules statement. It is possible to select a subset of the generated rules that verify a certain condition

```
SelectRules(R) where Body has { (Age=*), (Sex=*) } and
                  Consequent is { (Address=*) }
```

as well as to select the tuples of the input database that violate (satisfy) all (any of) the extracted rules:

```
Select * from Actor where VIOLATES ALL(
    GetRules(Actor)
    where Body is { (Age = *) }
    and Consequent is { (Sex = *) }
    and Confidence > 0.3
).
```

A limitation of the above proposals is their specificity and limited extensibility: in such a sense, they appear to be ad-hoc proposals. By this term we mean that they have been proposed on top of specific pattern languages or solvers. Most of the proposals only deal with association rules, and no support to different mining tasks is provided. The ATLaS System proposed in [81] overcomes such limitations by proposing minimal extensions to SQL that are particularly effective at expressing several different data mining tasks.

ATLaS adds to SQL the ability of defining new Table Functions and User Defined Aggregates. A table function can be used in a FROM clause and returns a table. For example, given a tuple made of four attributes, the dissemble function breaks down the tuple into four records. Here, each record represents one column in the original tuple and includes the column number, column value, and the value of column YorN:

```
FUNCTION dissemble (v1 Int, v2 Int, v3 Int, v4 Int, YorN Int):
        (Col Int, Val Int, YorN Int);
{   INSERT INTO RETURN VALUES
    (1, v1, YorN), (2, v2, YorN), (3, v3, YorN), (4, v4, YorN);
}.
```

In addition, ATLaS adopts the SQL-3 idea of specifying User Defined Aggregates by an initialize, an iterate, and a terminate computation, and expresses these three computations by a single procedure written in SQL. The following example defines an aggregate equivalent to the standard avg aggregate in SQL.

```
AGGREGATE myavg(Next Int)  : Real
{     TABLE state(sum Int, cnt Int);
    INITIALIZE : {
        INSERT INTO state VALUES (Next, 1);
    }
    ITERATE : {
        UPDATE state SET sum=sum+Next, cnt=cnt+1;
    }
    TERMINATE : {
        INSERT INTO RETURN SELECT sum/cnt FROM state;
    }
}.
```

The first line of this aggregate function declares a local table, state, to keep the sum and count of the values processed so far. While, for this particular example, state contains only one tuple, it is in fact a table that can be queried and updated using SQL statements and can contain any number of tuples. These SQL statements are grouped into the three blocks labeled, respectively, INITIALIZE, ITERATE, and TERMINATE. The principle is that, to compute the aggregate function for an input stream of values, a statement is executed for each value in the stream. In particular, the first value triggers the execution of the INITIALIZE block, while the remaining values issue the execution of the ITERATE block. The TERMINATE block is finally executed when the end of the stream is reached.

Thus, INITIALIZE inserts the value taken from the input stream and sets the count to one. The ITERATE statement updates the table by adding the new input value to the sum and 1 to the count. The TERMINATE statement returns the final result(s) of computation by INSERT INTO RETURN (to conform to SQL syntax, RETURN is treated as a virtual table; however, it is not a stored table and cannot be used in any other role).

User Defined Aggregates can call other aggregates or call themselves recursively. This approach to aggregate definition is very general and, combined with the possibility of defining table functions in SQL, plays a critical role in expressing data mining queries in ATLaS.

Let us consider, e.g., the well-known Play-tennis example detailed in [54]. In this example, we assume a table *PlayTennis (ID, Outlook, Temperature, Humidity, Wind, Play)*, which describes if, according to weather conditions, a tennis match was played or not. Suppose we want to build a classifier that predicts the value of the *Play* attribute on the basis of the values of the weather attributes. The following query relies on the classify recursive aggregate, and classifies the tuples in the *PlayTennis* table accordingly:

```
SELECT classify(0, p.ID, d.Col, d.Val, d.YorN)
FROM PlayTennis AS p,
  TABLE(dissemble(p.Outlook, p.Temp, p.Humidity, p.Wind, p.Play))
    AS d
```

The *classify* aggregate is detailed in [81] and implements a scalable decision-tree classifier by suitably defining the INITIALIZE, ITERATE, and TERMINATE statements. Other recursive aggregates have been defined, which allow the specification of clustering and frequent itemset mining tasks.

12.3.1.2 Logic-Based Approaches

Logic represents a significant opportunity for data mining query languages, as it can be profitably used as a unifying formalism for integrating input data sources with data mining tasks and discovered knowledge. Two alternative paradigms have been proposed in the literature, namely $\mathscr{LDL}_{Mine}$ and *RDM*. Both paradigms are logical query languages extended with data mining capabilities, which make them particularly attuned to represent complex reasoning tasks on the extracted knowledge.

$\mathcal{LDL}_{Mine}$ [30] is an extension of $\mathcal{LDL}++$, a logical query language that provides both the typical deductive features and advanced mechanisms for nondeterminism, stratified negation, and aggregation [85]. $\mathcal{LDL}_{Mine}$ implements an inductive engine that allows, by means of inductive clauses, the interaction between mining algorithms and deductive components. $\mathcal{LDL}_{Mine}$ distinguishes between inductive and deductive queries. An *inductive query* is a clause *Head* ← *Body*, where *Head* represents a pattern in $\mathcal{L}$ and *Body* represents the knowledge sources upon which patterns are defined. The evaluation of such clause corresponds to an exploration of the search space according to given search criteria.

Within $\mathcal{LDL}_{Mine}$, inductive queries are formally modeled in a structured way by means of clauses containing specific user-defined aggregates. For example, the *Patterns* aggregate defined in [30] can be used in inductive queries to extract frequent itemsets. We illustrate this by the following example. Suppose we are interested in the following tasks: find sets s of unpolluted places reached by at least three trajectories, which also traverse a square. In association rule terminology, the problem concerns finding frequent ("at least three") itemsets subject to constraints (places were never recorded as polluted, trajectories traversing them also traverse squares). The following $\mathcal{LDL}_{Mine}$ program performs this task:

$$Traverses(e, \langle p \rangle) \leftarrow Traverses(e, p, t)$$
$$ItemSets(Patterns\langle(3,s)\rangle) \leftarrow Traverses(e,s), p \in s,$$
$$Place(p, n, \text{Square})$$
$$ans(s,n) \leftarrow ItemSets(s,n), \neg NotPolluted(s)$$
$$NotPolluted(s) \leftarrow r \in s, \neg Pollution(r,d)$$

The first clause groups tuples in the *Traverses* relation. The second clause is an inductive query: the *Patterns*$\langle(3,s)\rangle$ aggregate extracts frequent itemsets from the grouped places; the details of this operation can be found in [30]. After the evaluation of the second clause, *ItemSets*(s,n) holds if s is a set of places traversed by $n \geq 3$ trajectories also traversing a square. Finally, the third clause restricts the result to sets of unpolluted places. This example shows three important issues:

- Primary knowledge is specified by means of $\mathcal{LDL}_{Mine}$ clauses. In the above example, the extension of the *Traverses* predicate constitutes the primary knowledge source. The use of $\mathcal{LDL}_{Mine}$ for specifying primary knowledge allows complex data manipulations, including recursion and stratified negation.
- The *Patterns* aggregate is then evaluated relative to the primary knowledge source. More precisely, the *Patterns*$\langle(3,s)\rangle$ aggregate acts as an interface to a specific data mining algorithm by explicitly defining the support threshold (the value 3) and the input transactions (the set of all possible s resulting from the evaluation of the query). The evaluation of such a data mining aggregate may benefit from the use of background knowledge. Moreover, user-defined aggregates allow defining search criteria in an ad-hoc way.
- The results of the mining phase can be queried and combined with further background knowledge, thus allowing the refinement of the results to specific

application needs. The adoption of a rule-based language in this phase allows the integration between induced and deduced knowledge, and supports the interoperability between different data mining packages.

The described approach can be generalized to a variety of mining tasks, provided that the patterns to be discovered can be conveniently encoded in $\mathscr{LDL}_{Mine}$. For certain tasks, like decision tree learning, such encoding may be intricate, thus impeding the interoperation between mining and querying.

Differently from $\mathscr{LDL}_{Mine}$, which is specifically designed for propositional knowledge sources, the logical language *RDM* proposed by De Raedt [69] focuses on multirelational knowledge sources. A feature of the approach is the use of terms for conjunctive queries. More precisely, a constant can be a conjunctive query, and a variable can be a placeholder for a conjunctive query. *RDM* queries can be regarded as higher-order queries that inquire standard conjunctive queries by means of certain constraints. Three possible constraints involving query variables are presented next.

- *Subsumption constraints* are introduced to restrict the range of query variables; they are of the form $Q \subseteq q_1$, $q_1 \subseteq Q$, or $q_1 \subseteq q_2$, where q_1, q_2 are query variables and Q represents a fixed conjunctive query. A satisfaction of a subsumption constraint is relative to a query variable assignment: By considering, e.g., the query

$$Q := ans(X) \leftarrow Traverses(X, A, \_), Intersects(A, B),$$
$$Pollution(B, \_).$$

  The constraint $q_1 \subseteq Q$, for example, restricts the range of q_1 to conjunctive queries that are contained in Q. This means that, for each assignment Q_1 of q_1, the head of Q_1 must be of the form $ans(X)$. In addition, each X that is an answer to Q_1 must also be an answer to Q. In other words, each assignment to q_1 must refer to queries that contain in their answer set solely trajectories traversing a place, which intersects a polluted area.
- *Frequency constraints* use $cnt(q_1)$, where q_1 is a query variable, to refer to the cardinality of the answer set of any assignment of q_1 w.r.t. a database of interest. Frequency constraints can then be defined as equalities and inequalities involving counts. For example, the constraint $q_1 \subseteq Q, cnt(q_1) \geq 3$ is satisfied by all queries containing in their answer set solely trajectories traversing a place that intersects a polluted area, and such that their answer set contains at least three trajectories.
- Finally, *Coverage constraints* are used to state that a target query must yield a specified answer on a given database. For example, a *positive* coverage constraint is of the form $t \in q_1$, where t is a ground fact and q_1 is a query variable, and is satisfied by all the assignments to q_1, which also contain t in their answer set.

On the basis of the above constraints, *RDM* queries are defined by extending standard rule-based conjunctive queries with query variables and the above subsumption, frequency, and coverage constraints. An example is the following clause:

$$out(q) \leftarrow q_1 \subseteq Q, cnt(q_1) \geq 3.$$

An example answer is the query

$$Q_1 := ans(X) \leftarrow Traverses(X,A,\_), Intersects(A,B),$$
$$Pollution(B,\_), Place(A,\_,Road).$$

Indeed, Q_1 contains, relative to the database exemplified in Fig. 12.1, the trajectories #1, #2, and #3 in their answer set.

RDM is powerful enough to ask for complex patterns. In addition, predicates used by *RDM* can be used for defining new predicates and for querying in the same style of $\mathscr{LDL}_{Mine}$.

12.3.1.3 Constraint-Based Approaches

Some orthogonal approaches proposed in the last decade consider the possibility of decoupling the specification of a query language in two components:

1. The definition of an expressive query language that allows users to interact with the pattern discovery system and specify declaratively how the desired patterns should look like and which conditions they should satisfy
2. The development of an efficient and scalable mining engine that keeps query response time as small as possible, giving frequent feedbacks to the user and thus allowing realistic human-guided exploration

The approaches shown in the previous sections essentially concentrate on the first point, i.e., the specification of the mining query. Other approaches, however, were proposed to consider, besides the specification issues, the possibility of equipping the mining engine with specific optimizations, which efficiently generate the answers to the query. This view of data mining as a declarative querying process shares many analogies with the theory of relational algebra (which may provide a useful source of inspiration, and consequently can serve as the basis for a theory of data mining).

This research theme was investigated from several perspectives, which essentially envisage constraint-based pattern mining as a query optimization problem, i.e., developing efficient, sound, and complete evaluation strategies for constraint-based mining queries. Under this perspective, several efforts were devoted to analyze the properties of some constraints comprehensively in order to exploit such properties to speed-up the computation of the patterns that satisfy them. Following this methodology, some classes of constraints that exhibit nice properties have been detected [57]. Specifically, the approaches assume that the pattern language $\mathscr{L}$ exhibits a partial order $\preceq$ among patterns. Then, a predicate Q is *monotone* if, for each $p,q \in \mathscr{L}$ and knowledge source Σ,

$$p \preceq q \text{ implies that } Q(p,\Sigma) \rightarrow Q(q,\Sigma).$$

Analogously, a predicate Q is *antimonotone* if, for each $p, q \in \mathscr{L}$ and knowledge source Σ,

$$p \preceq q \text{ implies that } Q(q, \Sigma) \rightarrow Q(p, \Sigma).$$

These classes of constraints are quite appealing, since efficient techniques can be employed for their solution: examples are the Apriori algorithm, which implements a levelwise bottom–up search, or FPGrowth, which by the converse implements a top–down strategy.

Interestingly, other classes of constraints can be effectively managed by means of specific engines: *succint* constraints [57], i.e., constraints allowing a succint characterization of the solution space that admits a member-generating function, or *convertible* constraints [65] for which there is no clear interplay between $\preceq$ relationship and constraint satisfiability, but an interplay can be found by arranging the components of the pattern in some order. These classes of constraints have been extensively studied and optimized solvers have been devised [4, 64, 65].

Approaches [47, 70] consider the possibility of combining constraints by defining an algebra for manipulating pattern sets. The framework is based on the notion of Boolean inductive query, i.e., a Boolean expression over monotonic and antimonotonic predicates. Boolean expressions are made of the conjunction, disjunction, and negation operators. Under this perspective, the approach represents a valuable generalization with respect to the standard approaches to constraint-based pattern mining. Indeed, the latter only consider "simple" queries (namely, either single constraints or conjunctions of monotone and antimonotone constraints), whereas the core of the proposal developed in [47, 70] is a divisive approach for solving arbitrary Boolean queries.

The proposal can be summarized as follows: given a pattern language $\mathscr{L}$ and a knowledge source Σ, find a decomposition of a generic query Q into a disjunction $Q_1 \vee Q_2 \vee \ldots \vee Q_k$ of k subqueries, such that each Q_i can be specified as a conjunction of a monotone and antimonotone constraint: $Q_i = Q_i^a \wedge Q_i^m$.

The solution of a query $Q_i = Q_i^a \wedge Q_i^m$ can be characterized in terms of *version spaces* [54]. A version space is a convex set, which can be represented by its border sets (its maximally general and maximally specific elements). Several effective algorithms exist for computing the solution sets of such convex sets. Thus, there are two features that characterize the approach:

- Each Q_i allows an efficient approach to its solution
- The solution to Q is simply the union of the solutions of all Q_i

As an example, let us consider the domain of sequences. Here, a pattern is simply any sequence of discrete symbols from a given alphabet, and Σ is a set of sequences observed according to a given phenomenon. Thus, a specific query could be

$$Q(s, \Sigma) = \text{superstring\_of}(s, ab) \wedge \text{superstring\_of}(s, uw) \wedge$$
$$(\text{length\_atleast}(s, 6) \vee \text{minimum\_frequency}(s, .3, \Sigma))$$

Then, Q allows the decomposition $Q = Q_1 \wedge Q_2$, where

$$Q_1(s, \Sigma) = \text{superstring\_of}(s, ab) \wedge \text{superstring\_of}(s, uw) \wedge \\ \text{minimum\_frequency}(s, .3, \Sigma)$$

$$Q_2(s, \Sigma) = \text{superstring\_of}(s, ab) \wedge \text{superstring\_of}(s, uw) \wedge \\ \text{length\_atleast}(s, 6) \wedge \neg\text{minimum\_frequency}(s, .3, \Sigma).$$

Interestingly, an inductive query Q can admit a variety of different decompositions. The question then arises as to which decomposition is optimal in terms of computing resources.

The outlined approach has a strong theoretical foundation, but it suffers from a number of limitations: its main drawback is a limited applicability since many data mining tasks are specified by constraints that are neither monotonic nor antimonotonic. The CONQUEST approach [2] implements a query language and a constraint solver capable of dealing with conjunctions of antimonotone, succinct, monotone, convertible, and even loose antimonotone constraints. The core of the CONQUEST approach is an efficient and scalable, levelwise, frequent pattern mining algorithm which, at each iteration of the mining process, prunes the data source by exploiting the independent data reduction properties of all user-specified constraints [3].

A different optimization perspective has been devised in [6]. The basic intuition is that, if the pattern language $\mathscr{L}$ were stored within relational tables, any constraint predicate Q could be specified by means of a relational algebra expression, and the DBMS could take care of implementing the best strategy for computing the solution space. Assume, for example, that sequences are stored within a relational engine by means of the following relations:

- *Sequences*(*sid, item, pos*), representing each sequence by means of a sequence identifier, an item, and its relative position within the sequence
- *Supports*(*sid, supp*) that specifies, for each sequence, its frequency

Then, the following SQL query asks for the sequences holding with frequency greater than 60%, or such that item a occurs before item b within the sequence:

```
SELECT Supports.sid
FROM Sequences S1, Sequences S2, Supports
WHERE S1.sid = Supports.sid AND S2.sid = S1.sid
  AND Supports.supp > 60
  OR (S1.item = a AND S2.item = b AND S1.pos < S2.pos).
```

Clearly, the pattern language can be extremely huge, and hence it is quite unpractical to effectively store the *Sequences* and *Supports* tables. And, indeed, the pattern language is represented as a *virtual* table, i.e., an empty table that has to be populated. In the above example, although the *Sequences* and *Supports* tables are exploited within the query, they are assumed to be virtual tables, i.e., no materialization actually exists for them within the DBMS. The idea here is that, whenever the user queries such pattern tables, an efficient data mining algorithm is triggered by the DBMS, which materializes those tuples needed to answer the query. Afterwards, the query can be effectively executed.

Thus, the core of the approach is a constraint extraction procedure, which analyzes a given SQL query and identifies the relevant constraints. The procedure builds, for each SQL query, the corresponding relational algebra tree. Since virtual tables appear in leaf nodes of the tree, a bottom-up traversal of the tree allows the detection of the necessary constraints. Finally, specific calls to a mining engine can be raised to populate those nodes representing virtual tables.

12.3.2 Support to KDD Process

The search for knowledge from data is a complex process-oriented task, typically including the combination of several mining tasks and ad-hoc interpretation and evaluation of the extracted knowledge. It is therefore natural to embody data mining query languages in a more general framework, where effective support to the whole knowledge discovery process is provided. Indeed, several data mining tools have been developed, which provide the user with process-oriented capabilities. Paradigmatic examples are Clementine [13] and Weka [82].

Clementine is a data mining workbench that uses a visual approach to model the process and provides a tangible way to work with data. Each data mining task in Clementine is represented by a node, which the user can connect to form a stream representing the flow of data through a variety of tasks. Streams embed some fixed mining models as well as preprocessing and evaluation tasks. Also, input data are represented by single tables. Figure 12.2 gives an idea of the "Clementine way" to express the process.

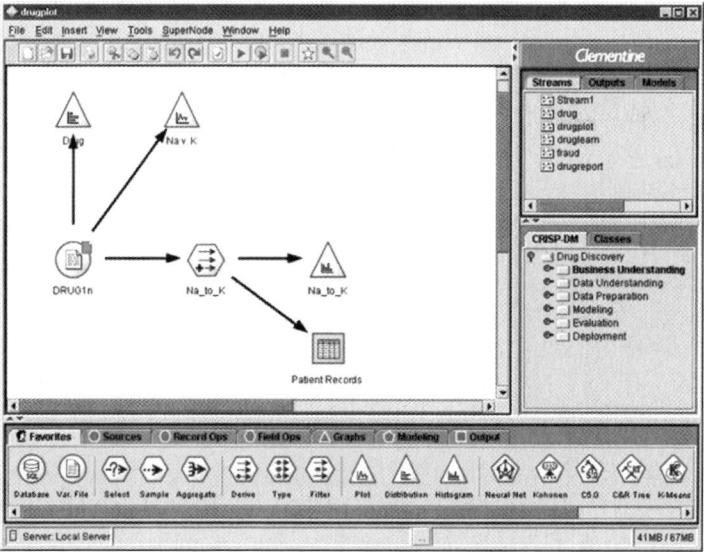

Fig. 12.2 *Clementine main window*

Weka is an open-source toolbench for machine learning and data mining, implemented in Java. The algorithms provide a standard interface that makes them directly available within custom Java code. Main features of Weka are the following:

- A comprehensive set of data preprocessing (filtering) tools
- Several learning algorithms for classification, regression, clustering, and association mining, together with model evaluation tools
- Standard interfaces for filters, algorithms, and evaluation methods, which can be hence customized to specific application needs

Its specification as a library allows to model complex tasks within Java code. For example, the following fragment specifies the choice of a suitable preprocessing task (among some of interest), which allows to obtain a misclassification error below 12%:

```
1.     reader = new FileReader(...some file containing data ...);
2.     Instances data = new Instances(reader);

3.     while (true){
4.         Filter filter = ...filter chosen among some of interest...
4.         Instance processed = applyFilter(filter);

5.         Filter sampler = new Resample();
6.         sampler.setSampleSizePercentage(70);
7.         Instances trainInstances = applyFilter(processed);
8.         sampler.setSampleSizePercentage(30);
9.         Instances testInstances = applyFilter(processed);
10.        Classifier scheme = ... scheme got from somewhere

11.        Evaluation evaluation = new Evaluation(trainInstances);
12.        evaluation.evaluateModel(scheme, testInstances);

13.        if (evaluation.errorRate() < 0.12)
24.            break;
15.    }
```

The main strength of Weka lies in its flexibility: a clean, highly customizable, object-oriented Java class hierarchy for each element of interest in a knowledge discovery process. In the above example, the classes Instance, Filter, Classifier, and Evaluation provide standard interfaces to data sources, preprocessing, classification, and evaluation algorithms, respectively.

Clearly, the high flexibility in Weka also represents its weakness: there is no standard way of encoding and exploiting background knowledge, which is on the contrary demanded to the user/programmer. In particular, there are no standard mechanisms for reasoning on the extracted knowledge, which should be explicitly encoded and programmed.

A proposal that concentrates on the formalization of the KDD process as a whole is MQL [1,76]. This formalism models the KDD process as a query process resulting in the composition of algebraic operators. Thus, a query in MQL specifies a complex knowledge extraction process, in which different data mining algorithms are combined. Consider for example the following query:

```
begin query
  let Input = runSQLQuery [select ID,Type from Traverses INNER
                           JOIN Place ON [TrajectoryID = ID]];
    Rules1 = createRules Input Apriori 0.1 0.6;
    Rules2 = filter Rules1 if !"Road" in Head & !"Square" in Body;
    Exception = RuleException Rules2 from Input
  in  Result = createRules Exception Apriori 0.1 0.6;
end query
```

Here, the result is specified as the application of a sequence of operations. Each operation acts on an MQL object, which can be either a data set or a mining model. For example, the `runSQLQuery` operator requires the specification of a query and returns an MQL object representing a data set, whereas the `filter` operator requires an object representing a model and returns a new (filtered) model.

MQL represents primary knowledge sources as propositional tables stored on relational databases, and manipulates them by means of external algorithms that implement specific search criteria. In the above example, the `createRules` operator acts on the `Input` object by exploiting the Apriori algorithm with support and confidence thresholds, respectively, 0.1 and 0.6.

Similar approaches based on different specifications of the available algebraic operators have been proposed. For example, OLE DB for data mining [59] is an extension to SQL defining primitives for the extraction and the application of a model. Again, the main drawback of such approaches is that no specific reasoning mechanisms can be integrated within a query, unless ad-hoc operators are added.

Some theoretical issues arising from the idea of modeling the KDD process as an algebraic process have been studied in [7,43]. The authors start from the observation that the essence of a knowledge discovery process is the interaction between two (apparently) separate worlds: the *Data* world and the *Model* world. Each world has its own entities, together with properties and relations: e.g., relational tables and operators in the data world, and conjunctions of linear inequality constraints in the model world. Now, since linear inequality constraints specify regions in high-dimensional spaces, such regions can be equivalently represented in an equivalent extensional way, as the set of all data points that satisfy those constraints. Thus, the resulting 3W model [43] can be specified as a set of three worlds: the D (Data) world, the I (Intensional) world, and the E (Extensional) world. Bridging operators can hence be specified for relating entities in these different worlds. In [7], the authors study the expressiveness of the underlying model subject to different choices for the operators. In particular, there are some specific operators that make the resulting algebra computationally complete.

A major limitation in the proposed approach is in the I-World, which is populated, as mentioned, by linear inequality constraints. This limitation means that the results of some data mining operations might not be expressible, as they require more complex mathematical objects.

12.3.3 A DMQL for Spatial Data

As pointed out in the introduction, to the best of our knowledge, in the literature there is no proposal for a DMQL for moving objects. However, there are some proposals concerning only spatial data, which are interesting to analyze since they highlight the peculiarities introduced in the data mining process due to the handling of such kind of data.

It is worth recalling that the first main difficulty concerns the huge amount of spatial data and the complexity of spatial data type. Spatial objects are characterized by a *geometrical representation*, a *position*, and a nonspatial description of these objects. Some spatial attributes, such as orientation, area, and some spatial relations like topological relations, are *implicit*. Hence the handling of spatial objects requires a higher level of reasoning with respect to the traditional data stored in databases. Indeed, in Sect. 12.4 we will present a brief overview of spatiotemporal query languages focusing on the reasoning support they provide.

The investigation of the relationships between spatial objects is typically a crucial issue, since many spatial data are interrelated, i.e., spatial objects are influenced by their neighboring objects. For instance, the presence of a highly polluting factory can reduce the price of the houses close to it. This is why in some spatial data mining approaches [49] there is distinction between the reference objects of analysis and other task-relevant spatial objects, which can have some attributes that affect the values of the unit of analysis. Besides distinguishing between these objects, it is also important to represent the interactions between them, and to do that it seems easier to use a multirelational approach, as already highlighted in Section 12.2. In this respect, Koperski et al. [44] suggest to adopt an Object-Oriented model to cope with the complexity of spatial objects. Finally, it is fundamental to have proper visualization techniques to present spatial data. In fact these tools can help in selecting data to be analyzed, in understanding the extracted knowledge, and in giving a quick feedback for the refinement of queries.

In this context, a data mining query language can be very profitable for a user because it provides a high-level interface hiding the problems related to the integration of different technologies, such as data mining, inference engines, and spatial data repositories. As an example we will briefly outline the *Spatial Data Mining Object Query Language* (SDMOQL) proposed by Malerba et al. [49], which is designed to support the interactive data mining process in INGENS [50]. INGENS is a prototype GIS, which integrates data mining tools to assist users in the task of topographical map interpretation. When a user wants to formulate a query concerning geographical objects not explicitly represented in the database, he/she can train the system giving examples and counterexamples of the concepts he/she wants the system to learn.

According to the classification presented in Sect. 12.3, SDMOQL belongs to the research line that considers a DMQL as an interface to data mining tasks. In fact it provides an object-based query language (OQL) to select the sources to mine, two data mining tasks, i.e., classification and association rules, to detect the patterns of interest, and it also allows one to specify some search criteria, such as interestingness measures for data patterns and their corresponding thresholds.

Typically, geographical data are organized according to an object-oriented model, stored in a commercial object-oriented DBMS and queried by a simplified version of OQL (Object Query Language) [58]. An example of query specifying the objects to be mined is the following:

SELECT x
FROM x in Cell
WHERE x->part_map->map_name = "Canosa"
AND x->log_incell->num_cell = 11,

which selects the cell 11 from the topographic map of Canosa.

The processing of this query requires first the selection of the involved object(s) from the *Map Repository*, which is the database instance containing the actual collection of maps in the GIS. Then there is a component, called *Map Descriptor*, which is responsible for the *automated* generation of first-order logic description of the selected objects. Such a representation is expressed as a conjunction of atoms of the kind $f(t_1,\ldots,t_n) = value$, where f is a function symbol called *descriptor*. It contains information about *geometrical* aspects (such as area, density, kind of shape); *topological* relationships (like which relation holds between two regions); *directional* values concerning orientation of the object (such as north, south-east); *locational* aspects, expressing the coordinates of the objects; and finally *nonspatial* features (such as color, the kind of object).

To give an idea of this transformation, a fragment of the symbolic description of cell 11 is

contain(c11,pc494_11) = true, ..., contain(c11,ss296_11) = true,
type_of(pc494_11) = parcel, ..., type_of(ss296_11) = street,
color(pc494_11) = black, ..., color(ss296_11) = black,
part_of(pc494_11,x1) = true, ..., part_of(ss296_11,x68) = true,
area(x1) = 99962, ..., part_of(ss296_11,x68) = true,
line_to_line(x1,x68) = almost_parallel, ...

This states that the cell $c11$ contains a parcel pc494_11, whose color is black. pc494_11 is a logical object composed by a physical object x1, having area 99962 and which is almost parallel to x68, which is a line part of the street ss296_11.

Thanks to this description and by using a *deductive inference engine*, the conditions in the **WHERE** part of the query are checked. This combination of high-level qualitative representation and deductive ability has many advantages. First it allows the user to express complex conditions, which can involve also mined rules. For instance, as shown in [49], one can first learn the concept of "system of farms" by using a classification task, which returns a set of clauses defining the predicate *class*. Then, inserting the condition *class(C) = system_of_farms*, one can use the extracted knowledge to formulate new queries in SDMOQL. Second, this first-order description is adequate as input for many relational data mining algorithms, which return spatial patterns expressed in a first-order language. Third, it eases the specification and integration of *Background knowledge*, since such knowledge is expressed in a declarative way as a set of definite clauses, directly defined by the user or imported by a deductive database. Finally, it could be a uniform framework to represent also temporal and spatiotemporal information. In particular, various approaches in the spatiotemporal field [10, 56] use constraints in order to model and to reason on such kind of data. In Sect. 12.4.2.4 we will deepen the proposal in [56].

The main construct of SDMOQL is the following:

⟨Object_Specification_Query⟩
mine⟨Kind_of_Pattern⟩
analyze⟨Primitive_descriptors⟩
with descriptors⟨Pattern_descriptors⟩
[⟨Background_knowledge⟩]
{⟨Hierarchy⟩}
[**with**⟨Interestingness_Measures⟩]
[⟨Result_Displaying⟩].

SDMOQL supports only two data mining tasks: classification and association rules, specified in the **mine** clause. For example,

mine classification as MorphologicalElements
for class(_)=system_of_farms, class(_)=fluvial_landscape,
    class(_)=royal_cattle_track, class(_)=system_of_cliffs

specifies the concepts to be learnt, which allows to classify the cells as system of farms, fluvial landscape, royal cattle track, or system of cliffs.

The **analyze** clause concerns data preparation for the mining task: it states what descriptors can be used in the first-order representation of the geographical objects satisfying the query. On the other hand, **with descriptors** establishes the descriptors to be used to describe the *generated* patterns.

The remaining part of the main construct allows the user

- To specify background knowledge expressed as a set of definite clauses or recalling rules already present in the deductive database
- To define hierarchies that permit knowledge mining at multiple abstraction levels
- To control the data mining process by specifying interestingness measures for data patterns and their corresponding thresholds

12.4 Querying Spatiotemporal Data

In view of the definition of a spatiotemporal data mining query language, it is important to analyze the kind of primitives that are provided by spatiotemporal query languages in the literature. In fact, spatiotemporal query languages are useful in many steps of the knowledge discovery process. In particular, they can serve as tools for specifying the data of interest and for interpreting the extracted patterns. In Sect. 12.4.1 we describe, in short with no claim to be exhaustive, the history of the field of spatiotemporal databases with the emphasis on spatiotemporal query languages, whereas in Sect. 12.4.2 we focus on approaches supporting qualitative spatiotemporal reasoning, which allow to specify highly expressive declarative queries.

12.4.1 Spatiotemporal Query Languages

A taxonomy of spatiotemporal applications ranging from those that rely on a stepwise constant geometry to applications that need a more complete integration of space and time (like for instance a continuous description of a trajectory) can be found in Erwig et al. [23].

In the early nineties, when research on spatial databases and GIS was flourishing, questions arose about the role of time in GIS. In 1993, the "Specialist Meeting on Time in Geographic Space" [21] was organized by Research Initiative 10 from the U.S. National Center for Geographic Information and Analysis (NCGIA) and its main goal was "to formulate a research agenda for spatiotemporal reasoning about geographic space" (a summary of the identified research questions can be found in a technical report of the NCGIA [21]). One of the participants of the Specialist Meeting was Worboys, who in 1994 proposed the first spatiotemporal database model [83]. He defined a spatiotemporal database as a collection of unified objects that have a spatial extent (a point, line segment, or triangle) and a temporal extent (a time interval). As the spatial and temporal extent are completely independent of each other, only piecewise constant movement can be represented. Chomicki and Revesz [11] propose an extension of this model, where objects also have a spatiotemporal component.

In 1996, the CHOROCHRONOS research network [27], a cooperation between ten European institutes, was started. The objective of this network was to work together on "Spatial and Temporal Databases." In 2003, an overview of the realizations of CHOROCHRONOS was published [45]. Here, we summarize the main achievements of the CHOROCHRONOS network with respect to the models and languages it proposed for spatiotemporal databases. Two approaches towards spatiotemporal data modeling were explored: an approach based on data types and one based on constraint databases.

We start with the data type approach [36]. A set of base, spatial, temporal, and spatiotemporal data types is proposed. The (two-dimensional) spatial data types are point, points (a finite set of points), line (a finite set of continuous curves), and region. Time is to be considered linear and continuous. A type constructor named *moving* exists that, given any type α, yields a mapping from time to α. Examples of types that can be constructed this way are *moving(point)* and *moving(region)*. Next to a set of data types, a set of spatial operations is proposed, which can be lifted to spatiotemporal operators. For example, the intersection operator defined on a region and a point can be lifted in such a way that it can compute the spatiotemporal intersection between a *moving(region)* and a *point*, a *moving(point)* and a *region*, or a *moving(region)* and a *moving(point)*. Those operations are embedded in an SQL-like language. The set of data types is fixed, but the user is allowed to define new operations using those data types. A discrete implementation of this data type approach has been proposed. Here, the data type line is implemented as a set of line segments, the data type region as a collection of polygons with polygonal holes, etc. A *moving(region)* is allowed to change in such a way that its three-dimensional representation is a polyhedron. As further work the construction of a set of spatiotemporal predicates [25], based on the well-known set of eight topological spatial predicates [20], is proposed. Also, the need for spatiotemporal partitions [24], spatial partitions that are preserved over time, is recognized.

We illustrate the query language of [36] with the example query "At what time and distance does flight 257 pass the Eiffel tower?" taken from [36]. It is assumed that a closest operator exists with signature $moving(point) \times point \rightarrow intime(point)$, which returns time and position when a moving point is closest to a given fixed point in the plane. This query would be expressed as

> *LET EiffelTower =*
> *ELEMENT(SELECT pos*
> *FROM site WHERE name = "Eiffel Tower");*
> *LET pass = closest(route257, EiffelTower);*
> *inst(pass); distance(EiffelTower, val(pass))*

The second approach of the CHOROCHRONOS network towards spatiotemporal data modeling is the constraint database approach. In this approach, the DEDALE data model, spatiotemporal data are represented using linear constraints. It extends the standard language of linear constraints with some additional primitives like *dist* for computing distances and connect for testing connectivity. An SQL-like query language for users is developed on top of the constraint algebra, hiding the data model from the user. The DEDALE model was implemented at INRIA [33]. The developers of DEDALE also introduced the concept of orthographic dimension of a constraint relation, which can speed up query evaluation on (spatiotemporal) constraint databases [34, 48].

In 1997, the MOST (Moving objects Spatiotemporal) data model was proposed by Sistla et al. [77]. In this model, objects can have dynamic attributes, having a value, an update time, and a function. This function can be any function f of time for which $f(0) = 0$. The value is the value of the function at the current time, and the update time indicates when the value has to be updated. When functions change, for example, when an object has a piecewise linear movement, the previous function can be kept by computing the database state before the change and store this state in the database history. A spatiotemporal query is a predicate over the database history. The FTL (Future Temporal Logic) query language is proposed to query MOST-data. This logic contains two basic future temporal operators **Nexttime** and **Until**. It also uses additional temporal operators such as **Eventually** and **Always**. As an example, we give the following query, taken from [77]. It retrieves all the objects o that enter the polygon P within three units of time and stay in P for another two units of time, and is expressed as

RETRIEVE o
WHERE Eventually_within_3 *INSIDE*(o,P) ∧
Always_for_2 *INSIDE*(o,P)

In 1998, an Esprit Working Group, called *DeduGIS* – Deductive constraint databases for intelligent Geographical Information Systems(GISs), started. The goal was to envisage a new generation of GISs, characterized by enhanced capabilities to support spatiotemporal reasoning and semantic integration of diverse data models. One of the proposals [51] constructed a programing environment in which logical reasoning and geographical information management are integrated. To this end, the system integrates a logic-based spatiotemporal knowledge representation language with a robust, commercial GIS. The main advantages are the following. First, the definition of a reasoning component where a logic-based layer connects to a GIS to perform the most resource-consuming operations on geographical data. The idea is to exploit the GIS ability to perform spatial operations and visualization in an efficient and user-oriented fashion. Second, the data stored in the GIS can be exported at the logical level in order to perform complex deductive reasoning, which standard GISs do not usually provide. Third, the logic-based representation language allows the user to add temporal annotations to the information stored in the GIS, which can then be used to perform temporal reasoning on the spatial data stored in the GIS. In Sect. 12.4.2.4 we will show a logic-based language that can be used as a reasoning component on top of a GIS.

In 1999, Chomicki and Revesz [11] proposed the *Parametric Data Model*, an extension of the object–data model of Worboys [83]. Spatiotemporal objects were represented by a spatial reference object and a time domain (analogous to Worboys' spatial and temporal extent [83]) and also a spatiotemporal component, namely, a time-dependent transformation function describing the movement of the spatiotemporal reference object throughout the time domain. Several classes of such objects were proposed depending on the type of spatial reference objects and the type of transformation functions. Revesz [5,75] further developed an algebra for the class of linearly moving rectangles. This algebra includes the specific spatiotemporal operators buffer, compose, and block. Both the general model and the query language for linearly moving rectangles are also included in the constraint database textbook by Revesz [74].

In the same year, the Tripod [32] project emerged, a joint collaboration between researchers in the computer science departments at Keele and Manchester Universities. An existing database system was extended with the spatial types proposed for the ROSE algebra [37] and the temporal types Instants and TimeIntervals. The Tripod data model can express only discrete changes. Previous states of an object are kept in histories.

In 2000, an European project called Multirepresentations and Multiple Resolutions in Geographic Databases (MurMur) [63] was started. The participants' goal was "enhancing GIS (or STDBMS) functionality so that, relying on more flexible representation schemes, users may easily manage information using multiple representations." The added functionality will support multiple coexisting representations of the same real-word phenomena (semantic flexibility), including representations of geographic data at multiple resolutions (cartographic flexibility). This will in particular make possible a semantically meaningful management of multiscale, integrated, and temporal geo-databases. MurMur started from an existing spatiotemporal data model, called MADS (Modeling of Application Data with Spatiotemporal features), proposed by Parent et al. [62]. This model contains objects, attributes, and relationships of several types. Special to MADS is the perception stamp, including the viewpoint of the user (public, manager, or technician) and the resolution or level of detail of a representation (e.g., 1:2000). These perception stamps allow users to define sub-schemas in a given schema, personalize data types, etc. A two-sorted algebra (the MADS algebra) and two visual query languages were developed to manipulate spatiotemporal data.

At the same time, Chen and Zaniolo [9] proposed SQL^{ST}, a spatiotemporal data model and query language. Here, a moving object is modeled as a series of snapshots containing directed triangles. The query language is based on SQL^T [8] for its temporal operators, and a set of spatial operations, like intersect, area, etc. No real spatiotemporal operations are needed because of the snapshot view.

Also in 2000, Kuijpers et al. [46] proposed a model and query language for Movie Databases. In this proposal, the constraint database approach is used. A movie is modeled as a two-dimensional semi-algebraic figure that can change in time. A number of computability results concerning movies are given, e.g., it can be decided whether a frame of a movie is only a topologically transformation of another frame, a movie has a finite number of scenes and cuts and these can be effectively computed, etc. Based on these computability results, an SQL-like query language for movie databases is developed. This query language supports common movie editing operations, like cutting, pasting, and selection of scenes.

Another constraint database approach to spatiotemporal databases was proposed in 2001 by Ibarra et al. [55, 79]. Logical properties of moving objects were considered in connection with queries over such objects using tools from differential geometry. An abstract model was proposed where object locations can be described as vectors of continuous functions of time. Using this conceptual model, logical relationships between moving objects, and between moving objects and (stationary) spatial objects in the database were examined. These relationships were then characterized in terms of position, velocity, and acceleration. Based on this theoretical foundation, a concrete data model for moving objects was developed, which is an extension of linear constraint databases. The authors also presented a preliminary version of a logical query language for moving object databases.

Recently, Pelekis et al. in [66, 67] proposed a moving object database developed on top of Oracle, where, in addition to the spatial data cartridge, a new temporal data cartridge has been defined. The resulting system provides an efficient database management system for moving objects. The associated query language is defined as an extension of PL-SQL and allows to express complex spatiotemporal queries.

For an overview of spatiotemporal database systems and query languages we refer to the PhD-thesis of Sofie Haesevoets [39] and for an overview of languages specifically directed towards moving objects we refer to the recent book by Güting and Schneider [38].

12.4.2 Spatiotemporal Qualitative Reasoning

As already pointed out in the previous sections, dealing with the complexity of spatial, and even more complex spatiotemporal objects, needs the ability to represent and reasoning on implicit information.

Let us take again as an example the query posed in Sect. 12.2. The user wants to find out patterns (moving objects) intersecting a spatial object (a river) within X time units from their start. The intersection property is implicit on the shape and location of objects and we need here to make it explicit. Furthermore, this relation is *qualitative* since we abstract away from the huge amount of numerical data, allowing the user to specify, in a synthetic way, all the possible intersections between objects.

Qualitative reasoning, as opposed to quantitative one, deals more with the way humans reason, whereas quantitative methods are closer to machine reasoning. Quantitative location information is presented as "76.87 N, 13.45 E" instead of "Shopping Mall," or "245.5°" instead of "West" for directions, or speed can be indicated as "fast" instead of "150 km h^{-1}" and so on. From these simple examples it becomes clear how having qualitative information about objects is important for the end user. Furthermore, qualitative reasoning deals with imprecision and uncertainty. When we say that "a car crosses a park," a degree of imprecision is involved, since there could be several possible paths that cross a park. When the qualitative relation involves a (set of) moving object(s) (i.e., the trajectories or the patterns), it is called *spatiotemporal* since it changes during the time.

Queries involving relations between objects have been defined as *comparison* question in Chap. 1. Here, the target can be the relations between given objects (i.e., "In which relations are the car and the park?"), or the objects for which a given relation holds ("Which cars are *inside* the park?").

Relations between spatiotemporal entities are typically expressed, in an end user query, in a qualitative manner. It is, therefore, fundamental to provide the query language with a set of qualitative spatiotemporal primitives. It is worth noticing that quantitative primitives are necessary too, and that usually these kinds of operations are provided by the spatiotemporal DBMS itself (such as computation of the area or the perimeter of a spatial object).

Among qualitative approaches, much research has been done in spatial qualitative reasoning, coming mainly from the artificial intelligence field. Only a few approaches are present in the literature about spatiotemporal qualitative reasoning, some of them presented later in this section.

Most spatial qualitative approaches focus on the description of the relationships between spatial entities. Prominent examples are *directional* and *topological* relations, that are, respectively, relations that represent the relative positions of objects and the spatial relations that are invariant under topological transformations like translation, rotation, scaling.

Many approaches to spatial topological relations can be found in the literature coming from both mathematics and philosophical logics: from the *RCC (Region Connection Calculus)* [73], originating from a proposal of Clarke [12], to the *9-intersection model* proposed by Egenhofer [22], or the *CBM (Calculus-based*

Method) [14], adopted by the OpenGIS standard [60]. Direction relations deal with cardinal points such as *north, south-west*. There are the *projection-based* approaches [61, 86] where the space is divided using horizontal and vertical lines passing through the reference point or delimiting the reference object, and the *cone-based* approaches [26, 68] where space around a reference object is partitioned into four (or eight) partitions of 90° or 45°. Finally, a recent approach [31] allows the representation of cardinal directions between objects by using their exact geometries.

In the following, we will briefly present some of the most recent spatiotemporal qualitative reasoning approaches proposed in the literature, ranging from double cross calculus, to the uncertainty predicates of Wolfson et al., to the abstract data type spatiotemporal predicates of Güting et al., to the logic-based approach STACLP.

12.4.2.1 The Double-Cross and the Qualitative-Trajectory Calculi

The *double-cross calculus* [28, 86] is an expressive way of qualitatively representing a configuration of two vectors in the plane $\mathbf{R}^2$ by means of a 4-tuple of elements of $\{+, 0, -\}$ that expresses the orientation of both vectors with respect to each other. Figure 12.3 gives the intuition behind this idea. If we have two vectors $\vec{\ell_1}$ and $\vec{\ell_2}$ in the plane, then the vector $\vec{u}$ between their starting points determines the three lines RL, PL_1, and PL_2, as shown in the figure. The double-cross formalism records in which of the four quadrants or on which of the four lines the vectors $\vec{\ell_1}$ and $\vec{\ell_2}$ are situated. In this example, the vectors $\vec{\ell_1}$ and $\vec{\ell_2}$, shown in Fig. 12.3, are qualitatively described by $(+---)$. Indeed, vector $\vec{\ell_1}$ is in the upper left quadrant $(+-)$ and $\vec{\ell_2}$ is in the lower left quadrant (see [15] for details).

The double-cross formalism is used in the *qualitative trajectory calculus* [15–18]. This calculus was introduced to describe relative changes between moving objects that do not change their topological relationships (i.e., that remain disconnected, for example). For this purpose different versions of the qualitative trajectory calculus were introduced as a theory for representing and reasoning about movements of

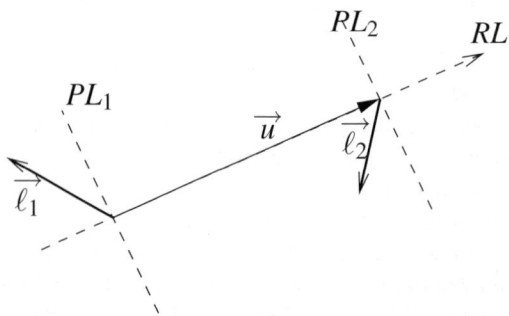

Fig. 12.3 The double-cross design with the lines RL, PL_1, and PL_2

objects in a qualitative framework, differentiating groups of disconnected objects. Van de Weghe et al. [16] illustrate their calculus by describing the evolution of the interaction between a predator, a lion, and its prey, a zebra, during the hunt of the lion. This evolution of movements is described by sets of transitions of tuples over $\{+,0,-\}$, which describe the resting, running away, running towards, and finally overtaking of the animals.

12.4.2.2 Wolfson's Range Query Predicates

Wolfson et al. proposal defines a number of Boolean predicates to express relations of trajectories [80] with respect to a spatial region. These predicates express whether a moving point is in a region R (a region is considered to be a polygonal figure without holes) in the time interval $[t_1,t_2]$.

This model provides an explicit support for dealing with uncertainty that comes from different sources. An *uncertainty trajectory*, represented as a couple $T = (\text{Tr}, u)$, where Tr is a trajectory in $\mathbf{R}^2$ and $u > 0$ is a buffer, is a trajectory with a cylindrical uncertainty buffer. A *possible motion curve* for a given uncertainty trajectory $T = (\text{Tr}, u)$, denoted by PMC^T (Possible Motion Curve), is a curve within this cylindrical uncertainty buffer.

In this model three types of uncertainty are considered:

- Sometimes or always in $[t_1,t_2]$ the object is in R (*sometime* ↔ *always*)
- The object is located somewhere or everywhere in R (*somewhere* ↔ *everywhere*)
- The object is possibly or definitely in R during $[t_1,t_2]$ (*possibly* ↔ *definitely*)

Every combination of these basic predicates is possible, for example, *Possibly_Sometime_Somewhere_Inside*. The number of possible predicates is $2^3 \times 3! = 48$, (3! because the order is relevant). Since it is useless to express *everywhere* in a region, thus there are only $2^2 \times 2! = 8$ different predicates. The possible configurations are shown in Fig. 12.4.

- *Possibly_Sometime_Inside*(T,R,t_1,t_2) is *true* if and only if there is a PMC^T and a $t \in [t_1,t_2]$ such that PMC^T at t is in R.
- *Sometime_Possibly_Inside*(T,R,t_1,t_2) is *true* if and only if there is a $t \in [t_1,t_2]$ and a PMC^T such that PMC^T at t is in R (Since existential quantifiers commute, this is semantically the same as the previous case).
- *Possibly_Always_Inside*(T,R,t_1,t_2) is *true* if and only if there is a PMC^T that for each $t \in [t_1,t_2]$ is in R.
- *Always_Possibly_Inside*(T,R,t_1,t_2) is *true* if and only if for every $t \in [t_1,t_2]$ there is a PMC^T that is in R at time t. Semantically this predicate is different from the previous one, as illustrated in Fig. 12.4.
- *Always_Definitely_Inside*(T,R,t_1,t_2) is *true* if and only if for every $t \in [t_1,t_2]$, every PMC^T is at moment t in R.
- *Definitely_Always_Inside*(T,R,t_1,t_2) is *true* if and only if for every PMC^T it holds that for every $t \in [t_1,t_2]$ it is in R (Since universal quantifiers commute, this is semantically the same as the previous case).

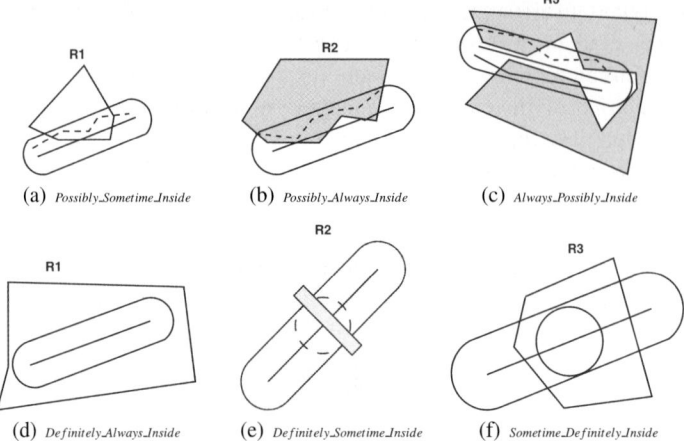

Fig. 12.4 The six predicates for range queries

- $Definitely\_Sometime\_Inside(T,R,t_1,t_2)$ is *true* if and only if for every PMC^T there is a moment $t \in [t_1,t_2]$ such that PMC^T at moment t is in R.
- $Sometime\_Definitely\_Inside(T,R,t_1,t_2)$ is *true* if there is a moment $t \in [t_1,t_2]$ such that every PMC^T is in R at that moment. Semantically this predicate is different from the previous one, as illustrated in Fig. 12.4.

An example of a query that can be expressed in this formalism is *"find all objects that are possibly always in R between the time instant at which object A is at location L_1 and the time instant where A is at location L_2."*

$$Possibly\_Always\_Inside(T,R,When\_At(T_A,L_1),When\_At(T_A,L_2)).$$

12.4.2.3 Güting et al. Spatiotemporal Predicates

As briefly explained in Chap. 5 and in Sect. 12.4.1, the spatiotemporal data model of Güting et al. aims at defining a set of abstract data types to express moving objects [36]. In this context, Erwig and Schneider introduced a wide range of *spatiotemporal predicates* that define qualitative spatiotemporal relationships between objects [25]. This work was inspired by Galton's pioneering work on qualitative theory of movements [29] and largely based on the Egenhofer qualitative relations [20], extended with time.

The authors distinguish between *base* and *complex* predicates and they define an algebra over these predicates. Essentially, basic predicates are defined by temporally lifting Egenhofer spatial primitives, thus obtaining a function from spatiotemporal objects to temporal Booleans. The essence is that they can be used to express conditions that can be both true and false during the time. For example, the *inside* predicate between a point and a spatial region is either true or false, whereas between a moving point and an evolving region yields true at the time points where the moving point is inside the moving region, false otherwise. Thus, the relationship between the moving object and the moving region can change as time flows. However, one can define a predicate *always inside*, yielding true only if the lifted version of inside is true for all time points of interest. In a similar way all the other 9-intersection predicates are redefined by temporal lifting.

Relationships between objects can be modeled by sequences of spatial and (base) spatiotemporal objects. For example, given a continuously moving point P and an evolving region R, the event *P Crosses R* can be represented by the following sequence of predicates:

Disjoint(P,R); meet(P,R); Inside(P,R); meet(P,R); Disjoint(P,R),

where the first spatiotemporal *Disjoint* predicate holds from the P location time point t_1 until the time point t_2, where P touches R; the first spatial predicate *meet* holds at t_2; *Inside* holds between $]t_2,t_3[$, where t_3 is the time point when the second *meet* predicate holds; after t_3 the last *Disjoint* spatiotemporal predicate holds. It is worth noticing that to describe the developments of spatiotemporal objects, the validity of spatiotemporal predicates has to be restricted to time intervals.

Spatiotemporal and spatial predicates can be combined by means of composition operators, described in [25] along with their properties.

12.4.2.4 A Logic-Based Approach to Reason on Spatiotemporal Data

A different promising research field concerns the integration of declarative paradigms and systems for dealing with spatial and/or temporal information, such as spatial databases and GISs. In the literature we can find several attempts to exploit the deductive capabilities of logics to *reason* on geographic data [35, 50, 51, 71, 78, 84]. Underlying these approaches there is the belief that the language for "programming" the extensions has to be a real knowledge representation language, or, better to say, a very high level query language. It should be able to handle not only data but also rules, and exhibit both deductive and inductive capabilities. Rules can be used to represent general knowledge about the collected data, and deductive capabilities can provide answers to queries that require some inference besides the crude manipulation of the data. Finally, induction can help in extracting implicit knowledge from data. In the following we briefly present the language STACLP [71] as an example of this research line.

STACLP (spatiotemporal annotated constraint logic programming) is a language based on constraint logic programing, extended with annotations. Constraint logic programing provides the deductive capabilities, and annotations allow a neat representation of temporal, spatial, and spatiotemporal knowledge. On this ground, knowledge extraction methods can be implemented, thus providing the required inductive capabilities [56].

STACLP offers three kinds of temporal and spatial annotations to represent spatiotemporal properties of objects:

- `atp`(X,Y) and `at` T specify that a property holds at a certain spatial/temporal point, respectively.
- `thr`$[(X_1,X_2),(Y_1,Y_2)]$ and `th`$[T_1,T_2]$ involve a region and an interval in which the property holds.
- `inr`$[(X_1,X_2),(Y_1,Y_2)]$ and `in`$[T_1,T_2]$ state that a property holds in some point(s) of the given spatial region or temporal interval, which ones may not be known.

In this way STACLP can support not only *definite* spatial and temporal information but also *indefinite* knowledge, thanks to the use of the `inr` and `in` annotations. Moreover, it can also provide primitives for qualitative spatial reasoning. In [72] it is shown how the topological 9-intersection model [22] and the direction relations based on projections [61] can be modeled in such a framework.

STACLP can be used to establish a dependency between space and time, thus permitting to model continuously moving points and regions. For instance, consider a car running on a straight road with speed v and assume that its initial position at time t_0 is (x_0, y_0). The position (X,Y) of the car at T can be computed as follows:

$$car\_position \, \text{atp}\,(X,Y)\,\text{at}\,T \leftarrow X = x_0 + v(T-t_0), Y = y_0 + v(T-t_0).$$

STACLP has been used to represent and reason on trajectories of moving points. As we have widely discussed in the previous chapters, object movements are given by means of a *finite* set of observations, i.e., a finite subset of points taken from the actual continuous trajectory of the object, together with some interpolation function(s) that reconstruct(s) the full trajectory. In STACLP the observations are represented by means of `atp`/`at` annotations, like `fix(o) atp (x1, y1) at t1` and the interpolation function is expressed as a constraint. For instance, by using a linear interpolation, the trajectory is *completed* by defining all the intermediate points by means of the following STACLP rules:

```
traj(O) atp (X, X) at T :- fix(O) atp (X, Y) at T.
traj(O) atp (X, Y) at T :- fix(O) atp (X1, Y1) at T1,
                           fix(O) atp (X2, Y2) at T2,
                           succ(T1,T2), T1 < T < T2,
                           X=(X1(T2-T)+X2(T-T1))/(T2-T1),
                           Y=(Y1(T2-T)+Y2(T-T1))/(T2-T1).
```

In the body of the second rule, approximate points (x,y) are computed by using the equation for the line passing through two given points. It is worth noticing that by changing the constraint one can implement different interpolation functions.

Given such a representation, some example queries can be the following:

- Where is the object *id* at noon?
  `traj(id) atp (X,Y) at 12.`
- Do the trajectories *id*1 and *id*2 meet? Where? At what time?
  `traj(id1) atp (X,Y) at T, traj(id2) atp(X,Y) at T.`
- Which car(s) cross(es) the square "Piazzale Roma" at 9 a.m.?
  `traj(O) inr R at 9, square(Roma) thr R.`

Deductive reasoning can be useful to solve analysis problems that essentially require to find entities and values having some, possibly complex, properties. However, when dealing with sophisticated analysis tasks, it is quite common to meet concepts and abstract entities whose definition through deductive rules can be extremely difficult. In many cases, a suitable solution to the problem at hand requires the *extrapolation* of new pieces of information from those already available. In other words, knowledge induction capabilities can be needed to properly tackle some difficult problems. This aspect has been introduced in [56], where a basic data mining tool, the *k-means* clustering algorithm, has been defined as STACLP rules, specifically tailored around trajectories. Hence, the user can ask the system queries such as *Which are the objects that are moving together in a certain time period?* or *Is there a group of objects greater than* 10 *inside a certain region?*. This offers the user a powerful tool to express complex spatiotemporal analysis tasks. We refer the reader to [56] for details.

12.5 Discussion

The overviews provided so far clarify and highlight the issues related to the design of a data mining query language specifically targeted for movement data, which also comprises reasoning capabilities. From a knowledge discovery perspective, the current literature has developed two orthogonal directions, which are not necessarily contrasting. On the one side, research has concentrated on how to interface data sources and data mining tasks. On the other side, the definition of procedural knowledge discovery workflows has been the main objective of investigation.

It is clear that both aspects should be addressed in the definition of a data mining query language. Under this perspective, an extension of the ideas proposed in [7,43] appears to be the most promising. The essence of a knowledge discovery process can be summarized as the interaction between two (apparently) different worlds: the *data* world and the *model* world. Whenever each world is populated by the appropriate entities, a set of operators can be defined and used, in compliance with the requirements described in Sect. 12.3.1, to specify a process specifically targeted to the extraction of actionable knowledge. Within such a model, accommodating reasoning capabilities is somehow straightforward: essentially, entities in the two worlds can be investigated within a general reasoning framework, where the focus is the study of their properties and relationships.

Starting from these assumptions, the main issues are concerned with the definition of the contours for the two worlds and their operators. One has to concentrate on which entities (which pattern language) are supported within the model world, how data entities relate to model entities, and how constraint solving takes place. One may ask what is so special about mining movement data in the above model. And indeed, the model is in a sense a sort of "meta-model": since we are dealing with movement data, the data world should be able to represent spatiotemporal entities, properties, and relationships. Similarly, the model world should concentrate on spatiotemporal patterns. In particular, we have seen that, when dealing with movement data and patterns, an essential requirement is that both the data and the model worlds should be able to cope with complex objects. Analogously, both worlds should be equipped with a calculus/algebra capable of dealing with such complex entities. These issues should be addressed, which are crucial for the specification of the model and provide a substantial differentiation of that from the approaches proposed in the literature.

Also, two further research lines arise from the definition of specific bridging operators, able to correlate data objects to model objects and viceversa.

- From an expressiveness viewpoint, one can be interested in investigating a minimal set of bridging operators supporting a vast majority of spatiotemporal data mining operations. This research line can be seen as an extension of the investigation in [7] to the case of movement data. In principle, the complexity of the underlying worlds increases with spatiotemporal data, and so one can expect that new minimal operators should be accommodated to the model.
- From a practical point of view, the identification of specific classes of constraints supported by efficient constraint-solving techniques is a critical aspect, which should be addressed in a real-life implementation of the model.

12.6 Conclusions

In this chapter, we investigated the research issues arising by the quest toward a language framework, capable of supporting the user in specifying and refining mining objectives, combining multiple strategies, and defining the quality of the extracted knowledge, in the specific context of movement data. The spatiotemporal domain, with its complexities and peculiarities, exacerbates the intrinsic difficulties underlying the design of a data mining query language, which were illustrated throughout the chapter.

Besides the traditional data mining issues, a suitable spatiotemporal DMQL should take advantage of spatiotemporal qualitative reasoning primitives, especially for trajectories and/or moving objects. In addition, support for reasoning on spatiotemporal objects is highly desirable. This would enable the formalization of implicit knowledge, which can be purely geographical or specific to the application domain.

Several issues arise: which formalism better represents all the peculiarities of the GeoPKDD domain; which pattern languages are of interest, and how to integrate them with qualitative reasoning; which coupling should be pursued among the reasoning and querying component. In this perspective, data mining query language for spatiotemporal data pose new and exciting challenges to research community. It is clear that research on data mining query language involves several orthogonal dimensions and a trade-off between all the requirements posed by the specific domain has to be found.

References

1. M. Baglioni and F. Turini. MQL: An algebraic query language for knowledge discovery. In *Proceedings of the 8th Congress of the Italian Association for Artificial Intelligence on Advances in Artificial Intelligence (AI*IA'03)*, pp. 225–236. Springer, 2003.
2. F. Bonchi, F. Giannotti, C. Lucchese, S. Orlando, R. Perego, and R. Trasarti. ConQueSt: A constraint-based querying system for exploratory pattern discovery. In *Proceedings of the International Conference on Data Engineering (ICDE'06)*, p. 159. IEEE, 2006.
3. F. Bonchi, F. Giannotti, A. Mazzanti, and D. Pedreschi. ExAMiner: Optimized level-wise frequent pattern mining with monotone constraints. In *Proceedings of the International Conference on Data Mining (ICDM'03)*, pp. 11–18, 2003.
4. F. Bonchi and C. Lucchese. Extending the state of the art of constraint-based frequent pattern discovery. *Data and Knowledge Engineering*, 60(2):377–399, 2007.
5. M. Cai, D. Keshwani, and P. Revesz. Parametric rectangles: A model for querying and animation of spatiotemporal databases. In *Proceedings of the 7th International Conference on Extending Database Technology (EDBT'00)*, pp. 430–444. Springer, 2000.
6. T. Calders, B. Goetals, and A. Prado. Integrating pattern mining in relational databases. In *Proceedings of the Conference on Principles and Practice of Knowledge Discovery in Databases (PKDD'06)*, pp. 454–461. Springer, 2006.
7. T. Calders, L.V.S. Lakshmanan, R.T. Ng, and J. Paredaens. Expressive power of an algebra for data mining. *ACM Transactions on Database Systems*, 31(4):1169–1214, 2006.
8. C.X. Chen and C. Zaniolo. Universal temporal extensions for database languages. In *Proceedings of the 15th International Conference on Data Engineering (ICDE'99)*, pp. 428–437. IEEE, 1999.
9. C.X. Chen and C. Zaniolo. SQLST: A spatiotemporal data model and query language. In *Proceedings of the 19th International Conference on Conceptual Modeling (ER'00)*, pp. 96–111. Springer, 2000.
10. J. Chomicki and P. Revesz. Constraint-based interoperability of spatiotemporal databases. *GeoInformatica*, 3(3):211–243, 1999.
11. J. Chomicki and P. Revesz. A geometric framework for specifying spatiotemporal objects. In *Proceedings of the 6th International Workshop on Temporal Representation and Reasoning (TIME'99)*, pp. 41–46. IEEE, 1999.
12. B. Clarke. A calculus of individuals based on 'connection'. *Notre Dame Journal of Formal Logic*, 22(3):204–218, 1981.
13. CLEMENTINE, http://www.spss.com/clementine/.
14. E. Clementini, P.D. Felice, and P. van Oosterom. A small set of formal topological relationships for end-user interaction. In *Proceedings of the 3rd International Symposium on Advances in Spatial Databases (SSD'93)*, pp. 277–295. Springer, 1993.
15. N.V. de Weghe. *Representing and Reasoning about Moving Objects: A Qualitative Approach*. PhD thesis, Ghent University, Belgium, 2004.
16. N.V. de Weghe, A. Cohn, G. de Tré, and P.D. Maeyer. A qualitative trajectory calculus as a basis for representing moving objects in geographical information systems. *Control and Cybernetics*, 35(1):97–120, 2006.
17. N.V. de Weghe, A. Cohn, P.D. Maeyer, and F. Witlox. Representing moving objects in computer based expert systems: The overtake event example. *Expert Systems with Applications*, 29(4):977–983, 2005.
18. N.V. de Weghe, G.D. Tré, B. Kuijpers, and P.D. Maeyer. The double-cross and the generalization concept as a basis for representing and comparing shapes of polylines. In *Proceedings of the International Workshop on Semantic-based Geographical Information Systems (SeBGIS'05)*, pp. 1087–1096. Springer, 2005.
19. S. Dzeroski. Multi-relational data mining: An introduction. *SIGKDD Exploration Newsletter*, 5(1):1–16, 2003.
20. M. Egenhofer and R. Franzosa. Point-set topological spatial relations. *International Journal of Geographical Information Systems*, 5(2):161–174, 1991.

21. M. Egenhofer and R. Golledge. *Time in Geographic Space, Report on the Specialist Meeting of Research Initiative 10*. Technical Report 94-9, National Center for Geographic Information and Analysis, Univeristy of California, 1994.
22. M.J. Egenhofer. Reasoning about binary topological relations. In *Proceedings of the International Symposium Advances in Spatial Databases (SSD'91)*, pp. 143–160. Springer, 1991.
23. M. Erwig, R.H. Güting, M. Schneider, and M. Vazirgiannis. Spatiotemporal data types: An approach to modeling and querying moving objects in databases. *GeoInformatica*, 3(3):269–296, 1999.
24. M. Erwig and M. Schneider. The honeycomb model of spatiotemporal partitions. In *Proceedings of the International Workshop on Spatiotemporal Database Management (STDBM99)*, pp. 39–59. Springer, 1999.
25. M. Erwig and M. Schneider. Spatiotemporal predicates. *IEEE Transactions on Knowledge and Data Engineering*, 14(4):881–901, 2002.
26. A. Frank. Qualitative spatial reasoning: Cardinal directions as an example. *International Journal of Geographic Information Systems*, 10(3):269–290, 1996.
27. A. Frank, S. Grumbach, R. Güting, C. Jensen, M. Koubarakis, N. Lorentzos, Y. Manopoulos, E. Nardelli, B. Pernici, H.-J. Schek, M. Scholl, T. Sellis, B. Theodoulidis, and P. Widmayer. CHOROCHRONOS: A research network for spatiotemporal database systems. *SIGMOD Record*, 28:12–21, 1999.
28. C. Freksa. Using orientation information for qualitative spatial reasoning. In *Spatiotemporal Reasoning*, Vol. 639. *Lecture Notes in Computer Science*, pp. 162–178. Springer, 1992.
29. A. Galton. Towards a qualitative theory of movement. In *Spatial Information Theory*, pp. 377–396, 1995.
30. F. Giannotti, G. Manco, and F. Turini. Specifying mining algorithms with iterative user-defined aggregates. *IEEE Transactions on Knowledge and Data Engineering*, 16(10):1232–1246, 2004.
31. R. Goyal. *Similarity Assessment for Cardinal Directions Between Extended Spatial Obejcts*. PhD thesis, The University of Maine, 2000.
32. T. Griffiths, A.A.A. Fernandes, N.W. Paton, and R. Barr. The tripod spatio-historical data model. *Data Knowledge and Engineering*, 49(1):23–65, 2004.
33. S. Grumbach, P. Rigaux, M. Scholl, and L. Segoufin. The DEDALE prototype. In *Constraint Databases*, pp. 365–382. Springer, 2000.
34. S. Grumbach, P. Rigaux, and L. Segoufin. On the orthographic dimension of constraint databases. In *Proceedings of the 7th International Conference on Database Theory (ICDT'99)*, pp. 199–216. Springer, 1999.
35. S. Grumbach, P. Rigaux, and L. Segoufin. Spatiotemporal Data Handling with Constraints. *GeoInformatica*, 5(1):95–115, 2001.
36. R. Güting, M. Böhlen, M. Erwig, C. Jensen, N. Lorentzos, M. Schneider, and M. Vazirgiannis. A foundation for representing and querying moving objects. *ACM Transactions on Database Systems*, 25(1):1–42, 2000.
37. R. Güting and M. Schneider. Realm-based spatial data types: The ROSE algebra. *The Very Large Data Bases Journal*, 4(2):243–286, 1995.
38. R. Güting and M. Schneider. *Moving Object Databases*. Morgan Kaufmann, 2005.
39. S. Haesevoets. *Modelling and Querying Spatiotemporal Data*. Doctor's thesis, Hasselt University, 2005.
40. J. Han, Y. Fu, W. Wang, K. Koperski, and O. Zaiane. DMQL: A data mining query language for relational databases. In *Proceedings of the Workshop on Research Issues in Data Mining and Knowledge Discovery (DMKD'96)*, 1996.
41. T. Imielinski and H. Mannila. A database perspective on knowledge discovery. *Communications ACM*, 39(11):58–64, 1996.
42. T. Imielinski and A. Virmani. MSQL: A query language for database mining. *Data Mining and Knowledge Discovery*, 3(4):373–408, 1999.
43. T. Johnson, L. Lakshmanan, and R. Ng. The 3W model and algebra for unified data mining. In *Proceedings of the International Conference on Very Large Data Bases (VLDB'00)*, pp. 21–32, 2000.

44. K. Koperski, J. Adhikary, and J. Han. Spatial data mining: Progress and challenges. Survey paper. In *Proceedings of the Workshop on Research Issues in Data Mining and Knowledge (DMKD'96)*, 1996.
45. M. Koubarakis, T.K. Sellis, A.U. Frank, S. Grumbach, R.H. Güting, C.S. Jensen, N.A. Lorentzos, Y. Manolopoulos, E. Nardelli, B. Pernici, H.-J. Schek, M. Scholl, B. Theodoulidis, and N. Tryfona (eds.). *Spatiotemporal Databases: The CHOROCHRONOS Approach*. Springer, 2003.
46. B. Kuijpers, J. Paredaens, and D.V. Gucht. Towards a theory of movie database queries. In *Proceedings of the 7th International Workshop on Temporal Representation and Reasoning (TIME'00)*, pp. 95–102. IEEE, 2000.
47. S. Lee and L.D. Raedt. An algebra for inductive query evaluation. In *Proceedings of the International Conference on Data Mining (ICDM'03)*, pp. 147–154. IEEE, 2003.
48. L. Libkin. Some remarks on variable independence, closure, and orthographic dimension in constraint databases. *SIGMOD Record*, 28(4):24–28, 1999.
49. D. Malerba, A. Appice, and M. Ceci. A data mining query language for knowledge discovery in a geographical information system. In *Database Support for Data Mining Applications*, pp. 95–116. 2004.
50. D. Malerba, F. Esposito, A. Lanza, F. Lisi, and A. Appice. Empowering a GIS with inductive learning capabilities: The case of INGENS. *Journal of Computers, Environment, and Urban Systems*, 27:265–281, 2003.
51. P. Mancarella, A. Raffaetà, C. Renso, and F. Turini. Integrating knowledge representation and reasoning in geographical information systems. *International Journal of Geographical Information Science*, 18(4):417–446, 2004.
52. H. Mannila and H. Toivonen. Levelwise search and border of theories in knowledge discovery. *Data Mining and Knowledge Discovery*, 1(3):241–258, 1997.
53. R. Meo, G. Psaila, and S. Ceri. An extension to SQL for mining association rules. *Data Mining and Knowledge Discovery*, 2(2):195–224, 1998.
54. T. Mitchell. *Machine Learning*. Mc Graw-Hill, 1997.
55. H. Mokhtar, J. Su, and O.H. Ibarra. On moving object queries. In *Proceedings of the 21st Symposium on Principles of Database Systems (PODS'02)*, pp. 188–198. ACM, 2002.
56. M. Nanni, F.T.A. Raffaetà, and C. Renso. A declarative framework for reasoning on spatiotemporal data. In *Spatiotemporal Databases. Flexible Querying and Reasonig*, pp. 75–104. Springer, 2004.
57. R. Ng, L. Lakshmanan, J. Han, and A. Pang. Exploratory mining and pruning optimization of constrained association rules. In *Proceedings of the Conference on Management of Data (SIGMOD'98)*, pp. 13–24, 1998.
58. Object Database Management Group. www.odmg.org.
59. OLE DB DM Specifications, http://www.microsoft.com/data/oledb/dm/.
60. OpenGIS Simple Features Specification For OLE/COM. The file can be downloaded at http://www.opengis.org/techno/specs/99-050.pdf.
61. D. Papadias and Y. Theodoridis. Spatial relations, minimum bounding rectangles, and spatial data structures. *International Journal of Geographic Information Science*, 11(2):111–138, 1997.
62. C. Parent, S. Spaccapietra, and E. Zimányi. Spatiotemporal conceptual models: Data structures + space + time. In C.B. Medeiros, (ed.), *Proceedings of the 7th International Workshop on Geographic Information Systems (GIS'99)*, pp. 26–33. ACM, 1999.
63. C. Parent, S. Spaccapietra, and E. Zimányi. The MurMur project: Modeling and querying multi-representation spatiotemporal databases. *Information Systems*, 31(8):733–769, 2006.
64. J. Pei and J. Han. Can we push more constraints into frequent pattern mining? In *Proceedings of the Conference on Knowedge Discovery and Data Mining (KDD'00)*, pp. 350–354. ACM, 2000.
65. J. Pei, J. Han, and L. Lakshmanan. Mining frequent itemsets with convertible constraints. In *Proceedings of the International Conference on Data Engineering (ICDE'01)*, pp. 433–442. IEEE, 2001.

66. N. Pelekis. *STAU: A Spatiotemporal Extension for the ORACLE DBMS*. PhD Thesis, UMIST, 2002.
67. N. Pelekis, Y. Theodoridis, S. Vosinakis, and T. Panayiotopoulos. Hermes – A framework for location-based data management. In *Proceedings of the International Conference on Extending Database Technology (EDBT'06)*, pp. 1130–1134. Springer, 2006.
68. D. Peuquet and Z. Ci-Xiang. An algorithm to determine the directional relationship between arbitrarily-shaped polygons in the plane. *Pattern Recognition*, 20(1):65–74, 1987.
69. L.D. Raedt. A logical database mining query language. In *Proceedings of the International Conference on Inductive Logic Programming (ILP'00)*, pp. 78–92. Springer, 2000.
70. L.D. Raedt, M. Jaeger, S. Lee, and H. Mannila. A theory of inductive query answering. In *Proceedings of the International Conference on Data Mining (ICDM'02)*, pp. 123–130. IEEE, 2002.
71. A. Raffaetà and T. Frühwirth. Spatiotemporal annotated constraint logic programming. In *Proceedings of the International Symposium on Practical Aspects of Declarative Languages (PADL'01)*, pp. 259–273. Springer, 2001.
72. A. Raffaetà, C. Renso, and F. Turini. Qualitative spatial reasoning in a logical framework. In *Proceedings of the 8th Congress of the Italian Association for Artificial Intelligence on Advances in Artificial Intelligence (AI*IA'03)*, pp. 78–90. Springer, 2003.
73. D. Randell, Z. Cui, and A. Cohn. A spatial logic based on regions and connection. In *Proceedings of the International Conference on Knowledge Representation and Reasoning (KR'92)*, pp. 165–176. Morgan Kaufmann, 1992.
74. P. Revesz. *Introduction to Constraint Databases*. Springer, 2002.
75. P. Revesz and M. Cai. Efficient querying and animation of periodic spatiotemporal databases. *Annals of Mathematics and Artificial Intelligence*, 36(4):437–457, 2002.
76. A. Romei, S. Ruggieri, and F. Turini. KDDML: A middleware language and system for knowledge discovery in databases. *Data Knowledge and Engineering*, 57(2):179–220, 2006.
77. A.P. Sistla, O. Wolfson, S. Chamberlain, and S. Dao. Modeling and querying moving objects. In *Proceedings of the 13th International Conference on Data Engineering*, pp. 422–432. IEEE, 1997.
78. S. Spaccapietra, (ed.). *Spatiotemporal Data Models and Languages (DEXA'99)*. IEEE, 1999.
79. J. Su, H. Xu, and O.H. Ibarra. Moving objects: Logical relationships and queries. In *Proceedings of the 7th International Symposium on Advances in Spatial and Temporal Databases (SSTD'01)*, pp. 3–19. Springer, 2001.
80. G. Trajcevski, O. Wolfson, S. Chamberlain, and F. Zhang. The geometry of uncertainty in moving objects databases. In *Proceedings of the International Conference on Extending Database Technology (EDBT'02)*, pp. 233–250. Springer, 2002.
81. H. Wang and C. Zaniolo. ATLaS: A native extension of sql for data mining. In *Proceedings of the SIAM Conference on Data Mining (SDM'03)*, 2003.
82. WEKA, http://www.cs.waikato.ac.nz/ml/weka/.
83. M. Worboys. A unified model for spatial and temporal information. *Computer Journal*, 37:26–34, 1994.
84. M.F. Worboys and M. Duckham. *GIS – A Computing Perspective, 2nd Edition*. CRC Press, 2004.
85. C. Zaniolo, N. Arni, and K. Ong. Negation and aggregates in recursive rules: The LDL++ Approach. In *Proceedings of International Conference on Deductive and Object-Oriented Databases (DOOD'93)*, pp. 204–221. Springer, 1993.
86. K. Zimmermann and C. Freksa. Qualitative spatial reasoning using orientation, distance, and path knowledge. *Applied Intelligence*, 6(1):49–58, 1996.

Chapter 13
Visual Analytics Methods for Movement Data

G. Andrienko, N. Andrienko, I. Kopanakis, A. Ligtenberg, and S. Wrobel

13.1 Introduction

All the power of computational techniques for data processing and analysis is worthless without human analysts choosing appropriate methods depending on data characteristics, setting parameters and controlling the work of the methods, interpreting results obtained, understanding what to do next, reasoning, and drawing conclusions. To enable effective work of human analysts, relevant information must be presented to them in an adequate way. Since visual representation of information greatly promotes man's perception and cognition, visual displays of data and results of computational processing play a very important role in analysis.

However, a simple combination of visualization with computational analysis is not sufficient. The challenge is to build analytical tools and environments where the power of computational methods is synergistically combined with man's background knowledge, flexible thinking, imagination, and capacity for insight. This is the main goal of the emerging multidisciplinary research field of Visual Analytics (Thomas and Cook [45]), which is defined as the science of analytical reasoning facilitated by interactive visual interfaces.

Analysis of movement data is an appropriate target for a synergy of diverse technologies, including visualization, computations, database queries, data transformations, and other computer-based operations. In this chapter, we try to define what combination of visual and computational techniques can support the analysis of massive movement data and how these techniques should interact. Before that, we shall briefly overview the existing computer-based tools and techniques for visual analysis of movement data.

G. Andrienko
Fraunhofer Institut Intelligente Analyse- und Informationssysteme, Sankt Augustin, Germany, e-mail: gennady.andrienko@iais.fraunhofer.de

13.2 State of the Art

13.2.1 Visualization Fundamentals

In a strict sense, visualization is representation of data in a visual form, i.e., creating various pictures from data: graphs, plots, diagrams, maps, etc. For this purpose, items of data are translated into graphical features, such as positions within a display, colors, sizes, or shapes. For the visualization to be effective, the translation is done according to the established principles and rules (see, for example, Bertin [6] or a summary in Andrienko and Andrienko [2], Sect. 4.3). Thus, numeric data should be encoded by positions or sizes, while color hues, shapes, and texture patterns are more suitable for qualitative data.

Ben Shneiderman [41] summarized the process of data exploration by means of visualization in the well-known information seeking mantra: "Overview first, zoom and filter, and then details-on-demand." For supporting an overall view of a data set, it is necessary to visualize the data so that all visual elements representing data items could be perceived together as a single image (Bertin [6]). For further data exploration, visual displays need to be complemented with interactive tools for zooming, filtering, and accessing various details, or "drilling down" into the data (e.g., Buja et al. [7]).

All interactive tools need to be carefully designed for maximum user convenience and effectiveness of the exploration process. Direct manipulation methods, when the user interacts directly with a visual display, are highly recommendable. Mouse-operated widgets such as sliders and switches are also appropriate. Response time may be a critical issue in implementing interactive tools. It is desirable that the computer responds to an interactive operation within 50 ms or at the most 100 ms; the user perceives this as an instantaneous response. However, in case of a very large data set, reaching such responsiveness may be extremely problematic.

When data have a complex structure (as, in particular, movement data, which involve space, time, population of entities, and a number of numeric and qualitative characteristics), they cannot be adequately visualized in a single display. Therefore, the use of multiple displays providing different perspectives into the data is important. The displays should be linked so that the information contained in individual views can be integrated into a coherent image of the data as a whole (Buja et al. [7]). The most popular method for linking parallel views is identical marking of corresponding parts of multiple displays, e.g., with the same color or some other form of highlighting. Usually, highlighting is applied to objects interactively selected by the user in one of the displays. This method, usually called "brushing," is a generalization of the "scatterplot brushing" technique first implemented by Newton [38] and later elaborated in various directions.

The idea of brushing is illustrated in Fig. 13.1. Five displays show different aspects of the same data: about movements of white storks in the course of their

13 Visual Analytics Methods for Movement Data

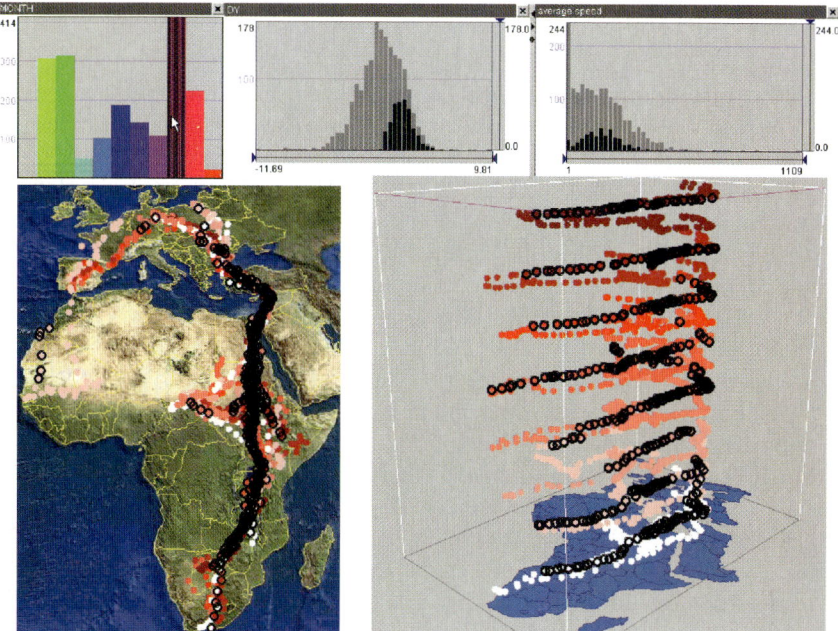

Fig. 13.1 An illustration of the technique of brushing between several parallel views of the same data. The illustration has been produced using the system CommonGIS. The satellite telemetry data about the seasonal migration of white storks have been collected in Vogelwarte Radolfzell, a department of the Max Planck Institute for Ornithology, Germany

seasonal migration during eight seasons from 1998 to 2006. The bar chart in the upper left corner represents the distribution of the movements by months of a year. The user has clicked on the highest bar, which corresponds to March. As a result, the data records about the movements that occurred in March have received a special status of selected records. All displays have reacted to this by marking the graphical elements corresponding to the selected records, as in the map and space–time cube in the lower part of the figure, or showing the positions and proportions of the selected data with regard to the whole data set, as in the two histograms on the top.

Current approaches to display linking are described, e.g., by North and Shneiderman [39], Roberts [40], Baldonado and Woodruff [5].

Animated displays are often considered as the first choice when data involve time (Eick [14]). However, psychological studies show that animation is not necessarily effective and superior to static displays (Tversky et al. [48]). It seems that animation is good for gaining an initial overview of a time-related phenomenon or process while the further, more comprehensive exploration requires combination of animation with other displays and rich facilities for user interaction.

13.2.2 Visualization of Individual Movement Data

The early visualizations of movements on maps or in space–time cubes were produced manually, which was a laborious and time-consuming process. Computers and graphical display facilities not only simplified and expedited the work but also provided new opportunities, in particular, dynamics and possibility of user interaction with a display. Nowadays, animated maps [3, 4] and interactive cubes [25, 29] are widely used to visualize movement data. Map and cube displays are complemented with graphs and diagrams exhibiting various aspects of the movement [13, 24, 29, 34–37]. One example is time–time plot or T–T plot [24], which has two time axes and represents changes of a certain characteristic of the movement, such as the speed, traveled distance, or direction, between the moments t_x and t_y by placing symbols in the positions corresponding to x and y or by coloring or shading the cells, in which the plot area is divided.

A comprehensive research on methods for exploration of individual movement behaviors has been conducted in London City University [13, 34–37]. A specific focus of the researchers is the very long trajectories, which require the use of data aggregation. Temporal aggregation occurs in a temporal histogram, which shows the number of visited locations by time intervals. Spatial aggregation is done by imposing a regular grid over the territory and counting trajectory points fitting in each cell. The resulting densities are visually represented by coloring or shading the grid cells on a map display. Densities counted for consecutive time intervals can be shown on an animated map display. A grid with densities can be treated as a surface, which may contain various features such as peaks (maxima), pits (minima), channels (linear minima), ridges (linear maxima), and saddles (channels crossing ridges). There are computational methods for detecting such features, which can then be visualized on a map.

In addition to the density surface, surfaces representing other movement-related characteristics may be built as suggested by Mountain [34]. Thus, an isochrone surface is a series of concentric polygons, centered on a selected location, representing the areas accessible from this location within specified "time budgets," e.g., 3, 6, 9 min, and so on. An accessibility surface is a grid where each cell represents the travel time from the selected location.

Laube et al. [32] represent movement behaviors of several entities such as football players in a matrix where the columns correspond to time intervals and the rows to the moving entities. Symbols or coloring in the cells of the matrix encode average characteristics of the movement of the entities, such as the speed or direction, on each of the intervals. Similarity of rows in such a matrix indicates that the respective entities have similar movement behaviors. The matrix is good for detecting certain types of patterns of collective movement, for example, "trend setting," when a group of entities repeats the movements of one entity after some time lag. However, with increase in the number of entities and the duration of the movement, the visual search for patterns becomes more and more difficult. It should be noted that visualization of movement data has not been the main research focus of Laube et al. The

13.2.3 Visualization of Movements of Multiple Entities

Analogous to long-time series of movement data, the visualization of movements of numerous entities requires data aggregation or other ways of summarization. Buliung and Kanaroglou [8] describe an approach where computational methods available in ArcGIS are applied to multiple trajectories. First, a convex hull containing all the trajectories is built. Then, the central tendency and the dispersion of the paths are computed and represented on a map. This method, however, is only suitable when the trajectories are sufficiently close to each other. When multiple entities synchronously move in the same direction, the visualization described by Wilkinson [49] may be appropriate, where the northerly migration of Monarch butterflies is shown on a map by "front lines" corresponding to different times. Again, this is a very special case.

Forer and Huismann [17] aggregate movement data into a surface by computing the total number of person-minutes spent in each cell of a regular grid. In a similar way, many other characteristics of multiple movements may be summarized and visualized. Kwan and Lee [31] build surfaces of summary characteristics of movements not in the geographical space but in an abstract space where the dimensions are the time of day and the distance from home. For this purpose, they use kernel density estimation methods. Such surfaces can be built for different groups of entities in order to compare their behaviors. Pairwise differences between surfaces can be computed and visualized.

Unfortunately, summarization of movement data into surfaces severely alters their nature so that one can no longer see the changes of the spatial positions of the entities, i.e., the very essence of movement. To preserve the information about changes of positions, the data need to be aggregated in a different way. A possible approach is to count for each pair of locations (points or areas) in space how many entities moved from the first to the second location between two time moments. The resulting counts may be visualized as a transition matrix where the rows and columns correspond to the locations and symbols in the cells, or cell coloring or shading encode the counts [20]. For more than one pair of time moments, one would need to build several transition matrices, which could be then compared. However, the limitations of this approach with respect to the length of the time series of movement data are evident. Another problem is that such a visualization lacks the spatial context. Some part of the spatial information may be preserved through ordering the spatial locations in the matrix in such a way that locations closer in space are also closer in the ordering. Guo and Gahegan [21] have made a survey of the existing methods applicable for this purpose.

Tobler [46, 47] suggests that numbers of entities or volumes of materials that moved from one place to another can be visualized by means of either discrete or

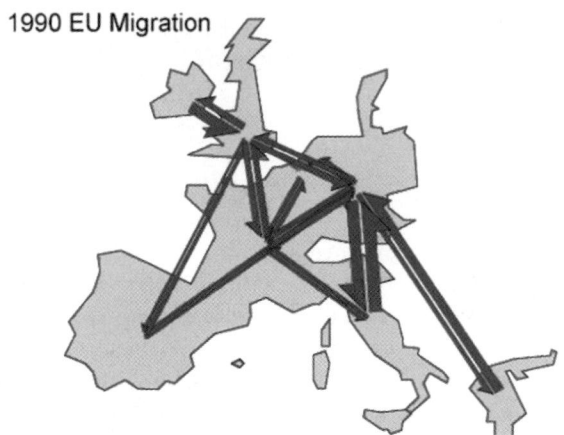

Fig. 13.2 A discrete flow map (Tobler [46, 47])

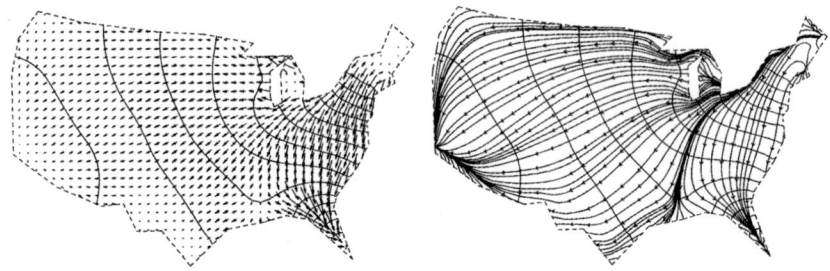

Fig. 13.3 Continuous flow maps (Tobler [46, 47])

continuous flow maps. A discrete map represents the movements by bands or arrows whose width is proportional to the volume moved (Fig. 13.2). Omitting minor flows increases map legibility when the number of locations is large. Continuous flow maps use vector fields or streamlines to show continuous flow patterns (Fig. 13.3). According to Tobler, the structure is immediately obvious in a vector field: adjacent vectors clearly being correlated in length and direction. Conversely, if this is not the case then it is also obvious. Continuous flow maps are, in principle, not limited with regard to the number of different locations present in the original data. However, producing such maps from discrete data is computationally intensive.

Tobler's flow maps do not reflect the temporal dimension of movement data but show cumulative movements that occurred during a certain time period. However, the concept can be extended to animated flow maps or to series of flow maps showing how the flows change over time.

Cartographers Drecki and Forer [12] have designed a very interesting visualization of aggregated movement data, specifically, movements of tourists coming to New Zealand (Fig. 13.4). They first transformed the travel times of the tourists from the absolute time scale (calendar dates) to a relative one, starting from the day of tourist's arrival to New Zealand. Then the cartographers built a diagram consisting

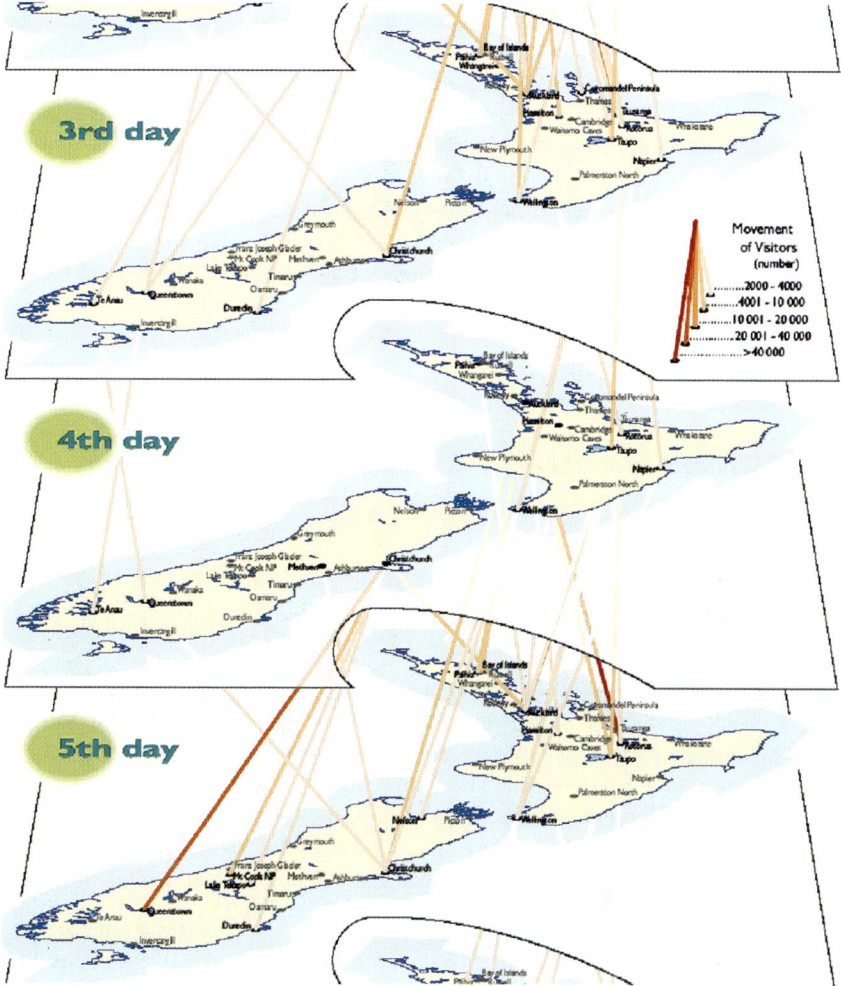

Fig. 13.4 A fragment of the visualization of the movements of tourists in New Zealand created by Drecki and Forer [12]

of six parallel planes, shown in a perspective view with a map of New Zealand depicted on each plane. The planes, from top to bottom, correspond to the first six days of the tourists' travel through New Zealand. The movements of the tourists are shown by lines connecting the locations of the major tourist destinations on successive planes. The brightness of a line corresponds to the number of people who moved from its origin location (on the upper plane) to the destination location (on the lower plane) between the days corresponding to the upper and lower planes. To make the view clearer, the authors have omitted minor flows. The visualization was designed for printing on paper. To our best knowledge, there are yet no software tools providing this kind of display for exploration of arbitrary movement data.

While software tools for data visualization are usually supplied with appropriate user interaction facilities for view manipulation (e.g., zooming or rotation), filtering, querying display elements, selection, etc., there are approaches supposing that the user interacts with data prior to the visualization. The idea is that the user selects a data subset of interest from a (possibly, very large) database, and only this subset is visualized. It is assumed that the size of the subset permits its visualization without the use of aggregation. Kapler and Wright [25] and Yu [50] apply this approach to movement data. Kapler and Wright suggest an ontology of movement data to support querying and search for information in the database. In particular, the user may consider the information on different levels of detail. In the system described by Yu, the user may formulate queries by referring to entities, their activities, and spatiotemporal relationships, specifically, colocation in space, colocation in time, and coexistence, i.e., colocation in both space and time.

It should be noted that the approaches based on selection and visualization of small data subsets do not support an overall view of the collective behavior of all entities and, hence, are insufficient for visual analysis of movement data.

13.2.4 Challenges

From the survey of the state of the art it may be seen that the research on methods for visual analysis of movement data did not reach yet its maturity. Most of the techniques and tools are not suitable for analyzing data about many entities moving during long time periods. The limitations, which are recognized by many researchers [31, 33], come both from the side of the hardware (much computation time required, low speed of rendering, insufficient display size and resolution, etc.) and from the side of the user (display illegibility, perceptual and cognitive overload, difficulties in interpretation of unfamiliar visualizations and in operating complex visualization environments). Hence, further research is required for finding ways to overcome these limitations.

Another problem is that each technique or tool allows one to consider movement data from a particular angle, while the data are multifaceted and influenced by numerous factors, from characteristics of the moving entities to properties of the environment and various phenomena and processes occurring in it (see Chap. 1 in this volume). For a comprehensive analysis, several tools need to be combined. The selection of appropriate tools and methods should be based on a careful consideration of the needs of potential users (i.e., analysts of movement data) as well as their capabilities and limitations.

In the following sections, we try to apply a systematic approach to the selection and design of visual analytics methods for movement data based on the consideration of possible analysis questions the users may have, on the one hand, and the established principles of visual presentation of information, on the other hand (coming from the best practices in visual representation of information, these principles take into account, explicitly or implicitly, the perceptual and cognitive capabilities of humans).

13.3 Patterns in Movement Data

Chapter 1 in this volume defines the types of possible analytical questions about moving entities and stresses the primacy of synoptic questions, which involve multiple time moments and/or multiple entities considered all together. It introduces a generic concept of behavior, the meaning of which embraces such notions as the trajectory of a single entity over a time period, the distribution of multiple entities in space at some time moment, and the collective movement of multiple entities in space over a time period. The primary objective in analyzing movement data is to understand and characterize the movement behavior of the entire population of moving entities over the whole time period the data refer to. On this basis, one can pursue further goals such as prediction of the future behavior or optimization of the movement.

Visual analytics mostly addresses the stage of gaining understanding and characterization of behaviors. The objective of visual analytics in application to movement data may be stated as follows:

Allow a human analyst (also referred to as "the user") to understand and characterize the movement behavior of a population of entities with the help of interactive visual displays, which are properly combined with other kinds of tools for analysis.

"Understand and characterize" a behavior means represent it by an appropriate *pattern*. A pattern may be viewed as a statement in some language [16]. The language may be chosen quite arbitrarily (e.g., natural language, mathematical formulas, graphical language); hence, the syntactic and morphological features of a pattern are irrelevant to data analysis. What is relevant is the meaning or semantics. It is natural to assume that representations of the same behavior in different languages have a common meaning. Hence, the constructs of the different languages refer to the same system of basic language-independent elements from which various meanings can be composed. By analogy with meanings of words in a natural language, we can posit that the basic semantic elements for building various patterns include *pattern types* and *pattern properties*. A specific pattern is an *instantiation* of one or more pattern types. This is analogous to the specialization of a general notion by means of appropriate qualifiers. In the case of patterns, the qualifiers are specific values of the pattern properties. For example, the pattern "entities e_1, e_2, ..., e_n moved together during the time period $\mathbf{T}$" instantiates the pattern type "joint movement" by specifying what entities and when moved in this manner.

It is quite reasonable to assume that the possible pattern types exist in the mind of a data analyst as mental schemata. Moreover, these schemata are likely to drive the process of visual data analysis, which is generally believed to be based on pattern recognition: the analyst looks for constructs that may be viewed as instantiations of the known pattern types. Therefore, for the design of proper visual analytics methods for movement data, it is important to define the pattern types relevant to such data.

13.3.1 Generic Pattern Types

On a very general level, pattern types are introduced in the book by Andrienko and Andrienko [2]. Descriptive patterns, which characterize behaviors, are distinguished from connectional patterns, which characterize relations between phenomena (see Chap. 1). The basic types of descriptive patterns are similarity, difference, and arrangement, where the latter type embraces such concepts as trend, sequence, periodicity, symmetry, etc. From instances of the basic pattern types, compound patterns are built as is shown graphically in Fig. 13.5.

The types of connectional patterns are *correlation* (which is treated in a more general sense than just statistical correlation between numeric variables and includes co-occurrence of qualitative characteristics and co-occurrence of behavioral patterns), *influence* (or dependency, if viewed in the opposite direction), and *structure*, i.e., composition of a complex behavior from simpler ones, like the visible movement of planets is a composition of their own movement and the movement of the Earth.

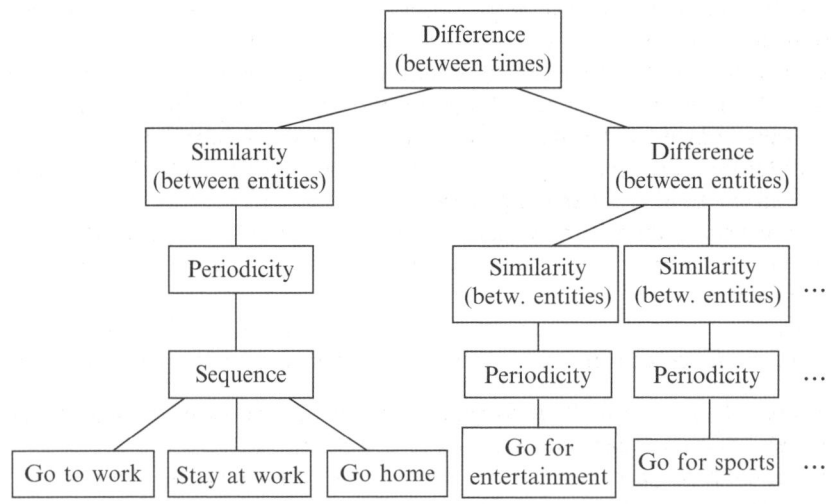

Fig. 13.5 An illustration of a compound pattern with several levels of nested sub-patterns

13.3.2 Descriptive Pattern Types for Movement Data

Let us now specialize these generic types of patterns for movement data. Our ultimate goal is to define pattern types for collective movement behaviors of multiple entities. To achieve this, it is necessary to consider the following "slices" or "projections" of this overall behavior:

- Individual movement behavior, i.e., movement of a single entity over time
- Distribution of movement characteristics (position, speed, direction, etc.) over the set of entities at a single time moment. For the sake of brevity, we shall call it "momentary collective behavior"

We shall use the following abbreviations: *IMB* for individual movement behavior, *MCB* for momentary collective behavior, and *DCB* for dynamic collective behavior, i.e., behavior of multiple entities during a time interval. A DCB can be viewed from two different perspectives:

- As a construct formed from the IMBs of all entities, i.e., the behavior (variation) of the IMB over the set of entities
- As a construct formed from the MCBs at all time moments, i.e., the behavior (variation) of the MCB over time

These two views may be called aspectual behaviors [2]. They are essentially different and need to be described in terms of different types of patterns.

The variation of the IMB over the set of entities can be described by means of similarity and difference patterns, i.e., as groups of entities having similar IMBs, which differ from the IMBs in other groups of entities. For example, the weekday movement patterns of working people may be considered as similar. At the same time, they differ from the movement patterns of housewives and pensioners. It may happen that some entities have quite peculiar IMBs, which differ from the IMBs of all other entities. For instance, the movement behavior of a tourist in a town may differ from the behaviors of the town residents. Such peculiar IMBs are also described by means of difference patterns.

Arrangement patterns are usually not relevant to the behavior of the IMB over the set of entities because this set has no natural ordering and no distances between the elements [2].

What does it mean that IMBs of several entities are similar? There are diverse possible meanings, and all of them may be relevant to analysis of movement data:

- Similarity of the overall characteristics: geometric shapes of the trajectories, traveled distances, durations, movement vectors, etc.
- Colocation in space, i.e., the trajectories of the entities consist of the same positions or have some positions in common
  - Ordered colocation: the common positions are attained in the same order

- Order-irrelevant colocation: the common positions may be attained in different orders
- Symmetry: the common positions are attained in the opposite orders

- Synchronization in time
  - Full synchronization: similar changes of movement characteristics occur at the same times
  - Lagged synchronization: changes of the movement characteristics of entity e_1 are similar to changes of the movement characteristics of entity e_0 but occur after a time delay Δt

- Coincidence in space and time
  - Full coincidence: same positions are attained at the same times
  - Lagged coincidence: entity e_1 attains the same positions as entity e_0 but after a time delay Δt

It should be noted that two or more IMBs may be similar during one time interval and dissimilar during another interval. Similarity and difference patterns may thus be applied not only to whole IMBs but also to their parts.

Let us now consider the other aspectual behavior, that is, the behavior of the MCB over time. Mathematically, time is a continuous set where ordering and distances exist between the elements, i.e., time moments. Hence, besides similarity and difference patterns, arrangement patterns are relevant. An arrangement pattern describes changes in the MCB with respect to the ordering and distances between the corresponding time moments. Here are the pattern types for describing the behavior of the MCB over time (we note in parentheses the basic pattern types that have been specialized):

- Constancy (similarity): the MCB was the same or changed insignificantly during a time interval. For example, massive traffic towards industrial areas is observed during a time interval from 7 a.m. till 9 a.m.
- Change (difference): the MCB significantly changed from moment t_1 to moment t_2. For instance, the movement in an area around a stadium after the beginning of a football game differs quite much from what could be observed before. Another abrupt change happens when the game is over
- Trend (arrangement): consistent changes of the MCB during a time interval. For example, the traffic in industrial areas tends to gradually decrease after 9 a.m. and tends to increase after 3 p.m.
- Fluctuation (arrangement): irregular changes of the MCB during an interval. Thus, the collective behavior of vehicle drivers on a busy highway may irregularly vary depending on emergence of obstacles such as a traffic accident or just a truck trying to overtake another truck
- Pattern change or pattern difference (difference): the behavior of the MCB during time interval T_1 differs from that during time interval T_2. The term "pattern change" applies when T_1 and T_2 are adjacent. For example, a decreasing traffic trend between 9 a.m. and 11 a.m. changes for constancy between 11 a.m. and

3 p.m., which, in turn, changes for an increasing trend after 3 p.m.. The term "pattern difference" applies to nonadjacent time intervals.
- Sequence (arrangement): patterns follow one another in a specific order, such as traffic increase – constant heavy traffic – traffic decrease – constant low traffic and so on
- Repetition (similarity): occurrences of the same patterns or pattern sequences on different time intervals. Thus, the traffic pattern sequences mentioned above occur every weekday
- Periodicity, or regular repetition (similarity and arrangement): occurrences of the same patterns or pattern sequences on regularly spaced time intervals, like the weekday traffic patterns
- Symmetry (similarity and arrangement): opposite trends like increase and decrease of traffic intensity; pattern sequences where the same patterns are arranged in opposite orders, for example, heavy traffic in the morning followed by low traffic at midday in industrial areas and low traffic in the morning followed by heavy traffic at midday in touristic and shopping areas

These pattern types are relevant not only for describing the variation of the MCB of the entire population but also to characterize movements of population subgroups. Thus, it seems reasonable to describe separately the variation of the MCB of car drivers and that of pedestrians.

13.3.3 Connectional Patterns

The types of *correlation* and *influence* patterns are similar since the relation of influence differs from the relation of correlation only in its being directed: from two related things, it is specified which one influences the other. Correlations or influences may exist between the following:

- Different movement characteristics, e.g., direction and speed
- Movement characteristics and supplementary characteristics (see Chap. 1), which include characteristics of entities, characteristics of time moments, and characteristics of spatial locations
- Individual or collective movement behaviors on different time intervals, e.g., after slow movement in a traffic jam, drivers tend to move faster than usual
- Collective movement behaviors of different subsets of entities, e.g., different teams in a football game
- Individual or collective movement behaviors and supplementary characteristics, e.g., properties of the surface
- Individual or collective movement behaviors and behaviors of external phenomena like weather, various types of events, etc.

What concerns *structure* patterns may be compositions of movement behaviors with regard to different temporal cycles like daily, weekly, and annual. For example, working people go to and from their work every day except weekends and go shopping on Saturdays. On Sundays, they usually stay at home in winter time and go to the countryside in summer time. Another example is the composition of the overall DCB in a city from the movements of the traffic, pedestrians, and cyclists, where each of the components is influenced by the others.

13.3.4 Pattern Properties

When a user detects, for example, a pattern of similarity of IMBs of multiple entities, he/she is interested to know how many entities have this common behavioral pattern. Likewise, when the user detects a pattern of synchronization or a trend, he/she measures the duration of the time interval when the entities moved synchronously or the trend lasted. These properties of the patterns taken as examples may be generalized as *support base*, that is, the size of the reference set on which this pattern takes place. Hence, for a pattern describing movements of multiple entities, the support base is the size of the subset of entities (i.e., the number of entities), and for a pattern describing movement on a time interval, the support base is the size (length) of the interval. Logically, the support base of a pattern describing movements of multiple entities during a time interval includes both the number of entities and the length of the interval. Besides the absolute support base, an important property is the *relative support base*, i.e., the size of the reference subset where the behavior corresponds to the pattern in relation to the size of the whole reference set.

Not only is the length of the time interval of a pattern interesting for a user but also the *temporal localization*, i.e., the position of the interval on the time scale. Likewise, it may be interesting to know which particular entities behave according to a pattern, in addition to the number of such entities. However, this pattern property may not always be accessible either because of a very large number of entities or because of privacy constraints.

Besides these general properties, there are also more specialized properties, which are relevant either to particular types of patterns or to characteristics involved in pattern definition. We shall not try to list exhaustively all these specialized properties but instead give a few examples. Thus, a change pattern may be characterized in terms of the magnitude and direction of the change, while a periodic pattern is characterized by the length of the period between the repetitions. An important property of a similarity pattern describing movement of multiple entities along the same route (i.e., visiting the same locations in space) is the spatial localization, i.e., where in space this common behavior takes place. The properties of a pattern describing some behavior in terms of the speed of movement include summarized speed characteristics such as average and maximum speed.

13.4 Helping Users to Detect Patterns: A Roadmap

It may be argued that the attempt to systemize and formalize movement patterns is more relevant to the development of computational analysis methods such as data mining than to visual analytics, where the detection and description of patterns involves such subjective processes as human perception and cognition. There are two major objections to this argument. First, depending on the presentation of material for human perception and cognition, these subjective processes can be either facilitated or impeded (they can even be purposefully manipulated, but this topic is out of the scope of our work). To facilitate the detection of patterns by appropriate presentation of data, we need to understand what types of patterns may exist in the data. On this basis, we can try to find visualization and interaction techniques increasing the probability of such patterns being noticed by a human viewer.

The second objection is that visual analytics is not solely data visualization but a synergistic work of human and computer supported by a synergy of visual and computational methods for data analysis. For achieving this synergy, it may be insufficient just to supply an analyst with independently developed visual and computational tools. It is more appropriate to design hybrid visuo-computational analysis methods and build corresponding tools. Knowing the types of patterns is essential for the design of such methods and tools. In particular, this may help to distribute the total analysis workload in an optimal way between the human and computer so that each side could apply its unique capabilities.

Let us give one example. A human analyst can effectively detect constancies, changes, trends, and other patterns in the variation of the MCB over time by viewing animated map displays or map series. Computer methods, at least currently existing, can hardly surpass humans in grasping characteristic features of spatial distributions and their temporal dynamics. The role of computers is to help humans with preparation of the data and with testing of hypotheses gained. However, the variation of the IMBs over a large population of entities cannot be effectively investigated without involving computational techniques such as cluster analysis. One of the reasons is that none of the known visualization techniques allows a legible representation of multiple IMBs. Another reason is that viewing of individual behaviors may be precluded for preserving the privacy of the individuals. Hence, the analyst needs to detect similarities and differences between IMBs as described in Sect. 13.3.2 without seeing the IMBs themselves. The only possibility is to develop appropriate computational methods and techniques for representing their results.

According to the focus of this chapter, we shall not further discuss the computational techniques, which are sufficiently covered in the rest of the book. We shall mainly consider visualization techniques as well as various data transformations, which may be required for effective visual exploration of movement data. In fact, the same or similar transformations may also be useful or necessary for preparing data to application of computational techniques. Moreover, it is important that visual and computational techniques are applied to the same data, either original or transformed, in order to extract complementary patterns contributing to comprehensive understanding of the data.

13.4.1 Data Manipulation

13.4.1.1 Need for Data Aggregation

In designing methods and tools for helping users to recognize various patterns, we must comply with a crucial constraint: detecting patterns must be done without seeing any information about individual entities, for preserving their privacy. In other words, only aggregated or otherwise generalized data should be available to the user. Data aggregation may be indispensable also for another reason: the number of different entities and/or time moments may be so large that the visualization of individual data becomes unfeasible because of the technical limitations (screen size and resolution) and/or impractical because of the human perceptual limitations. Hence, the role of data aggregation is both hiding individual information and reducing the amount of data.

While information reduction means substantial information loss, there is also a positive side, specifically, the possibility to omit "high-detail noise" and focus on characteristic features of the phenomenon under study. We may say that aggregation and generalization helps us to see forest for trees.

The degree of data aggregation and generalization matters a lot in data analysis. This is not only the matter of the size of the resulting data and the amount of information lost. This is also the matter of the scale on which the data are considered. Depending on the scale, the user sees the data differently and detects different patterns. Thus, in movement data, there may be local patterns like a flock (synchronous movement of several entities having close positions and same speed), or there may be larger scale patterns like massive movement towards industrial or commercial areas in mornings, or on a yet larger scale, the difference of collective movement patterns on weekdays and weekends, and so on.

Hence, the appropriate degree of data aggregation and generalization is not just a good trade-off between the simplification gained and the amount of information lost, but it must be adequate to analysis goals. When the interests of the user include patterns of different scales, it is necessary to consider the data on different levels of aggregation. The tools for visual analysis must thus enable the user to do this, as is illustrated in Fig. 13.6.

13.4.1.2 Approaches to Aggregation of Movement Data

Aggregation consists of two operations: (1) grouping of the individual data items or, in other words, division of the data into subsets and (2) deriving of characteristics of the subsets from the individual characteristics of their members. Typically, various statistical summaries are used as characteristics of the subsets: number of elements, mean, median, minimum, maximum values of characteristics, mode, percentiles, etc. It is also important to know the degree of variation of the characteristics within the aggregates. For this purposes, such statistical measures as variance (or standard deviation) or distance between the quartiles are computed. Aggregates with high

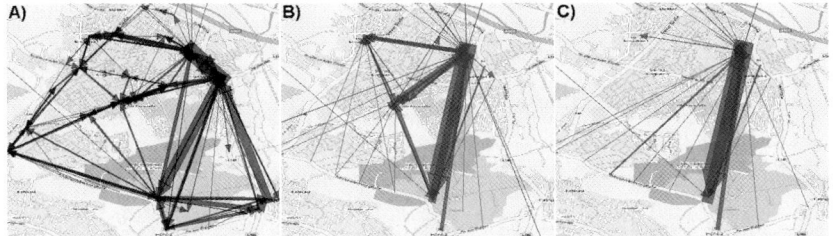

Fig. 13.6 Depending on the level of aggregation, different patterns can be observed in the same data about movements of a car. For the aggregation, the data have been divided into moves (trajectories) between stops of different duration: (**a**) for 10 s or more; (**b**) for 5 min or more; (**c**) for 2 h or more. Then, the numbers of moves between pairs of locations where the stops occurred have been counted and represented by vectors of proportional thickness. The illustration has been produced with the use of the system CommonGIS; the example data have been collected within the project GeoPKDD by GPS tracking of car positions

variation of characteristics of the members should be avoided since they may lead to wrong conclusions concerning the data. Grouping/division may be necessary not only for data aggregation but also for other kinds of data processing and analysis.

Primary items in movement data are usually tuples (records) consisting of entity identifiers, references to time moments, references to positions in space and, possibly, values of movement characteristics such as speed and direction. These microitems are typically combined into larger structures. The largest are the so-called *lifelines* of the entities where a lifeline includes all microitems referring to the same entity. Lifelines are often divided into trajectories or movement episodes. A *trajectory* is a sequence of items corresponding to a trip of an entity from one location (source) to another (destination), where the source and destination are defined semantically (e.g., home, work, shop, etc.) or according to the time the entity spends in a location. *Movement episodes* [13] are fragments of lifelines where the movement characteristics (speed, direction, sinuosity, etc.) are relatively constant, whereas a significant change indicates the beginning of the next episode. Movement episodes, trajectories, and lifelines can be viewed as macroitems of movement data. An analytical toolkit should enable the user to unite microitems into macroitems according to various criteria. Thus, in Fig. 13.6, microitems have been combined into trajectories. The sources and destinations have been chosen according to the time spent in a location. Depending on the choice of the minimum time threshold, shorter or longer trajectories are obtained.

There are also other possible methods of grouping, which can be applied either to microitems or to macroitems. Thus, microitems or short trajectories or movement episodes may be grouped according to the time of their occurrence. For this purpose, the whole time span is divided into intervals, and items occurring on the same intervals are grouped together, as in Fig. 13.9. Depending on the data and analysis goals, it may be useful to divide the time into equal intervals (e.g., 10 min, 1 h, or 1 week), or into slightly unequal intervals corresponding to calendar units such as months, quarters, or years, or to apply other division principles, for example, divide a year into semesters and holidays. Furthermore, it may be reasonable to divide the time into subsets consisting of noncontiguous intervals, in particular, according to one or more of the temporal cycles. Thus, the user may wish to group all Mondays, all Tuesdays, and so on. Hence, the data analytics toolkit should include a tool for time partitioning where the user can flexibly define the principles of division.

Another grouping principle applicable to microitems is according to the places where they occur. For this purpose, the space is divided into compartments, which may be cells of a regular grid, units of administrative or other existing territory division, or areas defined by the user according to any appropriate criteria such as surface type, way of use, accessibility, or other relevant properties of the space. The visual analytics tools should support such arbitrary divisions of the space. Thus, the user may define space compartments by interacting with a map display or by applying database search and computational operations like retrieving the locations of schools, shops, etc. and building buffer zones around them. Generally, it is not necessary that user-defined space compartments cover the whole territory, since there may be places never visited by the moving entities under analysis.

Space-based grouping is also applicable to movement episodes and trajectories, but in a different way: macroitems are grouped together if they start and/or end in the same compartments. This method of grouping has been applied in Figs. 13.6 and 13.9. The space compartments have been defined as circles encompassing the locations where the entities stopped.

It is also possible to group micro- and macroitems according to values of various attributes, including movement characteristics (speed, direction, transportation means, etc.) and characteristics of the entities (e.g., age or occupation in case of people). Since movement characteristics in macroitems are not constant but change over time, grouping can be done on the basis of values at selected time moments or on the basis of aggregated values on selected time intervals. Unfortunately, selection of each additional time moment or interval multiplies the number of groups and causes difficulties for the visualization and visual exploration of the results of the aggregation. Besides attribute values at selected time moments, macroitems can also be grouped on the basis of *changes* of the values that occurred between two time moments.

13.4.1.3 Other Data Transformations

Aggregation is not the only useful data transformation, and we shall briefly discuss some other data manipulation techniques that may increase the comprehensiveness of analysis and give additional insights into the data. One of them is the computation of the amounts or degrees and directions of changes, which is valuable not only for grouping of the entities by also by itself. Thus, it may be useful to look at change maps portraying (in a generalized manner) the changes of the MCB from one moment to another.

From other possible methods, especially useful may be transformations of space and time from absolute to relative. Thus, similarities between temporally and/or spatially separated behaviors represented by lifelines or trajectories can be more easily detected when these behaviors are somehow aligned in time and/or in space. To align behaviors in time, the "objective," absolute time of each behavior (i.e., the calendar dates and times) is ignored and only its "internal" time is considered, i.e., the time relative to the moment when this behavior began. Thus, in the representation of the tourist movement in New Zealand (Fig. 13.4), the analysts superposed the starting times of the IMBs of different tourists. It may also be useful to superpose both starting and ending times. In this case, the absolute time moments in each IMB are transformed into their distances from the starting moment divided by the durations of the behaviors (i.e., the lengths of the intervals between the starting and ending moments). This facilitates detecting similarities between movements performed with different speeds. Such an approach could be useful, for example, in comparing movements of cars and bicycles.

Analogous ideas can be applied for spatial alignment of trajectories or lifelines initially disjoint in space. A user may try to bring a set of trajectories to a common origin and search for coincidences between them. Furthermore, the user may be interested in disregarding the movement directions and considering only changes of the direction (turns). For this purpose, the trajectories are "rotated" until the initial movement directions coincide. Coincidences between further trajectory fragments indicate similarities. It may also be useful to "stretch" or "shrink" the trajectories to adjust their lengths.

In looking for colocations between trajectories where positions are specified as points in the space, it may be reasonable to apply a kind of "spatial coarsening," i.e., replace the original points by regions (areas), for example, circles with some chosen radius around the points. The resulting trajectories are treated as similar when there is an overlap between their "expanded" positions while there may be no sharp coincidence between the original positions.

In studying MCBs and their behaviors over time, it may be appropriate to treat the space as a discrete set of coarsely defined "places" rather as a continuous set consisting of dimensionless points. For this purpose, one uses the methods for space partitioning, which has been discussed before in relation to data aggregation. Such a transformation may be called "space discretization." Furthermore, it may be useful to transform the geographical space into a kind of "semantic" space consisting of such locations as home, working place, shopping site, sport facility, etc. Then, each trajectory is transformed into a sequence of movements between pairs of these locations, and the user looks for similar subsequences occurring in different trajectories.

13.4.2 Visualization and Interaction

As we have mentioned earlier, a DCB of a set of entities over a period of time involves two aspects: the variation of the IMB over the set of entities and the variation of the MCB over the time. Different types of patterns are relevant to each aspect and, hence, different tools are needed to support the detection of pattern instances.

The pattern types corresponding to the first aspect are patterns of similarity and difference between IMBs. As we have already noted, the challenge is that similarities and differences have to be detected without the user seeing the IMBs, for the reasons of privacy and data size. The only possible solution is computational search for similarities and differences, for example, with the use of clustering methods. It would be reasonable to develop a clustering tool that allows the user to specify the kind of similarity he/she is currently interested in (see Sect. 13.3.2) and uses an appropriate function for computing the degrees of similarity from a library of possible functions. The results of clustering need to be visualized so that the user could interpret and investigate them. A general approach is to display various statistics about each of the clusters, i.e., aggregated data obtained from individual characteristics of the members of the clusters.

The exploration of the second aspect of the DCB may be supported by visual displays that represent the MCB at different times. There are two generic ways to do this: display animation and display iteration. In display animation, the views of the MCB at different times are arranged temporally and presented one by one. In display iteration, these views are arranged spatially (within the space of the screen) and presented simultaneously. Animation or iteration may be applied to various types of displays, such as maps, diagrams, or graphs. In the context of this study, it is essential to use displays representing aggregated rather than individual data, for the reasons of data size and privacy.

Hence, the exploration and analysis of both aspects of the dynamic collective behavior require the visualization of aggregated movement data. The difference is only in the way the data are aggregated, through clustering or through interactive division. Therefore, the same visualization techniques may be applicable in both cases. Let us now review the methods suitable for the visualization of aggregated movement data.

13.4.2.1 Maps and Map-Based Displays

Map displays are best suited for the visualization of spatial data such as positions, directions, and trajectories as well as nonspatial data associated with positions, directions, and trajectories. In our case, the data must be displayed in an aggregated or generalized way. Thus, instead of the individual positions of entities, the densities of the entities at various places may be visualized. The densities may be computed from data referring to a single time moment or to a time interval. In the latter case, both spatial and temporal aggregation take place.

There are two principal approaches for displaying densities. One of them is to build a smooth density surface using appropriate computational methods for spatial interpolation between original positions (see, for example, [42], Chap. 14). In cartography, there are several methods for portraying such surfaces ([42], Chap. 15), in particular, contour lines, or isolines (projected intersections of the surface with horizontal planes corresponding to selected values), hypsometric tints (shaded areas between the contour lines), and continuous-tone map, in which each point of the surface is shaded with a grey tone or color proportional to the value of the surface at that point. The surface can also be given a three-dimensional look in a perspective view.

Another approach is to compute aggregated values for areas, for example, cells of a regular grid, and represent them on a map as characteristics of the areas. This can be done, for example, in the way of shading or coloring the areas according to the respective values or drawing inside the areas symbols or diagrams with the sizes proportional to the values (Fig. 13.7).

While smoothing may be good for exposing large-scale patterns of MCB, the display of nonsmoothed aggregated data may be equally effective for this purpose and at the same time serve better for detecting local peculiarities. Moreover, not only densities can be displayed in this way but also other aggregated movement characteristics, for example, average speeds or travel distances. Thus, the iterated maps in Fig. 13.7 show data about movements of storks that have been aggregated spatially by cells of a regular rectangular grid and temporally by months. The graduated circles in the cells represent the average speeds of the birds' movement within the cells during the corresponding months, from August 1999 (top left) to April 2000 (bottom right). It should be noted that smoothing would be hardly effective for data like these, where the movements are not spread over the whole territory. Hence, data aggregation on the basis of space discretization has a more general applicability than summarization of data into smooth surfaces.

To explore data about movement directions, the user may be suggested a map display showing the prevailing movement directions in different places, which may look like the vector map on the left of Fig. 13.3. A vector map may show not only the prevailing direction in each place (by vector orientation) but also how many entities moved in that direction (e.g., by vector length) and how much this direction prevails in relation to the other directions (e.g., by vector shade or color).

However, it is not always the case when one direction significantly prevails over others. Therefore, it may be reasonable to look also at more detailed information, e.g., how many entities moved in each direction in any place over the territory. For a single place, this information may be portrayed by means of diagrams as shown in Fig. 13.8. In the diagrams B and C, the size of the internal circle may encode the number of entities that stayed in the same place. Multiple diagrams may be overlaid on a map to show movement directions in different places. To avoid overlapping, the diagrams have to fit into the corresponding space compartments, but if the compartments are very small, the diagrams may be illegible.

To see not only the movement directions but also how far the entities moved, one can apply a discrete flow map technique. Such a map, as any other, may be animated or iterated to represent movements done over a period of time (Fig. 13.9). Another possibility is to use a three-dimensional view, such as the tiered maps in the visualization of the tourist movement in New Zealand (Fig. 13.4).

In any variety of the discrete flow map technique, there is a risk that a large number of flows and intersections between them can make the display illegible.

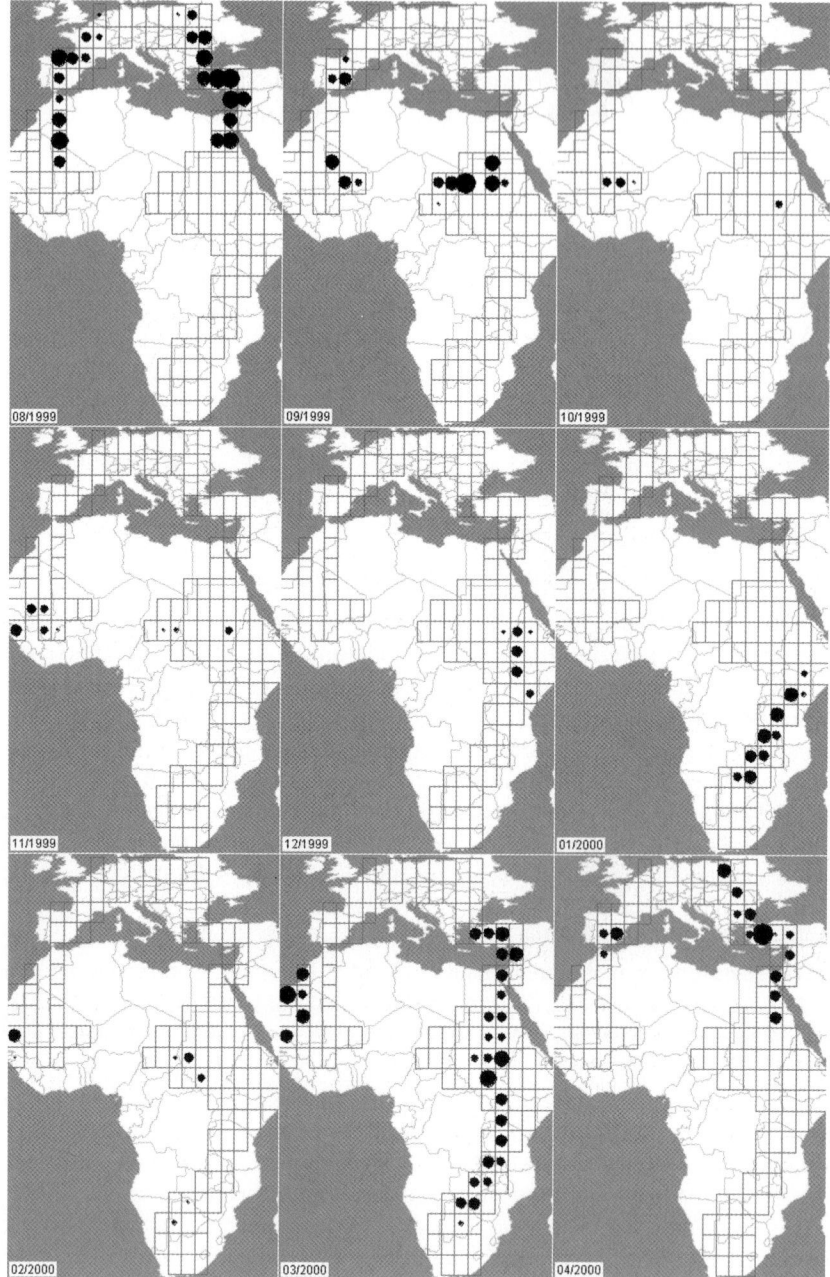

Fig. 13.7 A visualization of the data about the movements of storks (introduced in Fig. 13.1) aggregated spatially by cells of a regular rectangular grid and temporally by months

13 Visual Analytics Methods for Movement Data

Fig. 13.8 Possible methods for representing numbers of entities moving in different directions

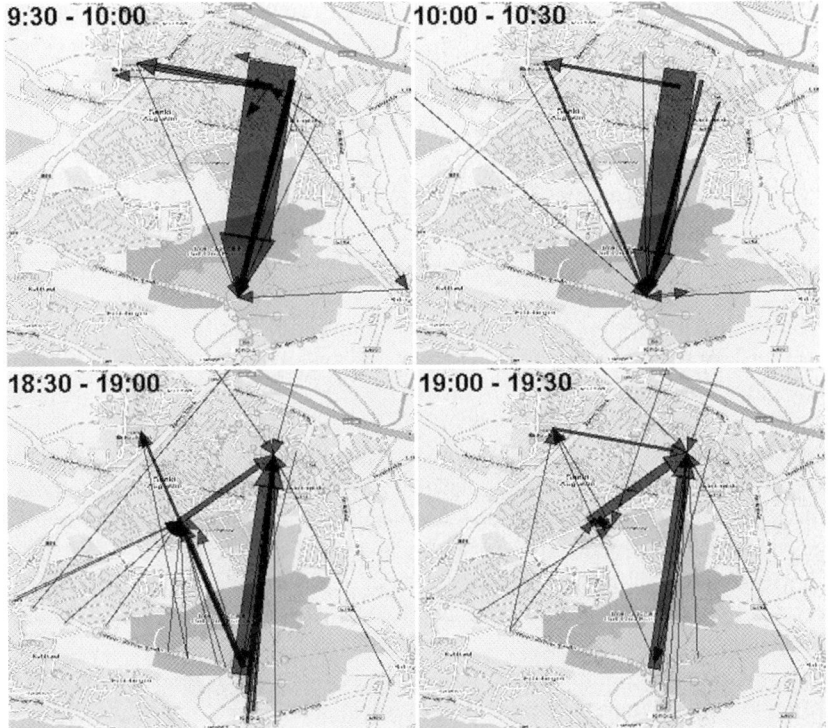

Fig. 13.9 An illustration of the technique of animated or iterated discrete flow map. The same data as in Fig. 13.6 have been divided into trajectories using the locations of the stops at least 5 min long as the sources and destinations. The space has been discretized by building circles around the source and destination locations. The time of the day has been divided into 30-min intervals. Each map in the series or each frame in the animation represents the movements having occurred during the corresponding intervals. Specifically, the widths of the arrows connecting the coarsened sources and destinations are proportional to the number of entities that moved between these locations

Additionally, iterated or tiered maps have limitations in the number of consecutive time moments that can be shown. Interactive techniques can compensate for these limitations, at least partly. Thus, interactive filtering can remove minor flows from the display. This reduces overlapping and allows the user to focus on major flows. Another useful operation is filtering according to the source or destination location. The user can also be given controls for "temporal scrolling," i.e., shifting the time period reflected in the iterated map display or map tiers. This alleviates the limitation concerning the number of time moments.

13.4.2.2 Noncartographic Displays

As maps and map-based displays cannot adequately reflect all relevant aspects of complex data such as movement, they need to be complemented with other types of display. One of them is frequency histogram known from statistics. It shows the distribution of numeric or qualitative characteristics over some population, which may be, in particular, a population of moving entities (see Fig. 13.1 top right). Histograms may be used for the exploration of the frequencies of different values of speed, direction, acceleration, etc. at selected time moments or over the whole time and for comparison of the distributions at different time moments. Histograms representing the frequencies of different movement directions may have polar or star-like layout rather than linear (see Fig. 13.8a). Histogram displays may allow interactive brushing as described by Spence and Tweedy [43].

There is an extension of the histogram technique known as two-dimensional histogram or binned scatterplot [11]. The plot has two axes corresponding to value ranges of two selected attributes. The area of the plot is divided into regular compartments (bins), in which the frequencies of the corresponding value combinations are shown by symbol sizes, shading, or coloring. Analogously, other aggregated characteristics may be represented, for instance, average or median values of another attribute. In the exploration of time-referenced data, it may be useful to apply a binned scatterplot where the axes correspond to temporal cycles. Thus, in Fig. 13.10, the horizontal axes of the two plots correspond to the time of a day divided into 1-h intervals, while the vertical axes correspond to the days of a week, from 1 (Monday) to 7 (Sunday).

A variation of two-dimensional histogram is a transition matrix (Fig. 13.11), where rows and columns correspond to different spatial locations while the symbol or color in each cell shows how many entities moved from one of the respective locations to the other between the selected time moments.

Figure 13.12 demonstrates how aggregated data may be visualized by means of segmented bars. One of the dimensions of the diagram may represent time or the

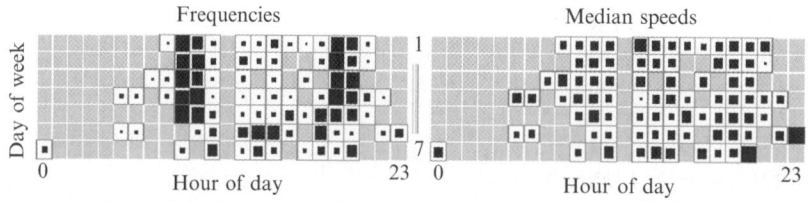

Fig. 13.10 Illustration of the technique of binned scatterplot. The car movement data introduced in Fig. 13.6 have been aggregated by days of week and hours of day. It may be noted that no movements occurred from hour 1 till 5 and from 13 till 14

13 Visual Analytics Methods for Movement Data

	E x h i b i t s	C i t r u s h	R i v e r s	P l a i t o	H e a t h g i b e r	G a i n a	L e o n a B	D o l m e	H e l m e h r	C o p s c a r	D e s . T e	S C a r m e	C o l u m b
Braun and Co		■	■	■	■		■	■		■			
Albert College		■	■		■	■		■	■		■	■		■
ABC mall	■	■	■	■			■			■		■	■
Beethoven Gymnasium	■		■	■	■	■				■		■	
XYZ school	■	■										■	
Kindergarten	■			■	■					■		■	
Frings Gymnasium		■	■			■	■						
Real school	■						■					■	
St.Joseph,s basic school		■							■	■			
Kindergarten C	■		■										
Kindergarten D						■							■

Fig. 13.11 A transition matrix built from simulated data about transportation of people. The display has been produced using the system CommonGIS

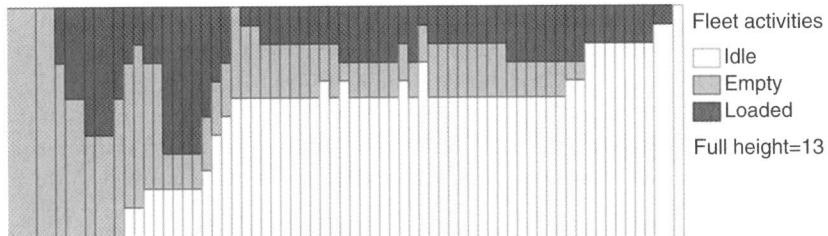

Fig. 13.12 Illustration of the technique of segmented bars. Each bar corresponds to a time interval. The full height of a bar represents the number of moving entities (in this case vehicles transporting people or goods) and is divided into segments according to certain movement characteristics (in this case, activity type: absence of motion, movement without load, or movement with load). The time span has been divided into intervals according to moments when relevant changes occurred in the situation. For this reason, the bars differ in their widths. The display has been built in the system CommonGIS from simulated transportation data

value range of any selected attribute. The other dimension represents the numbers of corresponding data items or entities divided into subsets according to values of another attribute.

As mentioned in Sect. 13.2.2, a useful technique supporting the exploration of time-related data is T–T plot [24]. However, T–T plots represent information about individual entities, which is inappropriate in our case. Therefore, the technique needs to be modified to show aggregated information. Instead of the changes of the individual values, the plot may show statistical summaries of the changes over the whole population or a group of entities, such as the means or medians.

13.4.2.3 Multiple Linked Views

When data have complex structure, there is no way to represent all the information in a single display. As may be seen from the discussion of various visualization techniques suitable for movement data, each of them shows only a certain aspect of the data. Therefore, different techniques need to be combined for a comprehensive visual exploration of the data. Moreover, these different techniques need to be used in parallel; otherwise, the user will not be able to relate, for example, the distribution of the entities in space at a particular moment to changes of their positions and other movement characteristics. The user should be provided interactive facilities to support establishing connections between different views, in particular, finding elements in different displays corresponding to same spatial positions, same times, same groups of entities, and/or same values of movement characteristics.

Brushing has been mentioned several times as a technique that supports establishing links between displays. Besides brushing, there are other methods of display linking such as propagation of a division of the set of entities into classes (each class is assigned its specific color, which is consistently used in all displays) and simultaneous reaction of all displays to interactive filtering of the data; a review may be found in [2]. It should be noted that most of the currently existing techniques for coordination of multiple displays involve dealing with individual data items and are therefore not scalable to very large data sets. There is a need in new technical solutions, which could properly work in a situation when all displays show only aggregated data (moreover, differently aggregated data!) while the original individual data are not present in the computer memory.

13.4.3 Supporting Search for Connectional Patterns

Linking of two or more displays providing complementary information may be helpful in a search for connectional patterns, i.e., for correlations, influences, and structural links between characteristics, phenomena, processes, events, etc. For example, linking can help the user to relate speeds represented on a histogram to positions on a map. Another approach is to represent the different phenomena or characteristics on the same display. The most common techniques suitable for this purpose are (binned) scatterplots and maps. Scatterplots expose correlations between values of two numeric characteristics. Maps combine representations of two or more characteristics or phenomena by overlaying several information layers. For example, movement data may be represented as flows or vectors on top of a layer representing weather or land cover information by area painting and a layer representing rivers and other waters by lines or shapes of a specific color.

For other display types, ways to incorporate additional information can sometimes be found while there are no general approaches. Thus, in a time series display like in Fig. 13.12, moments of various events may be represented as ticks on the temporal axis, which may be differently colored to indicate event types.

To investigate movement data for structure patterns related to various temporal cycles, it may be helpful to look at iterated displays arranged according to the cycles. For example, maps of city traffic aggregated by days may be arranged on the screen into a matrix with seven columns corresponding to the days of the week and the rows corresponding to different weeks. This arrangement facilitates noticing commonalities and differences within and between the cycles.

The user may also benefit from a temporal query tool capable of extracting data that refer to the same relative positions or subintervals in different cycles and aggregating the data across the cycles. For example, the user may be interested to extract all people movements made from 6 a.m. to 9 a.m. in all days and have the extracted data aggregated by the days of the week. Then, the aggregated morning hours movements on Mondays, Tuesdays, Wednesdays, and so on should be appropriately presented to the user so that the user could see the differences between the movements on weekdays and weekends. Moreover, the movements on Monday mornings may differ from the movements in the mornings of other weekdays and movements on Saturday mornings may differ from those on Sunday mornings. Similar queries can be applied to other hours of the day in order to understand in the result how the daily and weekly cycles interact in people movement.

13.5 Visualization of Patterns

13.5.1 Need for Pattern Visualization

We have discussed a number of visual and interactive techniques intended to help users in detecting patterns in movement data. According to the philosophy of visual analytics, visual search for patterns by a human can and should be complemented by automated search, or "mining," since computers may be able to discover such types of patterns that are hard to notice by a human, and vice versa. The use of automatic methods requires the results to be presented in a way allowing the user to interpret and evaluate them. In other words, automatically detected patterns need to be made perceptible to human mind. However, this requirement is also valid for patterns detected by visual methods: a person who has observed a pattern needs to represent it in such a way that it could be perceptible to other people as well as to this person after some time. Hence, irrespective of the method used for finding patterns, there is a need in their explicit *visualization* (let us recall that the word "visualize" is defined in a dictionary as "to make perceptible to the mind or imagination").

To our knowledge, the research on the visualization of patterns extracted from data is currently in its infancy and consists mainly of a few ad hoc methods devised for particular types of data mining results. In the area of data visualization, the researchers are focused on the task of enabling users to detect patterns and do not consider the problem of how these patterns can be explicitly represented. There is a more general research on knowledge visualization [44] conducted in the fields of knowledge management and education. Most methods suggested for knowledge visualization are based on the use of node-link structures or graphs. This includes semantic networks, also known as concept maps or cognitive maps, mind maps, argumentation maps, storyboards, etc. Generally, graphs are quite powerful instruments for representing various relationships and are therefore used for the visualization of some types of data mining results such as association rules and decision trees.

There are yet no specific methods for the visualization of patterns extracted from movement data. To find approaches to creating such methods, it seems reasonable to start with reviewing the existing methods used for pattern visualization in the area of data mining and knowledge discovery in databases.

It should be noted that the ideas of visual analytics are quite similar to the concept of *visual data mining* [26, 28], which means involvement of visual representations in all stages of the data-mining life cycle, including data preparation, model derivation, and validation [18]. The goal is to achieve a synergy of visualization and data mining and to enhance the effectiveness of the overall data mining process. Despite the similarity, visual analytics has a broader scope embracing visualization, interaction, various data transformations (not only as preparation to data mining), computational analysis methods, support of analytical reasoning, collaborative deliberation, and visual communication.

Presented below is a very brief overview of the most common forms of data mining results and the approaches to their visualization. For a more detailed survey, see [27]. Along with the descriptions of the existing approaches, we try to speculate how they can be adapted to the specific pattern types that may be discovered in movement data. This is not an easy task, first of all because of the necessity to deal with space and time, which substantially differ by their nature from numeric and nominal variables and therefore require special visualization techniques.

13.5.2 Visualization of Clusters

Clustering algorithms group various kinds of objects according to similarity (closeness) of their characteristics, and the user needs to understand what the objects in each cluster have in common. Unfortunately, clustering algorithms do not provide any general description of the clusters built. The clusters are defined extensionally, i.e., by listing the elements they consist of. Hence, any information about the common features of the objects in each cluster has to be extracted from the data that were used as the input of the clustering method. A realistic way to do this is to obtain various statistics about the characteristics of the members of a cluster and to visualize these statistics. By comparing the statistics for different clusters, the user can understand what is in common between the members of each cluster and how they differ from the members of the other clusters.

The general approach to the visualization of clustering results is illustrated in Fig. 13.13, which represents nine clusters, with their relative sizes (in percentage to the size of the whole data set) shown by bar segments and numbers on the left of the picture. The pie charts and bar charts represent the distributions of categorical and numeric characteristics, respectively, within each cluster in comparison to the distributions of these characteristics in the entire data set.

13 Visual Analytics Methods for Movement Data

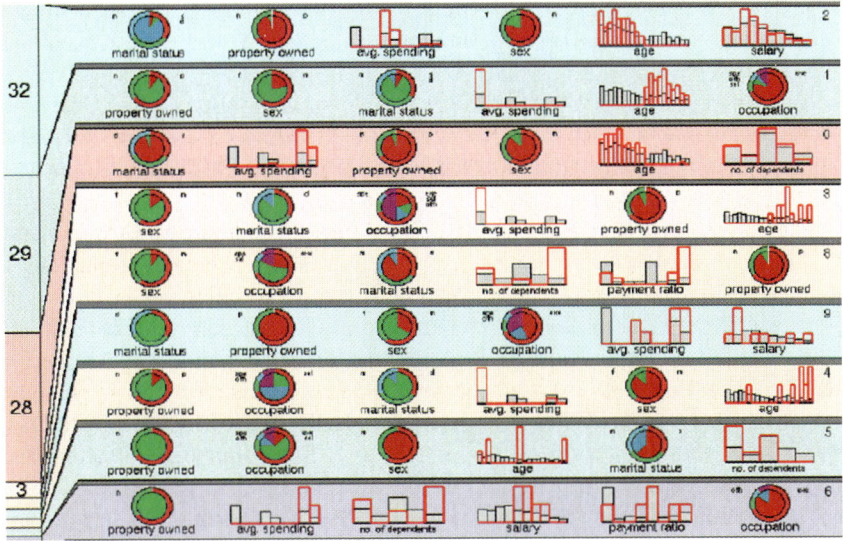

Fig. 13.13 Representation of clustering results in IBM Intelligent Miner

As has been mentioned, clustering is an appropriate instrument for the analysis of the variation of individual movement behaviors (IMBs) and their parts (trajectories or movement episodes). A clustering method divides these macroitems into groups according to a selected definition of similarity (see Sect. 13.3.2). The user then needs to see the common features of the items in each cluster as well as the degree of variation. As in the general case, this can be done by computing and visualizing statistics about the items in the clusters. Appropriate statistics and visualization techniques depend on the chosen definition of similarity.

Thus, when the similarity is defined as colocation of the trajectories, a suitable visualization of a cluster would be a map showing for each location (resulting from space coarsening or original, if there are not too many different locations in the source data) in how many trajectories it appears. Graduated symbols or graduated shading are suitable for this purpose. A separate map is built for each cluster, which enables comparison of the clusters. For ordered colocation and for spatiotemporal coincidence, it is reasonable to compute for each pair of locations x and y and time interval T how many cluster members moved from x to y during the interval T, where T results from an appropriate partitioning of the time (which may be previously transformed as discussed earlier). A possible way to visualize these statistics is an animated or iterated flow map, as in Fig. 13.9, or a three-dimensional map-based display (tiered maps), as in Fig. 13.4.

Unlike the pie charts and histograms in Fig. 13.13, map-based displays representing different clusters of trajectories are quite sizable and cannot be easily put together on a single screen. This means that the user will not be able to see the information about all clusters simultaneously and will need tools for browsing through the set of clusters and selecting pairs for comparison.

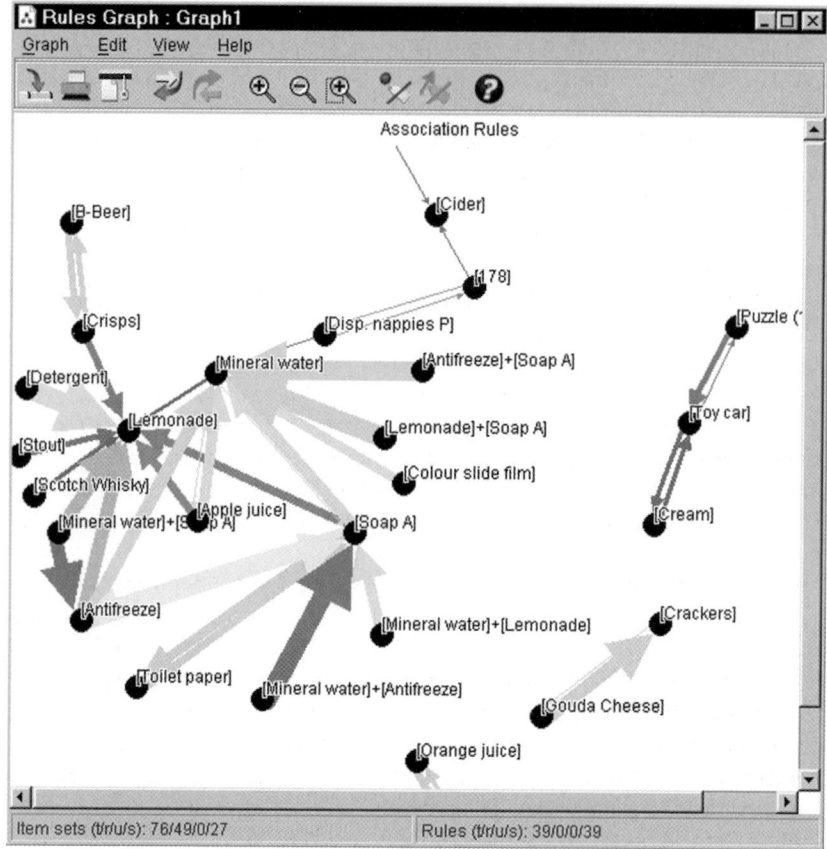

Fig. 13.14 A set of association rules visualized in IBM Intelligent Miner software by means of Rule Graph technique

13.5.3 Visualization of Association Rules

While there is extended research concerning association rules, the main focus of it is how to extract rules efficiently. Limited work has been done on how to enhance the comprehension of the discovered rules. The major problems that need to be addressed are the large number of association rules often generated, the difficulty of comprehending the output format that they have, and the difficulty of interpreting their specific semantic information [51]. Association rules that may be mined from movement data will pose additional problems, as indicated in Sect. 13.5.1.

At the present time, there are several approaches to the visualization of association rules. Some of them, like VisualMine [22], focus primarily on relationships among the items occurring in the rules but do not provide any techniques for the user to explore relationships among different rules. Other tools (e.g., [23]) allow the user to analyze individual rules in detail while loosing the overview of the entire rule set. Buono [9] introduces a graph-based technique that allows the user to visualize a great number of association rules. Interaction tools allow the user to manipulate the graph and explore the rules.

Figure 13.14 gives an illustration of a possible visualization of association rules, specifically, the Rules Graph technique from the commercially available data-mining package IBM DB2 Intelligent Miner for Data.[1] The nodes of the graph represent the item sets occurring in the bodies and heads of the rules and the arrows represent the rules. The color of an arrow represents the important measure of the corresponding rule (more precisely, the lift, which is defined as the confidence of the rule divided by the support of the rule head) and the width shows the confidence of the rule. Colors of the nodes represent the supports of the corresponding item sets.

This visualization demonstrates the use of node-link structures, which is, perhaps, the most fundamental and widely used method for knowledge representation [44]. Note that arrow symbols are paramount for visual communication; they are used multipurposely to represent directions, movements, orders, relations, interactions, and so forth [30]. This makes node-link structures quite suitable for the visualization of some types of patterns that can be extracted from movement data, in particular, temporally annotated sequential patterns [19], which may be related to locations or regions in space. An example of such a pattern is a frequently appearing sequence of places A, B, C with the transition time from A to B being t_1 and the transition time from B to C being t_2 (t_1 and t_2 may be average times or intervals). Nodes may be used to represent the places and links (arrows) may indicate the temporal order in which these places were visited. The arrow symbols may differ in width, color, brightness, and/or texture to represent the transition times and other characteristics. Additionally, text labels may be used. The graph may be drawn on top of a map to allow the user to recognize the places, see their relative positions, and relate the patterns to various geographical information.

However, simultaneous visualization of all sequential patterns extracted from movement data may be impracticable not only because of their potentially great number but also because of possible overlaps between the patterns when one and the same place appears in two or more patterns. Therefore, additional tools are required for navigation through the set of patterns and selection of patterns for more detailed examination and comparison. The user should be able to select subsets of the patterns according to various criteria, in particular, spatial (e.g., patterns involving place A or patterns where the movement direction is outwards from the center) and temporal (e.g., patterns occurring in the morning).

Besides sequential patterns, node-link drawings superimposed on a map can also represent rules referring to specific places, for example:

- trafficJam (Pisa, 7:30 a.m.) $\implies$ trafficJam (Lucca, 8:30 a.m.)
- trafficJam (Pisa, t) $\implies$ trafficJam (Lucca, t+1h)

(see Kuijpers et al., this volume). However, this approach will not work for rules involving more general spatial concepts such as "city center" and "outskirts,"

---

[1] http://www-306.ibm.com/software/data/iminer/fordata/.

"pedestrian area" and "major thoroughfare," etc., which have no precise localization and/or crisp boundaries, and hence cannot be adequately represented on a map display. Instead, such concepts can be represented verbally or symbolically, for example, with the use of the system of signs, the so-called "choremes," suggested by Roger Brunet for the representation of spatial objects and relations (cited in [15]). It seems reasonable to study which form, symbolic or verbal, is more effective and convenient for users.

13.5.4 Visualization of Classification Trees

Classification is applied to a set of records that contain class labels in order to create a profile for a member of each class from the values of available attributes. A typical result is a decision tree, which can be represented as a flow chart structure (as in Fig. 13.15) consisting of internal nodes, leaf nodes, and branches. Each internal node represents a test on an attribute and each branch represents one of the results of that test. Each leaf node represents, ideally, a single class, but in practice leaf nodes often represent several classes that could not be completely separated on the basis of available attribute values.

In movement data, class labels can be attached to records according to the types of the moving entities (e.g., pedestrians or vehicles) and their properties (e.g., people may be divided by age or occupation). The purpose of applying classification methods may be to find out how the classes of entities so defined differ by their movement characteristics. A user may also wish to divide trajectories or movement episodes

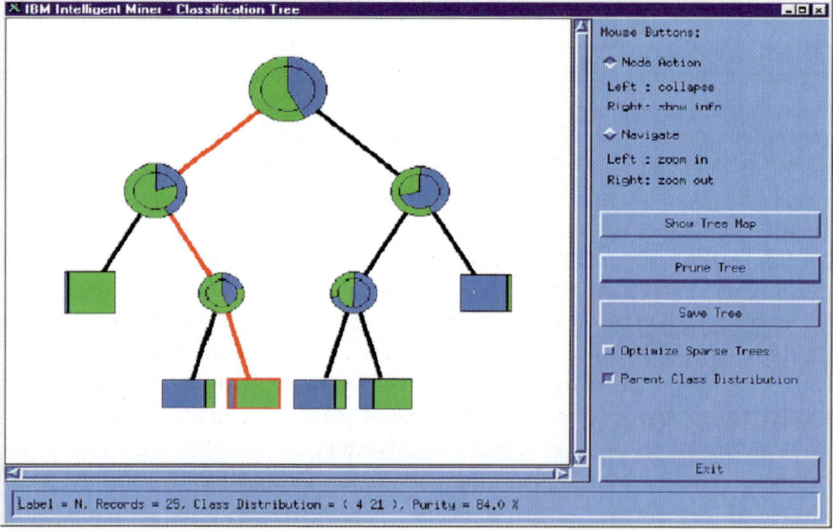

Fig. 13.15 A visualization of a classification tree in the IBM Intelligent Miner software

into classes according to the movement direction, speed, length, or geographic position and to see how these classes differ in terms of other characteristics. In any case, the use of tree (i.e., node-link structure with specific properties) seems to be the most natural approach to the visualization of the results. It is useful to extend this representation technique: where nodes refer to specific geographical locations or regions, interactive links to appropriate map displays should be included. This is analogous to combining concept maps with multimedia displays [1, 10].

13.5.5 General Notes

While specific research on visualization of patterns that can be found in movement data is yet to be done, the following general considerations may provide some guidance:

1. Node-link structures are widely used and are therefore familiar to users and easily understood. Such structures are well suited to representation of patterns involving various kinds of relationships. In particular, they can represent sequence patterns in movement data as well as correlation and dependency patterns.
2. Node-link structures may incorporate various media, in particular, maps.
3. The use of maps is reasonable when patterns refer to specific geographical locations or regions.
4. Besides techniques for pattern visualization, it is necessary to design tools for navigation through the set of patterns and for management of the patterns, which includes filtering, rearrangement, and establishing of links.

13.6 Conclusion

Current state-of-the-art methods and tools for visual and interactive exploration of movement data have significant limitations regarding the volumes of data they can be applied to. In this chapter, we have outlined a road map to developing methods for visual analysis of massive data sets, with numerous moving entities and long time series of measurements. The methods are based on data aggregation, which is performed prior to the visualization, as well as the use of computational analysis techniques. A number of technical problems need to be solved; in particular, effective linking between several displays presenting differently aggregated data.

The main goal of data exploration is detecting patterns and relationships in the data. We have considered the possible types of patterns an analyst may seek for in movement data. The role of interactive visual techniques is to allow the user to detect these patterns. We have also pointed out the need in tools for recording discovered patterns and in methods for the visualization of patterns. Visualization is necessary for a joint analysis of all detected patterns in order to gain an overall understanding of the data. This applies both to patterns detected by a human analyst and to patterns derived automatically, in particular, by means of data mining algorithms. Visualization of patterns is also required when the analyst wishes to communicate his/her discoveries to others. Currently, the problem of pattern visualization, in particular, visualization of movement patterns, is far from being solved and requires further research efforts.

References

1. S.R. Alpert. Comprehensive mapping of knowledge and information resources: The case of webster. In *Knowledge and Information Visualization*, pp. 220–237, 2005.
2. N. Andrienko and G. Andrienko. *Exploratory Analysis of Spatial and Temporal Data: A Systematic Approach*. Springer, 2006.
3. N. Andrienko, G. Andrienko, and P. Gatalsky. Impact of data and task characteristics on design of spatiotemporal data visualization tools. In *Exploring Geovisualization*, pp. 201–222. Elsevier, 2005.
4. N.V. Andrienko, G.L. Andrienko, and P. Gatalsky. Supporting visual exploration of object movement. In *Advanced Visual Interfaces*, pp. 217–220, 2000.
5. M.Q.W. Baldonado, A. Woodruff, and A. Kuchinsky. Guidelines for using multiple views in information visualization. In *Advanced Visual Interfaces*, pp. 110–119, 2000.
6. J. Bertin. *Semiology of Graphics. Diagrams, Networks, Maps*. University of Wisconsin Press, 1983.
7. A. Buja, J.A. McDonald, J. Michalak, and W. Stuetzle. Interactive data visualization using focusing and linking. In *Proceedings of the 2nd Conference on Visualization (VIS'91)*, pp. 156–163. IEEE Computer Society Press, 1991.
8. R.N. Buliung and P.S. Kanaroglou. An exploratory spatial data analysis (esda) toolkit for the analysis of activity/travel data. In *Proceedings of Computational Science and Its Applications (ICCSA'04)*, Vol. 3044. *Lecture Notes in Computer Science*, pp. 1016–1025. Springer, 2004.
9. P. Buono. Analysing association rules with an interactive graph-based technique. In *Proceedings of the International Conference on Human Computer Interaction (HCI'03)*, pp. 675–679, 2003.
10. A.J. Cañas, R. Carff, G. Hill, M.M. Carvalho, M. Arguedas, T.C. Eskridge, J. Lott, and R. Carvajal. Concept maps: Integrating knowledge and information visualization. In *Knowledge and Information Visualization*, pp. 205–219, 2005.
11. D. Carr. Looking at large data sets using binned data plots. In *Computing and Graphics in Statistics*, pp. 7–39. Springer, 1991.
12. I. Drecki and P. Forer. *Tourism in New Zealand – International Visitors on the Move (A1 Cartographic Plate)*. Tourism, Recreation Research and Education Centre (TRREC), Lincoln University, 2000.
13. J.A. Dykes and D.M. Mountain. Seeking structure in records of spatiotemporal behaviour: Visualization issues, efforts and applications. *Computational Statistics and Data Analysis*, 43(4):581–603, 2003.
14. S. Eick. Engineering perceptually affective visualizations for abstract data. In *Scientific Visualization Overviews, Methodologies and Techniques*, pp. 191–210. IEEE Computer Science Press, 1997.
15. C. Elzakker. *The Use of Maps in the Exploration of Geographic Data*. Doctoris Dissertation, University of Utrecht (Netherlands Geographical Studies 326), 2004.
16. U.M. Fayyad, G. Piatetsky-Shapiro, and P. Smyth. From data mining to knowledge discovery in databases. *AI Magazine*, 17(3):37–54, 1996.

17. P. Forer and O. Huisman. Space, time and sequencing: Substitution at the physical/virtual interface. In *Information, Place and Cyberspace: Issues in Accessibility*, pp. 73–90. Springer, 2000.
18. M. Ganesh, E. Han, V. Kumar, S. Shekhar, and J. Srivastava. *Visual Data Mining: Framework and Algorithm Development*. Technical report.
19. F. Giannotti, M. Nanni, and D. Pedreschi. Efficient mining of temporally annotated sequences. In *Proceedings of the Sixth International Conference on Data Mining (SDM'06)*, pp. 346–357.
20. D. Guo, J. Chen, A.M. MacEachren, and K. Liao. A visualization system for space-time and multivariate patterns (vis-stamp). *IEEE Transactions Visualization and Computing Graphics*, 12(6):1461–1474, 2006.
21. D. Guo and M. Gahegan. Spatial ordering and encoding for geographic data mining and visualization. *Journal of Intelligent Information Systems*, 27(3):243–266, 2006.
22. M. Hao, M. Hsu, U. Dayal, S. Wei, T. Sprenger, and T. Holenstein. *Market Basket Analysis Visualization on a Spherical Surface*. Technical Report. http://www.hpl.hp.com/techreports/2001/HPL-2001-3.pdf, 2001.
23. B. Hetzler, W. Harris, S. Havre, and P. Whitney. Visualizing the full spectrum of document relationships, 1998.
24. S. Imfeld. *Time, Points and Space: Analysis of Wildlife Data in GIS*. Dissertation, University of Zurich, Department of Geography, Zurich, 2000.
25. T. Kapler and W. Wright. Geotime information visualization. *Information Visualization*, 4(2):136–146, 2005.
26. D.A. Keim. Information visualization and visual data mining. *IEEE Transactions Visualization and Computer Graphics*, 8(1):1–8, 2002.
27. I. Kopanakis. *Visualization of Data Mining Outcomes*. http://www.csd.uoc.gr/kopanak/Sources/pattern_vis_review.pdf, 2006.
28. I. Kopanakis and B. Theodoulidis. Visual data mining modeling techniques for the visualization of mining outcomes. *Journal of Visual Languages and Computing*, 14(6):543–589, 2003.
29. M.-J. Kraak. The space-time cube revisited from a geovisualization perspective. In *Proceedings of the 21st International Cartographic Conference (ICC'03)*, pp. 1988–1995, 2003.
30. Y. Kurata and M.J. Egenhofer. Structure and semantics of arrow diagrams. In *Proceedings of Conference On Spatial Information Theory (COSIT'05)*, pp. 232–250, 2005.
31. M.-P. Kwan and J. Lee. Geovisualization of human activity patterns using 3D GIS: A time-geographic approach. In *Spatially Integrated Social Science*. Oxford University Press, 2004.
32. P. Laube, S. Imfeld, and R. Weibel. Discovering relative motion patterns in groups of moving point objects. *International Journal of Geographical Information Science*, 19(6):639–668, 2005.
33. H. Miller. Modeling accessibility using space-time prism concepts within geographical information systems: Fourteen years. In *Classics of International Journal of Geographical Information Science*, pp. 177–182. CRC Press, 2006.
34. D. Mountain. *Exploring Mobile Trajectories: An Investigation of Individual Spatial Behavior and Geographic Filters for Information Retrieval*. Dissertation, City University, London, 2005.
35. D. Mountain. Visualizing, querying and summarizing individual spatio-temporal behaviour. In *Exploring Geovisualization*, pp. 181–200. Elsevier, 2005.
36. D. Mountain and J. Dykes. What I did on my vacation: Spatio-temporal log analysis with interactive graphics and morphometric surface derivatives. In *Proceedings of The GIS Research UK (GISRUK'02)*, 2002.
37. D. Mountain and J. Raper. Modelling human spatio-temporal behaviour: A challenge for location-based services. In *Proceedings of 6th International Conference on Geocomputation*, 2001.
38. C. Newton. Graphics: from alpha to omega in data analysis. In *Graphical Representation of Multivariate Data*, pp. 59–92. Academic Press, 1978.

39. C. North and B. Schneiderman. *A Taxonomy of Multiple Window Coordinations*. Technical Report CS-TR-3854, 1997.
40. J.C. Roberts. On encouraging multiple views for visualisation. In *Information Visualization*. IEEE Computer Society, 1998.
41. B. Shneiderman. The eyes have it: A task by data type taxonomy for information visualizations. In *IEEE Visual Languages*, Number UMCP-CSD CS-TR-3665, pp. 336–343, 1996.
42. T. Slocum, R. MacMaster, F. Kessler, and H. Howard. *Thematic Cartography and Geographic Visualization*. Prentice Hall, 2005.
43. R. Spence and L. Tweedie. The attribute explorer: Information synthesis via exploration. *Interacting with Computers*, 11(2):137–146, 1998.
44. S.-O. Tergan and T. Keller. *Knowledge and Information Visualization: Searching for Synergies*. Springer, 2005.
45. J. Thomas and K. Cook. *Illuminating the Path. The Research and Development Agenda for Visual Analytics*. IEEE Computer Society, 1983.
46. W. Tobler. Experiments in migration mapping by computer. *The American Cartographer*, 14(2):155–163, 1987.
47. W. Tobler. *Display and Analysis of Migration Tables*. http://www.geog.ucsb.edu/tobler/presentations/shows/A_Flow_talk.htm, 2005.
48. B. Tversky, J.B. Morrison, and M. Bétrancourt. Animation: Can it facilitate? *International Journal Human-Computer Studies*, 57(4):247–262, 2002.
49. L. Wilkinson. *The grammar of graphics*. Springer-Verlag, 1999.
50. H. Yu. Spatial-temporal gis design for exploring interactions of human activities. *Cartography and Geographic Information Science*, 33(1):3–19, 2006.
51. K. Zhao and B. Liu. Visual analysis of the behavior of discovered rules. In *Proceeding of Workshop on Visual Data Mining (VDM'01)*, pp. 59–64, 2001.